ULTRASONICS

The Low- and High-Intensity Applications

ULTRASONICS

The Low- and High-Intensity Applications

DALE ENSMINGER
Applied Dynamics and Acoustics Division
Battelle Columbus Laboratories
Columbus, Ohio

MARCEL DEKKER, INC. New York 1973

MARCEL DEKKER, INC.

95 Madison Avenue, New York, New York 10016

LIBRARY OF CONGRESS CATALOG CARD NUMBER 72-90963 ISBN 0-8247-1175-0

PRINTED IN THE UNITED STATES OF AMERICA

PREFACE

The fact that ultrasonic energy has applications in so many areas makes it a most fascinating subject for study. This book is intended to highlight the scope of these applications and to provide others with information that the writer has found of considerable help in his research. With a subject so broad, it has been necessary to omit many topics and much information, but what has been feasible to include should help the individual reader, whether he is a novice or an experienced ultrasonics engineer or physicist. Perhaps this book will even stimulate the development of new applications.

It is impossible to present the complete fundamental physics of ultrasonics, describe in detail equipment and procedures, and discuss thoroughly all applications of ultrasound in one small volume. On the other hand, it is not sufficient to name the potential and practical applications of ultrasound without giving at least some theoretical background to show that these applications are indeed possible. Hopefully, the reader will find the necessary information for developing his own specific area of interest.

For those who may not be familar with the terminology used in ultrasonics, a glossary is included.

To use ultrasonic energy to the best advantage, whether in designing transducers or applying the energy to some process, it is necessary to understand the basic principles of waves and wave propagation. Three chapters are devoted to the presentation and derivation of many useful equations to this end. The derivations are carried out in a manner that can be followed by the average engineering student or graduate. These chapters alone could be the basis for a separate book, but with the limitation imposed by a volume devoted to applications, it was necessary to abbreviate the material. However, the material is presented in such a manner that the reader can expand it if necessary.

Presented in another chapter are a few basic principles of transducer design. No proprietary information has been included, however. The material should be useful to one starting in the field of ultrasonics, but readers should consult other reference works, including the literature of manufacturers of piezoelectric materials, for more thorough treatments of transducer design. To avoid publishing proprietary information and because oscillator and amplifier circuits are common knowledge among electronics engineers no circuit diagrams are offered. The matching circuitry for ultrasonic devices varies with the transducer, and the number of circuit arrangements for the various applications can be as varied as the number of applications.

A few block diagrams which give the functions of elements of the circuit at various points are presented in the chapters devoted to low-intensity applications of ultrasonics. This topic too, could be the basis of several books. For example, books devoted to the use of ultrasonics in nondestructive testing, or physical acoustics and the properties of matter, or acoustical holography are available. The applications of microwave acoustics have been bypassed , and surface-wave applications are treated with brevity, although both are receiving considerable attention at the present time. It was believed that these subjects could not be covered adequately in the time and space available. Tables of acoustical properties are included.

Originally, the plan was to include a chapter on underwater sound. However, several good books on this subject are available, and certain restrictions make whatever information could be included with the necessary brevity of little value. Those who desire this type of information may find it in references such as Fundamentals of Sonar by J. W. Horton, a Naval Institute publication, Annapolis, Maryland; Principles of Underwater Sound for Engineers by Robert J. Urick, published by McGraw-Hill Book

Company, and Underwater Acoustics Handbook--II by Vernon M. Albers, published by The Pennsylvania State University Press. Anyone who has the need to design equipment for underwater sound will have available the necessary material for reference.

Other chapters are presented which are devoted to high-intensity applications of ultrasonics. A table of solvents used in ultrasonic cleaning is also presented.

The final chapter is devoted to medical applications, which includes both low- and high-intensity applications.

The reader should find much useful information in this book.

The author wishes to express his appreciation to his employer and many colleagues for their encouragement and assistance in preparing the manuscript and to the industrial organizations that contributed photographs as acknowledged throughout the text. He is especially grateful to H. R. Ball for the many hours he spent drawing the illustrations, to Miss Dolores Landreman for her help in the organization, and to Mrs. Dorothy Wallace for her careful and detailed editing of the draft of the manuscript. Finally, the author wishes to pay a special tribute to a patient wife who was forced to assume many trying responsibilities during the time the manuscript was being written.

Dale Ensminger

CONTENTS

GLOSSARY

The following glossary includes terms that are commonly used in ultrasonics and are used throughout this book.

ABSORPTION: (of sound) the dissipation or conversion of sound energy into other forms of energy, such as heat, either within the medium or attendant upon a reflection.

ACOUSTICAL ATTENUATION CONSTANT: the real part of the acoustical propagation constant.

ACOUSTICAL PROPAGATION CONSTANT: the natural logarithm of the complex ratio of the steady-state particle velocity, volume velocities, or pressures at two points separated by unit distance in a uniform system (assumed to be of infinite length), or at two successive corresponding points in a system of recurrent structures (assumed to be of infinite length). The ratio is determined by dividing the value at the point nearer the transmitting end by the corresponding value at the more remote point.

ACOUSTIC IMPEDANCE: referred to a surface lying in a wavefront, the complex quotient of a sound pressure (force per unit area) on that surface by the flux (volume velocity or linear velocity multiplied by the area) through the surface. When concentrated rather than distributed impedances are considered, the impedance of a portion of the medium is defined by the complex quotient of the pressure difference effective in driving that portion by the flux (volume velocity).

ACOUSTICS: the science of sound including its production, transmission, and effects.

ACTIVE ACOUSTIC DEVICE: device which generates sound for the purpose of detecting an object.

ATTENUATION: loss of intensity in an ultrasonic wave with distance or time.

BEATS: periodic reinforcement of a sound produced by the interference of sound waves of slightly different frequencies.

COUPLANT: material through which an ultrasonic wave may pass from a source to another material with minimum loss of energy.

CYLINDRICAL WAVE: a wave wherein the wave fronts are coaxial cylinders.

DECIBEL: a term used in designating relative intensity. The number of decibels corresponding to a given intensity, I, is determined by $10 \log_{10}(I/I_o)$, where I_o is a reference intensity.

DOPPLER EFFECT: the phenomenon evidenced by the change in observed frequency of a wave in a transmission system caused by a time rate of change in the effective length of the path of travel between the source and the point of observation.

DOPPLER SHIFT: the magnitude of the change in the observed frequency of a wave due to the Doppler effect.

ECHOLOCATION: a process for locating distant or invisible objects by means of sound waves reflected back to the emitter by the object.

ELASTIC WAVES: a wave sustained by the mass and elasticity of a medium.

ELECTROMYOGRAPHY: the process of measuring electrical responses of nerves to various stimuli.

FLEXURAL WAVE: wave characterized by a bending, or flexing, type of motion.

FREE PROGRESSIVE WAVE: a wave in a medium free from boundary effects.

FREQUENCY: for a periodic quantity, in which time is the independent variable, the number of periods occurring in a unit of time.

HERTZ: cycle per second.

INTENSITY: the average rate of flow of energy through a unit area normal to the direction of wave propagation.

KILOHERTZ: one thousand Hertz.

LAMB WAVE: a type of wave associated with plates which propagates in a direction parallel to the surfaces of the plate at velocities which depend on the elastic properties of the plate and its thickness and on the frequency of the wave. There are two types of Lamb waves: symmetrical and asymmetrical. In a symmetrical wave, the particle motion along the center of the plate is longitudinal, thus producing a series of bulges in the surface of the plate at regular intervals. An asymmetrical wave is a flexural wave.

LONGITUDINAL WAVE: a wave in which the direction of displacement at each point of the medium is normal to the wave front.

LOVE WAVE: a surface wave in which the direction of displacement at each point of the surface is parallel to the wave front and to the surface of the medium.

MAGNETOSTRICTION: a change in dimensions in ferromagnetic materials, such as iron, nickel, cobalt, and alloys of these metals, which occurs when these materials are subjected to a change in magnetic state.

MEGAHERTZ: one million Hertz.

MODE: the vibration pattern of waves.

OSCILLATOR: a device for producing alternating current or motion. In ultrasonics, the radiofrequency or audiofrequency generator is usually referred to when this term is used. Occasionally it is used to refer to an oscillating mechanical device such as a tuning fork, pendulum, or active transducer.

PASSIVE ACOUSTIC DEVICE: a listening device for detecting sounds generated by a source under investigation.

PIEZOELECTRIC EFFECT: appearance of electric charges on the parallel faces of a slab cut from a crystal having an axis of nonsymmetry when these faces are lying normal to an axis of nonsymmetry and when the slab is subjected to a mechanical stress.

PIEZOELECTRICITY: electric charges that appear on the parallel surfaces of a piezoelectric crystal when the crystal is subjected to a mechanical stress.

PLANE WAVE: a wave in which the wave fronts are everywhere parallel planes normal to the direction of propagation.

PLATE WAVE: a Lamb wave.

RADAR: the art of detecting by means of radio echoes the presence of objects, determining their direction and range, recognizing their character, and employing the data thus obtained in the performance of various functions such as military, naval, or meterological observations.

RAYLEIGH WAVE: type of surface wave in which the particle motion is elliptical about axes that are parallel to the wave front.

REFLECTION COEFFICIENT: the ratio of acoustic intensity of the reflected wave to that of the incident wave at a boundary between two media of different acoustic impedances.

RESONANCE: condition of vibration at which the absolute value of the driving-point impedance is a minimum. Velocity resonance exists between a body, or system, and an applied sinusoidal force if any small change in the frequency of the applied force causes a decrease in velocity at the driving point. Except for simple systems, such as pendulums, it is usually necessary to specify the type of resonance or resonant frequency.

RESPONSE: ratio of the electric output to the acoustic input of a transducer (microphone or hydrophone).

SENSITIVITY: response of a transducer for a sine wave at a prescribed frequency.

SHEAR WAVE: transverse wave; a wave in which the particle motion is normal to the direction of propagation.

SONAR: acoustical analogue of radar; the art of detecting by acoustical means the presence of objects, determining their direction and range, recognizing their character, and employing the data thus obtained in the performance of various functions--submarine detection, navigation, distance, and depth measurements, etc. Underwater, sonar refers to man's use of sound for a variety of purposes in the use and exploration of the seas. Both active and passive sonar are used.

SONICS: the general subject of sound.

SPHERICAL WAVE: a wave in which the wave fronts are concentric spheres.

STRESS WAVE: any wave in a medium which is dependent upon the elastic nature of the medium.

SURFACE WAVES: waves that propagate along the surface of a medium and penetrate the medium only to shallow depths.

TORSIONAL WAVE: a type of shear wave in which, within any given plane parallel to the wave front, particle motion is normal to the direction of propagation and the plane itself oscillates through an angle about an axis that is parallel to the direction of propagation.

TRANSDUCER: an instrument that converts one energy form into a second energy form. The types most commonly used in ultrasonics convert electrical energy to ultrasonic energy and vice versa.

TRANSMISSION COEFFICIENT: the ratio of the intensity of the transmitted wave to that of the incident wave at a boundary between two media of different acoustic impedances.

ULTRASONICS: the general subject of sound in the frequency range above the average audible range of human beings.

ULTRASOUND: sound that is pitched too high for a human to hear.

WAVE: a disturbance which is propagated within a medium in such a manner that at any point in the medium the displacement is a function of the time, while at any instant the displacement at a point is a function of the position of the point.

WAVE FRONT: (a) for a progressive wave in space, a continuous surface which is a locus of points having the same phase at a given instant; (b) for a progressive surface wave, a continuous line which is a locus of points having the same phase at a given instant.

ULTRASONICS

The Low- and High-Intensity Applications

Chapter 1

ULTRASONICS--A BROAD FIELD

I. INTRODUCTION

The study of ultrasonics is the study of a form of energy. Ultrasonics has been a subject of interest to researchers since World War I, but its use in industry has grown to proportions worthy of note only since World War II.

This book deals primarily with industrial and medical applications of both low-intensity and high-intensity ultrasonic energy and with fundamental principles pertaining to these applications.

Ultrasonics, which is a branch of acoustics, deals with vibratory waves at frequencies above those within the hearing range of the average person, i.e., at frequencies above 16 kHz (16,000 cylces per second). Often, the hearing range of young persons extends beyond 20 kHz, so setting the lower limit of the ultrasonic range is somewhat arbitrary. Ultrasonics is sometimes used loosely and erroneously to refer to applications of vibratory waves at lower frequencies for producing results that are similar to those which ultrasonic energy is capable of producing. Recently, the term macrosonics has been applied to high-intensity applications of sonic energy, regardless of frequency.

Ultrasonic waves are stress waves, and for this reason they can exist only within mass media. They are transmitted from one mass to another by direct and intimate contact between the masses. In this respect, they differ from light and other forms of electromagnetic radiation which travel freely through vacuum. In other respects, these two forms of energy obey similar laws of propagation.

Ultrasonic waves also are termed elastic waves since it is the elastic property of the medium which is responsible for the sustained vibrations required for ultrasonic wave propagation. Fundamentals of elastic-wave phenomena are discussed in Chap. 2.

Division of the applications of ultrasonics into two categories -low intensity and high intensity-is natural. Low-intensity applications are those wherein the primary purpose is transmitting the energy through a medium. The objective may be to learn something about the medium or to pass information through the medium. The objective is never to change the state of the medium. Typical low-intensity applications are nondestructive testing of materials or devices, measurement of the elastic properties of materials, medical diagnosis, and livestock judging. Although many underwater sound generators require enough electrical power to supply a small city, marine applications such as depth sounding, echo ranging, communication, and submarine detection may be included in this category.

High-intensity applications are those wherein the purpose is to produce an effect on a medium, or its contents, through which the wave propagates. Typical high-intensity applications of ultrasonics are medical therapy, atomization of liquids, machining of brittle materials, cleaning, disruption of biological cells, welding of plastics and metals, and homogenization or mixing of materials.

II. HISTORY

The history of ultrasonics is part of the history of acoustics. The earliest record of a scientific examination of music is probably that of Pythagoras who, in the 6th century B.C., discovered that the shorter of two similar stretched strings of unequal length emits a higher note than the longer one and that if one string is twice the length of the other, their pitch differs by one octave. However, records of the use of stringed instruments predate Pythagoras by several centuries.

A second notable contribution to the science of acoustics is a published observation by Galileo in 1638 that pitch is associated with the frequency of vibration. A Franciscan friar named Mersenne, who was a contemporary of Galileo, was the first to actually measure the frequency of a long vibrating string and to calculate the frequency of shorter ones from his observations.

A number of physicists and mathematicians laid the groundwork for developing the science of acoustics during the 17th and 18th centuries. Among the familiar names are Robert Hooke (1635-1703), the author of Hooke's law; Joseph Saveur (1653-1716), who first suggested the name acoustics for the science of sound; Brook Taylor (1685-1731), the author of Taylor's theorem on infinite series; Isaac Newton (1642-1727), whose law of motion is basic to the wave-equations used in ultrasonics; Jean D'Alembert (1717-1783); Joseph LaGrange (1736-1813); and Jean Fourier (1768-1830). The elementary wave equation of D'Alembert and the more elaborate wave equation of LaGrange are basic to an understanding of wave motion. The mathematical series proposed by Fourier, which is a method of expressing an arbitrary function as a series of sine and cosine terms, is ideally suited to the analysis of ultrasonic waves.

In 1877 Lord Rayleigh published his famous work The Theory of Sound. (1) This was a milestone in the development of the science of acoustics. Rayleigh had worked in his own home with apparatus that was crude according to modern standards, but much of his treatise that resulted from this research is valid today. His work still is available in two volumes.

Two 19th Century discoveries were significant in the development of methods of generating and detecting ultrasonic energy. The first, magnetostriction, was revealed by Joule in a paper published in 1847. (2) Magnetostriction involves a change in the dimensions of a magnetic material under the influence of a magnetic field. The second, piezoelectricity, was discovered by the Curie brothers in 1880. (3) Piezoelectricity relates to the electric charges developed on the surfaces of certain types of crystals when these crystals are subjected to pressure or tension. The piezoelectric effect is commonly used to detect ultrasonic signals. The inverse effect, in which a voltage impressed across two surfaces of a piezoelectric crystal induces stresses in the material, is presently the most common method of generating ultrasonic energy in commercial systems.

Ultrasonics as a specific branch of the science of acoustics had its birth in the study of underwater sound. Prior to the development of ultrasonics, as a specific discipline, ships were warned of dangerous obstacles by bells submerged from lightships. Trained crew members of passing ships could detect these warning signals by means of microphones or stethoscopes pressed against the hulls. The invention, in 1912, of the Fessenden oscillator was a decided improvement over the system of bells. The Fessenden oscillator operates electromagnetically at frequencies between 500 and 1000 Hz. It is capable of sending Morse-code messages between ships over distances to 10 miles.

When submarines became a menace to the Allies during World War I, Wood and Gerrard in England and Langevin in France were assigned the task of developing methods of detecting enemy vessels.

Langevin was the first to suggest using a piezoelectric receiver. Wood made the first directional hydrophone for the location of submarines. Ultrasonics was used in the initial devices to obtain directional beams of energy.

During their underwater sound investigations, Wood and a co-worker, Loomis, and Langevin observed that small water animals can be stunned, lamed, or killed in intense ultrasonic fields. Consequently, Wood and Loomis conducted some interesting experiments with high-intensity ultrasonics in addition to their studies of the propagation of sound in water.

Since 1940, interest in ultrasonics has increased. Rapid developments in other technologies, such as electronics and piezoelectric ceramics, have made possible the production of practical ultrasonic systems for domestic, industrial, and military use. The interesting effects often obtained by the use of ultrasonic energy have sometimes resulted in its being considered as though it were some form of magic. Many impractical uses have been suggested, which reveals that there is a lack of knowledge of the fundamental principles of ultrasonics. This lack of understanding has, in turn, led to discouragement and false conclusions concerning practical applications of ultrasound. Both misconceptions can be avoided, or at least minimized, by sound understanding of the fundamental principles of ultrasonics.

Presented in the remaining chapters of this book are basic equations, design principles, and other material which demonstrate that there are broad applications of ultrasonics in many areas.

III. ULTRASONICS IN NATURE

The use of ultrasonics is not limited to human beings. Some of the world's greatest experts in the use of sound are found in the wild--in the oceans and the open fields. Certain animals use ultrasonics for locating and identifying food, for navigational purposes, and for detecting danger. Obviously, such animals must be capable of interpreting sonic signals instantly.

The study of ultrasonic signals used by animals--the types of signals that are emitted, their frequencies, their harmonic content, etc.--can help man improve his own techniques in the application of ultrasonic energy. Many studies have already been made of bats, moths, porpoises, birds, and dogs.

III.A. BATS

Perhaps the animal that is best known for its use of ultrasonics is the bat. Many scientists have studied these interesting animals and their use of sound to navigate and locate food. According to Pye (4), in 1794 Lazzaro Spallanzani demonstrated that the location of obstacles and prey by insectivorous bats depends on hearing rather than on vision. Although many of Spallanzani's friends confirmed his findings without knowing the true mechanism, other influential scientists disputed their contentions. Not until 1920 was any further effort made to investigate the theory that bats can produce ultrasonic calls by which they navigate. The revival of interest can be credited to Hartridge (5) who observed bats in flight and pointed out the advantages of high frequencies for producing echoes from small objects. Pye (4, 6-8) and his co-workers in England have been among those doing active research relative to bat sonar. On occasion he has cooperated with D. R. Griffin of Harvard University, another scientist who has contributed much to our knowledge of echolocation by bats and other animals.

The possibility that a bat may modulate the frequency of the sonic pulses he transmits to obtain a beat note between his voice and an echo and to permit inference of the distance to the echoing object by the frequency of the beat note was investigated by McGue (9). He hypothesized that bats in the Boston area judge range by sensing the time delay between an outgoing pulse and the echo. Pye agrees that bats may occasionally make such use of beats. He reports that a characteristic feature of bat sounds is that the frequency falls steadily by nearly an octave throughout each pulse (7). The frequency at the start of each pulse varies with the species and falls within the range 30 to 120 kHz. When the bat is flying within close range of an object, frequencies become lower, pulses are shorter, and pulse-repetition rates are higher than when flying at a distance. The pulse-repetition rate varies from less than 10/sec when the bat is cruising freely to over 100/sec when he is within close range of objects.

Pye discusses in particular the Rhinolophidae. This bat makes longer pulses of constant frequency, which suggests that the time-difference mechanism is used for location of prey or obstacles at a distance and that beats are used at closer range. He has proposed that Doppler shift also may be a factor in echolocation.

Small bats can fly at full speed through barriers of vertical wires only 0.4 mm in diameter and spaced only one wingspan apart. They can catch small insects, such as mosquitoes, at the rate of one every 10 sec for up to 30 min. Their ability to discriminate between such things as food and raindrops or foliage and hedges is phenomenal. Their sonar appears to be nearly "jam proof." When flying in close proximity to other bats, background noise could be expected to be confusing; nevertheless, they continue to locate prey and avoid collisions. The bat Noctilio of Trinidad catches small fish by dipping its feet under the surface of the water. It appears to locate its prey ultrasonically despite the difficulty that the sound must twice cross a barrier with a very high reflection coefficient. Prior to dipping, this bat markedly raises its pulse repetition rate. It is possible that characteristic ripples caused by the fish are being detected rather than echoes from an object beneath the surface. However, such bats, even when blinded, were able to locate not only fish, but also small air-filled bladders completely submerged in a pond. This was done more consistently than would be expected with a random sweeping of the pond. These bats would also approach objects with acoustic impedances similar to that of water.

All bats are completely disoriented if both of the external ears are carefully plugged. In most cases, this also is true if only one ear is plugged. The Rhinolophid bat is an exception. This bat can fly well with one ear plugged and avoids obstacles without difficulty. The reason appears to be that the Rhinolophid bat makes fewer but longer pulses of constant frequency, concentrating the sound in a beam for scanning an area. The ears move in a complex manner, independent of the direction of the beam.

As would be expected in an animal which is so dependent upon sound, the auditory regions of a bat's tiny brain are extraordinarily well developed.

III.B. MOTHS

Just as bats use ultrasonics for detecting prey, one of their prime targets, the moth, uses ultrasonics for self-defense. When the moth detects the bat sonic pulse, he immediately takes evasive action in zigzag flight and power dives. Roeder of Tufts University and Treat of City College of New York have used electromyographical techniques in studying the response of the moth to bat sounds (10). By attaching an electrolyzed silver probe to the tympanic nerve of the moth, they were able to detect the nerve response to various sounds such as hisses, bat cries, pure tones, and jingling keys. They displayed the amplified signals on an oscilloscope screen.

By thus "listening" through the moths' ears, they learned that moths can detect bats as far away as 100 ft. The moths' ears registered the cries of approaching bats at this distance, but when the bats were moving away, the ears stopped registering at about 40 ft. When a bat made a straight-on approach, the sensors recorded an uneven, sporadic signal. This was interpreted as an indication that the bat was maneuvering by varying the intensity of his sonar pings.

III.C. BOTTLE-NOSED DOLPHINS OR PORPOISES

Porpoises or dolphins are most interesting creatures, particularly from the standpoint of ultrasonics. Certainly, all who have had the privilege of observing the playfulness, the speed, and the gracefulness of these swimmers can scarcely forget this fascinating experience. They can put their sonar to good use for foraging even in murky water. These animals are highly intelligent, and their expert use of sonar has captured the interest of scientists. It is claimed that a small vocabulary has been discovered in use by porpoises.

Several books and papers on the habits of porpoises and their use of sonar have been published. Porpoises and Sonar by Kellogg (11) is especially interesting and easy reading.

Porpoise sounds have been studied by various methods including recording sounds on tape and observing the waveforms "live" on oscilloscopes. Their noises have been variously described as mewing, barking, whistling, rasping, and repetitive clicks or pulses. Kellogg (11) believes that all of the sounds except whistling are actually the same basic series of pulses which occur so rapidly that they take on a tonal quality of their own. The easiest way to detect porpoise sounds is by means of a hydrophone located in the vicinity of the animal. One technique for recording porpoise echolocation signals during target-discrimination tests is to attach sensors directly to the animal and to observe the detected signals as the animal maneuvers.

As is true of bats, porpoises vary the frequency, repetition rate, and amplitude of their echolocation signals. In fact, there are many similarities in the way the two animals use their sonar signals.

The vocabulary use has been observed, for instance, when free-swimming porpoises encounter a barrier. A scout may be sent to investigate and when he locates an opening he returns to the others and leads them through the opening.

The porpoise noises that are heard most often are the clicks and the whistle, the clicks being the more common. The repetition rate of these clicks varies from approximately 5 per sec to several hundred per sec; their frequencies may be as high as 170 kHz, but usually become fairly weak above 120 kHz. The whistle is approximately 0.5 sec in duration and appears in several pitch patterns. The frequency range is approximately 7-15 kHz.

As is true of bats, the echolocation of porpoises appears to be unaffected by jamming or interfering noises. Kellogg (11) tells of beaming sound, intended to interfere with their sonar, at porpoises as they swam through obstacles. Even broadcasting recordings of their own sonar signals failed to produce any observable disturbing effects on their swimming or avoidance behavior.

III.D. BIRDS

At least two families of birds, Steatornis and Collocalia, are echolocators. It is interesting to note that calls by various birds may be understood only by members of their local flocks. Gulls or crows from one area may or may not respond to distress or assembly calls of gulls or crows from another area, although they will always respond to the calls from birds in their own flocks.

III.E. DOGS

The silent dog-whistle makes use of the fact that the dog's hearing range extends to frequencies considerably above the human audible range. These animals are illustrative of the fact that the ultrasonic range is defined only by the response of the human ear. There is no evidence that dogs make use of echolocation. The dog is equipped with a keen sense of smell, sight, and hearing. It appears to have no need to be equipped with echolocation mechanisms. This is true of most animals.

IV. ULTRASONIC TRANSMITTERS

An ultrasonic transmitter is an instrument designed to generate the disturbance from which the ultrasonic energy emanates. Therefore, any device capable of generating ultrasound is an ultrasonic transmitter. It does not imply any specific type of transducer. It may be a whistle, a piezoelectric plate, a magnetostrictive stack driving a piston, a diaphragm driven electromagnetically, a siren, or any of various mechanical devices such as rotating eccentrics. All of these methods have been used to generate ultrasonic energy.

The ultrasonic transmitters most frequently used are either piezoelectric devices or magnetostrictive devices. Piezoelectric transducers may be used throughout the ultrasonic range of frequencies. Magnetostrictive transducers are useful for generating high-intensity ultrasonic energy to frequencies of approximately 50 kHz. Whistles for high-intensity applications of ultrasonic energy are difficult to fabricate for operation much above 20 kHz.

The design of a transmitter depends upon the application for which it is to be used and, therefore, it may take on many forms.

V. ULTRASONIC RECEIVERS

The ear is a passive device, responding to sounds and making the information available for recording in the brain. All forms of animal life either have ears or the capability of detecting vibrations. Sonics and ultrasonics in nature serve many purposes: defense, assembly calls, mating calls, distress calls, to name the obvious. Hunters are familiar with the use of duck and moose calls. Attempts have been made to frighten off pests by duplicating their distress calls. Many animals respond to frequencies exceeding 150 kHz.

The device that detects the ultrasound is called an ultrasonic receiver. It serves the same function as the ear, receiving the signal and hopefully transforming it faithfully into another form of energy (usually electrical) where it can be processed and analyzed as desired. Microphones are typical receivers. The transducers most often used as receivers of ultrasonic energy are the piezoelectric types. Magnetostrictive devices are used occasionally. In air, capacitive-type microphones are often used at frequencies well into the low end of the ultrasonic spectrum.

Basic design of ultrasonic transmitters and receivers is discussed in Chap. 5.

VI. LOW-INTENSITY APPLICATIONS

There are numerous low-intensity applications of ultrasonic energy, including process control, nondestructive testing, intrusion detection, measurement of elastic properties, medical diagnosis, delay lines, and appliance controls (such as television-channel selectors). Low-intensity applications are discussed beginning with Chap. 6.

In the first attempt at nondestructive testing continuous waves were used. A quartz transducer coupled to a specimen by means of a liquid emitted the continuous wave which was transmitted through the specimen. A second transducer scanned the opposite side searching for shadow areas indicative of flaws in the specimen. This was not a satisfactory method.

In 1942, Firestone (12) received a patent for his Reflectoscope. This was another milestone in the industrial use of ultrasonic energy. The Firestone instrument is a pulse-echo type of instrument incorporating some of the principles employed by bats and porpoises for echolocation. A piezoelectric crystal produces a short pulse of ultrasonic energy which is transmitted through the test object in a very narrow beam. A discontinuity reflects the energy to the same crystal which now serves as a microphone, and the pulses and travel time are displayed on the screen of an oscilloscope for interpretation. Until 1950, this industry was still in its infancy, but by now, ultrasonic nondestructive testing is a very important industry and, perhaps, the largest of the low-intensity applications. Ultrasound can penetrate some materials deeper than can any other form of energy--to 40 ft in some metals.

A second basic type of ultrasonic inspection device correlates the thickness of a piece of material with its resonance frequency in a thickness mode. The frequency of the tank circuit of the oscillator, which also includes the transducer, is varied either manually or automatically by electrical means. The increased output at resonance points is either displayed on an oscilloscope screen or processed and recorded. The useful thickness range is on the order of 0.010-8 in. of steel, the greater thickness being calculated from harmonic measurements. This method of measuring thickness is especially useful for locating areas of corrosion and nonuniformities in wall thicknesses when only one surface is accessible.

Ultrasonics has been used in nondestructive testing (a) to determine grain size of metals and atomic fuel elements by the attenuation of a pulse as it travels through the material; (b) to locate holes, cracks, lack of bond between layers, or any other undesirable discontinuity in materials; (c) to inspect the quality of welds; (d) to determine elastic properties of materials; and (e) to determine the quality of concrete.

Ultrasonic devices which operate on low-intensity principles also include door openers, height detectors, liquid-level gages, detectors of web breaks in paper or textile mills, and process controls. The amount of solids in suspension in a liquid is determined on the basis of attenuation of ultrasonic energy.

VII. HIGH-INTENSITY APPLICATIONS

High-intensity applications of ultrasonics are also numerous. These are discussed beginning with Chap. 10. The leading commercial high-intensity applications of ultrasonics are cleaning, plastic bonding, machining, and soldering.

The ultrasonic soldering iron was available prior to 1950. It was used for the fluxless soldering of aluminum by eroding the oxide film from the surface of the aluminum beneath the solder to promote intimate contact between the molten solder and the base metal.

The ultrasonic machine tool was in production by 1952 when Balamuth's patent (13) was issued. Conventional ultrasonic machining is done by a transducer resonating a tool in a longitudinal mode. An abrasive slurry is washed across the face of the tool. The action of the bit sets up cavitation in the slurry which produces a high-intensity scouring action on the workpiece, thus abrading the material. Recently, a rotary-type drill has been developed which uses commercially available diamond-tipped drills and needs no slurry.

Cleaning probably provides the largest market for high-intensity ultrasonic equipment. Applications range from very small denture cleaners to very large industrial production-line installations. Prior to 1952, only a few small units were being sold for ultrasonic cleaning, including equipment manufactured by a company in Switzerland selling to watch manufacturers. Development of this industry increased rapidly in the years following 1955.

A device has been manufactured for preparing liquid emulsions utilizing a high-pressure jet of liquid forced across the edge of a reed which is set into high-frequency, high-intensity vibration. The result is a system which produces stable liquid-in-liquid emulsions.

Gas-phase whistles are used to control or eliminate foam which otherwise creates a storage problem, to dry heat-sensitive materials, and to disperse cleansing fluids for cleaning large storage tanks. The most popular of these is the stem-jet whistle, a modified Hartmann whistle, which has a high energy-conversion efficiency.

VIII. MEDICAL APPLICATIONS-LOW- AND HIGH-INTENSITY

Medicine is a promising field for the applications of ultrasonic energy. Applications of ultrasonics for medical purposes include both low-intensity and high-intensity applications. This subject is treated as a unit in the final chapter.

Low intensities are used for diagnostic purposes, high intensities for therapeutical and surgical purposes.

Diagnostic applications incorporate many of the principles of nondestructive testing--reflection of ultrasonic energy from interfaces and selective absorption. Synchronizing the oscilloscope sweep trigger and position with the position of a scanning probe results in a picture that outlines various interfaces and from which anomalies can be identified.

In another application, a passive probe has been used for locating calculi in the urinary tract. The probe consists of a catheter to which is attached a magnetostrictive transducer. When the catheter encounters the calculus within the tract, an impulse is transmitted through the catheter to the transducer. The signal is amplified and indicated by an ammeter.

A device based upon the Doppler principle is used to indicate blood flow and motions of internal organs such as the heart and heart valves. This is a popular instrument and many makes are available.

There are many therapeutical applications of ultrasonics. Ultrasonic therapy is administered to athletes for treating various sorenesses and injuries. It is used after injections to disperse the injected fluids. It has been used effectively for treating warts and laryngeal papilloma. It provides relief for various types of rheumatic diseases. Ultrasonics has a relaxing and sedative effect which is useful in relieving cramps or "Charley horses".

Blood clots exposed to high-intensity ultrasonic energy dissolve very rapidly. One method of introducing the energy into the clot is to bring a wirelike tool into contact with the clot through a catheter.

Atomization of water and inhalants by ultrasonic means has clinical value. At high frequencies, the nebulized particles may be small enough to be transported into the alveoli where they are absorbed. The liquid is fed to the atomizer by gravity, and little heat is produced.

Surgical applications include ear surgery, pituitary surgery, and neurosurgery. Ultrasonics has been used to treat diseases such as Parkinson's disease, and this treatment has resulted in the elimination of the tremors.

For a short period of time, ultrasonics was used in the preparation of dental cavities. Modern dental equipment has forced an end to this application, but the same equipment is used for removing dental calculus.

REFERENCES

1. Lord Rayleigh, The Theory of Sound, 1st Am. Ed., Vols. 1 and 2, Dover, New York, 1945.
2. J. P. Joule, Phil Mag., [III], 30, 76 (1847).
3. J. Curie and P. Curie, Compt. Rend. Acad. Sci., Paris, 91, 294 (1880).
4. J. D. Pye, Inst. Laryngol. and Otol., 10, 114-125 (1959-1960).
5. H. Hartridge, J. Physiol., 54, 54 (1920).
6. J. D. Pye and M. Flinn, Ultrasonics, 2, 23-28 (1964).
7. J. D. Pye, Endeavour, 20, 101-111 (1961)
8. J. D. Pye, Nature, 190, 362-363 (1961).
9. J. J. G. McGue, Ind. Photog., 10, 48-54 (1961).
10. K. D. Roeder and A. E. Treat, J. Exptl. Zoo., 134, 127-158 (1957).
11. W. N. Kellogg, Porpoises and Sonar, Phoenix Books, University of Chicago Press, Chicago and London, 1965.
12. F. A. Firestone, U.S. Pat. 2,280,226 (April 21, 1942).
13. L. Balamuth, U.S. Pat. 2,580,716 (January 1, 1952).

Chapter 2

ELASTIC WAVE PROPAGATION AND ASSOCIATED PHENOMENA

I. INTRODUCTION

Designing for the effective use of ultrasonic energy requires an understanding of the basic principles of wave propagation and associated phenomena. Because of their importance in the applications of ultrasonics, these principles are presented in some detail. This chapter is devoted to describing these phenomena. The succeeding chapter is devoted to derivations and solutions of those equations that describe propagation characteristics of various wave types and which are used in ultrasonic designs and analyses. Chapter 4 includes equations and procedures for designing certain components, particularly solid horns, or mechanical amplifiers, for high-intensity applications. The reader should find the derivations of the equations as well as the equations themselves of considerable value in ultrasonic designs and in clarifying various statements and equations which he may encounter whether in this book or elsewhere.

Ultrasonics touches on nearly every aspect of human endeavor. The phenomena that make this wide range of interest possible have logical scientific explanations. The factors that affect the propagation of ultrasonic energy are basic to designing systems that will produce the phenomena desired and will do this efficiently. A study of the propagation characteristics themselves yields much information about the media through which the wave travels. This chapter is intended to acquaint the reader with some of the fundamental principles involved in the various applications of ultrasonic energy discussed in later chapters.

In many respects, ultrasound is similar to light and other forms of electromagnetic radiation in that it is a wave motion and obeys a general wave equation. Each wave type in a given homogeneous medium travels at a velocity that is dependent upon the properties of the medium. Similarly to light, ultrasonics is reflected from surfaces, refracted when going from one medium into another which effects change in the velocity of sound, and diffracted at the edges of surfaces or around obstacles. Energy is scattered from particles or rough surfaces just as light is scattered from similar surfaces or dust particles.

In many respects, ultrasonics is like electricity. Forces acting across an area at a given point in the wave are analogous to electrical voltage and the velocity potential at that point is analogous to current. Just as the ratio of voltage to current is an electrical impedance, the ratio of force to velocity is an acoustical impedance. The impedances in the two cases are used in similar ways, for example, to calculate reflections and transmissions at an impedance discontinuity or to match components for effective energy transfer from one element to another. When a stationary impedance in a transmission line produces a reflection that is everywhere in phase with an incident, steady-state wave, the two waves reinforce each other to produce a standing wave. This is the condition at resonance which is important in certain interferometer types of measurements and in some types of high-intensity applications.

At a given frequency, the particle motion at any point in an ultrasonic wave is sinusoidal if the stresses developed in the waves remain in the linear, elastic range of the medium. If two sinusoidal waves of slightly differing frequencies are superimposed, their amplitudes are alternately added and subracted so that the overall effect is a wave with an amplitude equal to the sum of the amplitudes of the individual waves and a frequency equal to the difference in the frequencies of the same two waves. The effect is called beats and the new frequency is called a beat frequency. The beat frequency is used effectively with the Doppler effect in many applications such as measuring velocity of blood flow, measuring the velocity of heart actions, or detecting the movement of intruders. The Doppler effect is a change in frequency between an initial wave and a received wave occurring as a result of a changing path length between a source of sound and an observer and can be caused by a moving reflector even when the source and the receiver are stationary. The beats are produced between the initial outgoing wave and the received wave in which the frequency has been shifted by the Doppler effect.

As a wave propagates through a medium, its amplitude decreases, or attenuates. There are several causes of this attenuation, such as spreading of the wave front, change of the acoustical energy to heat (absorption), and scattering from irregular surfaces. One factor that affects attenuation is relaxation, a term that describes the lag between an initiating disturbance and a readjustment of energy distribution induced by the disturbance, for example, heat flow from a region under compression to a lower pressure region. A finite time is involved and the exchange of energy approaches equilibrium value exponentially. A study of relaxation phenomena yields considerable information regarding the nature of the solid, liquid, and gaseous states of matter.

At high intensities of ultrasound, the absorbed energy can produce considerable heat; in fact, at extremely high intensities it can produce enough heat to rapidly melt glass or steel. Another phenomenon associated with high intensities in liquids is cavitation, to which many of the effects produced are related. Cavitation is associated with the stresses generated in the ultrasonic waves and their rate of change. Free chemical radicals are produced with cavitation which promotes certain chemical reactions. The stresses developed are sufficiently strong to erode hard and tough materials. Cavitation is especially helpful in ultrasonic cleaning.

These principles are discussed more fully in later sections of this chapter.

II. POWER DELIVERED TO AN OSCILLATING SYSTEM

An ultrasonic wave may be considered to consist of an infinite number of oscillating masses, or elements, connected by means of elastic springs so that each element is influenced by the motion of its nearest neighbor. To induce oscillation in each mass requires an initial power input, i.e., introduction of energy at a given rate.

Consider a simple oscillating system, driven by an external source of power. The instantaneous power delivered to the system by a driving force is given by

$$P_i = F_i v_i \tag{2.1}$$

where F_i is the instantaneous driving force and v_i is the resulting velocity. The instantaneous power P, dissipated, is the product of the real parts of F_i and v_i, or, for the simple oscillator of Eq. (2.1)

$$\begin{aligned} P &= (F \cos \omega t)[F \cos (\omega t - \phi)] \\ &= F^2 \cos \omega t \cos (\omega t - \phi) \end{aligned} \tag{2.2}$$

where ω is $2\pi f$, ϕ is the phase angle between the driving force F and the velocity v, and f is the frequency of oscillations.

The average power P_{ave} dissipated in the system is equal to the total work done per cycle divided by the period T, or

$$\begin{aligned} P_{ave} &= \frac{\int_0^T P\, dt}{T} \\ &= (F^2/TZ_m)\int_0^T \cos \omega t \cos(\omega t - \phi)dt \end{aligned}$$

$$= (F^2/2Z_m) \cos \phi \tag{2.3}$$

where Z_m is the complex mechanical impedance of the system.

When a vibratory system is driven by a constant oscillatory force, the initial energy is partitioned between a component that is used to overcome the losses of the system and a second component that is used to change the vibratory state of the system. Energy from the latter component is stored in the system until the system reaches an equilibrium where the energy supplied exactly equals the energy dissipated.

In Eq. (2.3), the term $\cos \phi$ ($= R_m/Z_m$) is the mechanical power factor.

III. THE VELOCITY OF SOUND

In keeping with the analogy of a series of an infinite number of mass elements in an ultrasonic wave, when a disturbance is induced in a mass, the first element affected transfers energy to the next one in line, the second influences the third in a similar manner, and so on as the energy introduced passes from element to element (propagates) through the medium until it is dissipated. The elements of the entire mass do not move in unison. The reasons for this are (1) that a mass has inertia and accelerates at a rate corresponding to the applied force according to Newton's laws of motion and (2) that it is elastic and deforms under stress. Thus the disturbance propagates through the medium at the velocity of sound.

The term "velocity of sound" requires some qualification. The nature of the disturbance is a factor in this velocity; i.e., the type of oscillatory motion that the disturbance produces is a factor in the speed at which the disturbance travels through the medium. The rate of propagation depends upon the type of wave, the elastic properties of the medium, the density of the medium, and, in some cases, the frequency. Since sound, or ultrasound, represents energy transmitted as stress waves, the velocity of sound also depends upon the mode of vibration with which it is associated. Gases are capable of transmitting only compressional (or longitudinal) waves. Most liquids transmit only longitudinal waves and surface waves. Solids may transmit a large variety of wave modes.

III.A. VELOCITY OF SOUND IN SOLIDS

The velocity of sound in a long, slender bar in which the particle motion is parallel to the axis of the bar (longitudinal thin-bar velocity) is given by

$$c_o = (Y/\rho)^{1/2} \tag{2.4}$$

where Y is Young's modulus of elasticity (dyn/cm^2) and ρ is the density of the bar (g/cm^3). The dimensions of c_o are in centimeters per second.

When an element in a slender bar is compressed axially it is free to expand radially, and when it is in tension it contracts radially. If a small element in a large mass of large cross-sectional area and length is compressed in one direction, its expansion in all other directions is limited by the surrounding mass. The effect is to increase the bulk modulus over that of Young's, or bar, modulus by a function of Poisson's ratio. There is a similar effect on the velocity of sound, which becomes

$$c_B = c_o \; [(1 - \sigma)/(1 + \sigma)(1 - 2\sigma)]^{1/2} \tag{2.5}$$

where c_B is longitudinal bulk velocity and σ is Poisson's ratio.

When the particle motion in the wave is normal to the direction of propagation, the energy passes from element to element by shear stresses and the wave is called a shear wave. In this case, the shear modulus, μ, is a factor in the velocity. Shear modulus may be related to Young's or bulk modulus as functions of Poisson's ratio. The relationships result in the following equation for shear-wave velocity, c_s:

$$c_s = c_o[2(1 + \sigma)]^{-1/2} = (\mu/\rho)^{1/2} \tag{2.6}$$

Equation (2.6) also applies to torsional waves, since these are shear waves.

Some ultrasonic waves follow the contour of the surface of a medium and penetrate the medium only to shallow depths. These waves are known as surface waves. Two types are of interest in the application of ultrasonics. These are Love waves and Rayleigh waves. The particle motion in Love waves is normal to the direction of propagation and is polarized in the plane of the surface. Rayleigh wave motion combines a longitudinal component with a shear component which is normal to the surface and the particle motion is elliptical. Since the restraints differ somewhat from those within the medium, the velocity of surface waves also differs from that of either shear or longitudinal waves. The velocity of Rayleigh waves, which are of the type that is most commonly used in applications of ultrasonic energy, is slightly lower than that of shear waves in the same medium. The velocity of Rayleigh waves is given by

$$c_R = K_R c_s \tag{2.7}$$

where K_R is a constant determined by the relationship between the longitudinal velocity and the shear velocity, and, therefore, Poisson's ratio. The relationship between c_B, c_s, and c_R is given by

$$\left(\frac{c_R}{c_s}\right)^2 \left\{\frac{c_R^6}{c_s^6} - \frac{8c_R^4}{c_s^4} + c_R^2\left(\frac{24}{c_s^2} - \frac{16}{c_B^2}\right) - 16\left(1 - \frac{c_s^2}{c_B^2}\right)\right\} = 0 \tag{2.8}$$

where c_R is the particular solution for which its value is slightly less than c_s. The effective depth of penetration of surface waves is approximately one wavelength.

Also used in ultrasonic applications are Lamb waves, which propagate in thin plates. There are an infinite number of velocities for Lamb waves, since the velocity of propagation of these waves depends upon the relationship between the wavelength and the thickness of the material. These waves are discussed in greater detail in Chap. 3.

III.B. VELOCITY OF SOUND IN LIQUIDS

Generally, shear waves are rapidly attenuated in liquids. Surface waves at ultrasonic frequencies in liquids are seldom of interest to the ultrasonics engineer. Therefore, the velocity of sound in a liquid usually refers to a longitudinal type of wave mode. The velocity of sound, c (cm/sec) in liquids is

$$c = (\gamma/\rho\beta_i)^{\frac{1}{2}} = (\rho\beta_s)^{-\frac{1}{2}} \tag{2.9}$$

where $\gamma = C_p/C_v$ is the ratio of specific heats, C_p is specific heat at constant pressure, C_v is specific heat at constant volume, β_s is the adiabatic compressibility (cm^2/dyn).

Equation (2.9) applies at low pressure amplitudes (acoustic intensities). At high pressure levels (finite amplitude wave intensities), propagation characteristics become more complex. Finite amplitude wave theory is applied to shock waves and the velocity of these waves is always higher than the acoustic velocity. In general, only acoustic velocities will be considered in this book.

Measurement of the velocity of sound in a liquid provides a means of determining the adiabatic compressibility of a liquid and the ratio of specific heats of the liquid.

III.C. VELOCITY OF SOUND IN GASES

The equation for the velocity of sound in gas is derived by the equation of state. All fluids (gases and liquids) and solids are characterized by an equation of state of the form

$$f\ (p,V,T) = 0 \tag{2.10}$$

where p is the pressure exerted by the system (gas, liquid, or solid), V is its volume, and T is its temperature on some suitable scale. It can be shown from Eq. (2.10) that

$$\left(\frac{\partial p}{\partial V}\right)_T \left(\frac{\partial V}{\partial T}\right)_p \left(\frac{\partial T}{\partial p}\right)_V = -1 \tag{2.11}$$

Generally, the propagation of ultrasonic energy in gases (or in liquids) is adiabatic because the variations in pressure and corresponding volume changes are so rapid that heat does not have time to flow from high-pressure regions to rarefaction regions, and therefore, no heat is lost in the process.

For an ideal gas, the equation of state is

$$pV = \frac{N}{N_o} RT \tag{2.12}$$

where p is the pressure (g/cm^2), V is the volume of the gas (cm^3), N is the number of molecules in V, N_o is the number of molecules in a gram-molecular weight of the gas, R is the gas constant, T is absolute temperature, (°K). The ratio R/N_o is called Boltzmann's constant.

The velocity of sound, c, in a gas is given by the relation

$$c^2 = \gamma \left(\frac{\partial p}{\partial \rho} \right)_T \tag{2.13}$$

where c is velocity in centimeters per second.

Therefore, from Eq. (2.12), the equation for the velocity of sound in an ideal (perfect) gas is

$$c^2 = \gamma \; RT/M \tag{2.14}$$

where M is the molecular weight.

Roberts and Miller (1) give the following relationship for the velocity of sound, c, in a real gas

$$c^2 = \gamma \frac{RT}{M} \left\{ 1 + \frac{9}{64} \frac{p}{p_c} \frac{T_c}{T} \left(1 - 6 \frac{T_c^2}{T^2} \right) \right\} \tag{2.15}$$

where T_c is the critical temperature of the gas (°K) and p_c is the critical pressure of the gas (g/cm^2). Measurement of velocity of sound in a gas is one of the most convenient and accurate methods of measuring specific heats.

IV. IMPINGEMENT OF AN ULTRASONIC WAVE ON A BOUNDARY BETWEEN TWO MEDIA

When an ultrasonic wave encounters an interface between two media, the energy of the wave is partitioned in a manner that depends upon the type of incident wave, upon how the wave approaches the interface, and upon the acoustic properties of the two media. The condition is generally more complicated than the optical analogue of a light beam incident on the surface of a transparent object. As with light, Snell's law is used to determine angles of reflection and refraction, but the acoustic problem is complicated by the greater number of wave modes and by the longer wavelengths usually associated with ultrasonic energy. Reflection and refraction influence all aspects of the propagation and application of ultrasonic energy and, thus, are an important factor in the applications discussed in this book.

The characteristic that leads to Snell's law is that, along a nonslip boundary, the phase velocity along the interface must be the same for each wave in order to fulfill the boundary conditions on displacements and stresses (see Fig. 2.1). The phase velocity is the rate at which a point of constant phase travels along the boundary and is

$$c_p = \frac{c_a}{\sin \alpha_a} \tag{2.16}$$

where c_a is the velocity of propagation of the particular wave under consideration and α_a is the angle between the wave front and the boundary surface.

According to Eq. (2.16),

$$c_p \geq c_a \tag{2.17}$$

Mode conversion occurs in solids because solids can sustain both shear and dilatational stresses. The types of waves that can be produced by a wave incident on an interface may be determined by resolving the stress and displacement components of the incident wave along the coordinates of the boundary. These stress and displacement requirements fix the boundary conditions for the particular solutions of the general wave equation by which it is possible to determine the parameters of the waves reflected from the interface and of the waves transmitted across the interface. Four separate boundary conditions exist for the general case. On both sides of the interface, the following quantities must be equal:

1. the normal displacements
2. the tangential displacements
3. the normal stresses
4. the tangential stresses.

IV.A. SIMPLE REFLECTION AND TRANSMISSION AT NORMAL INCIDENCE

When a plane wave impinges at normal incidence on an interface between two semi-infinite media, as in Fig. 2.2, the energy of the incident wave I is partitioned between a reflected wave R and a transmitted wave T. At the boundary, the total displacement is the same in either medium. Then

$$\zeta_T = \zeta_i + \zeta_R \tag{2.18}$$

where ζ_T, ζ_i, and ζ_R are potential functions defined as

$\zeta_T = \zeta_{To} \sin \omega \left(t - \frac{x}{c_2}\right)$, the displacement of the transmitted wave in Medium 2;

$\zeta_i = \zeta_o \sin \omega \left(t - \frac{x}{c_1}\right)$, the displacement of the incident wave;

$\zeta_R = \zeta_{Ro} \sin \omega \left(t + \frac{x}{c_1}\right)$, the displacement of the reflected wave.

Similarly, the stresses on one side of the interface are equal to those on the other side, i.e.,

$$p_T = p_i + p_R \tag{2.19}$$

where p_T, p_i, and p_R also are potential functions of stress and are related to the displacements according to the following equation:

$$p = -j \frac{\rho c^2}{\omega} \frac{\partial \dot{\zeta}}{\partial x} \tag{2.20}$$

where $\dot{\zeta} = \frac{\partial \zeta}{\partial t}$. (2.21)

From Eqs. (2.19) through (2.21), we obtain

$$\left.\begin{aligned} p_T &= -\,j\,\omega\rho_2 c_2 \zeta_T \\ p_i &= -j\,\omega\rho_1 c_1 \zeta_i \\ p_R &= j\,\omega\rho_i c_1 \zeta_R \end{aligned}\right\} \quad (2.22)$$

Substituting Eqs. 2.22 in Eq. 2.19 gives

$$\rho_2 c_2 \zeta_T = \rho_1 c_1 (\zeta_i - \zeta_R \quad (2.23)$$

where $\rho_1 c_1$ and $\rho_2 c_2$ are the specific acoustic impedances of Medium 1 and Medium 2, respectively, and c_1 and c_2 are the velocities of wave propagation in Medium 1 and Medium 2, respectively.

At the interface, x = 0, and from Eq. (2.18),

$$\zeta_{To} = \zeta_o + \zeta_{Ro} \quad (2.24)$$

Then

$$\frac{\zeta_T}{\zeta_i} = [1 - \frac{\zeta_R}{\zeta_i}] \frac{\rho_1 c_1}{\rho_2 c_2} \quad (2.25)$$

From Eq. (2.18)

$$\frac{\zeta_R}{\zeta_i} = \frac{\zeta_T}{\zeta_i} - 1 \quad (2.18a)$$

Therefore

$$\frac{\zeta_T}{\zeta_i} = [\,2 - \frac{\zeta_T}{\zeta_i}\,] \frac{\rho_1 c_1}{\rho_2 c_2} \quad (2.25a)$$

Substituting for

$\frac{\zeta_T}{\zeta_i}$ and simplifying gives

$$\frac{\zeta_T}{\zeta_i} = \frac{2\rho_1 c_1}{\rho_1 c_1 + \rho_2 c_2} \quad (2.26)$$

Again, from Eq. (2.18a)

$$\frac{\zeta_R}{\zeta_i} = \frac{\zeta_T}{\zeta_i} - 1 = \frac{2\rho_1 c_1}{\rho_1 c_1 + \rho_2 c_2} - 1 = \frac{\rho_1 c_1 - \rho_2 c_2}{\rho_1 c_1 + \rho_2 c_2} \tag{2.27}$$

The relationships between the stresses in the waves are found by equating the stresses to their corresponding displacements and impedances and substituting in Eqs. (2.23) and (2.24). Then:

$$\frac{p_T}{p_i} = \frac{\zeta_T \rho_2 c_2}{\zeta_i \rho_1 c_1} = \frac{2\rho_2 c_2}{\rho_1 c_1 + \rho_2 c_2} \tag{2.28}$$

and

$$\frac{p_R}{p_i} = \frac{-\zeta_R}{\zeta_i} \frac{\rho_1 c_1}{\rho_1 c_1} = \frac{\rho_2 c_2 - \rho_1 c_1}{\rho_1 c_1 + \rho_2 c_2} \tag{2.29}$$

According to Eqs. (2.26) through (2.29), no phase shift occurs between the incident wave and the transmitted wave either in the displacement or in the stress, regardless of which medium has the higher acoustic impedance. However, when the acoustic impedance of Medium 2 is greater than that of Medium 1, the displacement of the reflected wave is 180° out of phase with that of the incident wave, but the stresses are in phase. The reverse is true if the acoustic impedance of Medium 2 is less than that of Medium 1.

More general equations which apply in the case of finite media are

$$\frac{\zeta_R}{\zeta_c} = \frac{z_1 - z_2}{z_1 + z_2} \tag{2.27a}$$

$$\frac{p_T}{p_i} = \frac{2z_2}{z_1 + z_2} \tag{2.28a}$$

$$\frac{p_R}{p_i} = \frac{z_2 - z_1}{z_1 + z_2} \tag{2.29a}$$

where z_1 and z_2 are the complex acoustic impedances of the respective media.

In practice, the actual situations are often considerably more complicated than the ideal conditions described in this section. Reflections and refraction at interfaces between media enter into every aspect of the applications of ultrasonic energy. For this reason this topic is given extra consideration here.

IV.B. SOME BASIC MECHANICS

As a preface to the analysis of stress waves incident at any angle on an interface between two media, consider the mechanics of stresses in a block. A block of width ℓ, of height dy and of length dx (see Fig. 2.3) is subjected to a compressive stress p on faces AB and CD. The resultant stresses on the plane AC are p' and s'. The total applied force is $p\ell dx$. To maintain equilibrium, the force $p\ell dx$ is resisted by a force $(p'\ell ds \cos\theta + s'\ell ds \sin\theta)$ neglecting Poisson's ratio, or

$$p'\ell ds \cos\theta + s'\ell ds \sin\theta = p\ell dx \qquad (2.30)$$

where p' is the compressive stress normal to the plane AC and s' is the shear stress parallel to the plane. By substituting $(dx/ds) = \cos\theta$, Eq. (2.30) can be rewritten

$$p' = p - s'(\sin\theta/\cos\theta) \qquad (2.30a)$$

In the horizontal direction, normal to the direction of p

$$p'\ell ds \sin\theta = s'\ell ds \cos\theta \qquad (2.31)$$

from which

$$p' = s'(\cos\theta/\sin\theta) \qquad (2.31a)$$

Substituting Eq. (2.31a) into Eq. (2.30a) and solving for s' in terms of p gives

$$p = s'(\sin\theta/\cos\theta + \cos\theta/\sin\theta) = s'\left(\frac{2}{\sin 2\theta}\right) \qquad (2.32)$$

or

$$s' = \frac{p \sin 2\theta}{2}$$

Substituting Eq. (2.32) into Eq. (2.31a) and solving for p' in terms of p, we obtain

$$p' = p \cos^2 \theta = p\left(\frac{1 + \cos 2\theta}{2}\right) \tag{2.33}$$

Now assume that the same two faces AB and CD are subjected to a pure shear stress, s, instead of p, as shown in Fig. 2.4. The horizontal forces acting on the diagonal AC at equilibrium are

$$(p' \sin \theta + s' \cos \theta)\ell ds = s\ell dx \tag{2.34}$$

and the vertical forces are

$$p'\ell ds \cos \theta = s'\ell ds \sin \theta \tag{2.35}$$

From Eqs. (2.34) and (2.35)

$$\left.\begin{aligned} s' &= s \cos^2 \theta = s\left(\frac{1 + \cos 2\theta}{2}\right) \\ &\text{and} \\ p' &= \frac{s}{2} \sin 2\theta \end{aligned}\right\} \tag{2.36}$$

The same procedure can be used to determine the stresses produced on any plane by stresses acting on any of the other faces of the block.

If both p and s were acting on the faces AB and CD simultaneously, the resultant stresses acting on the plane AC would be the sum of those caused by p and by s, as though each were applied independently of the other. However, it is essential that the directions of the stresses be identified by proper signs. For instance, combining p and s of Figs. 2.3 and 2.4 and defining the positive directions above the diagonal AC as being downward and toward the left, we see that the normal stress on AC is the sum of the stresses p' given by Eqs. (2.33) and (2.36) or

$$p'_{(sum)} = p\left(\frac{1 + \cos 2\theta}{2}\right) + \frac{s}{2} \sin 2\theta \tag{2.37}$$

The total shear stress on diagonal AC is s' calculated by Eq. (2.36) minus s' calculated by Eq. (2.32), since these are in opposite directions. Then

$$s'_{(sum)} = s\left(\frac{1 + \cos 2\theta}{2}\right) - p\,\frac{\sin 2\theta}{2} \tag{2.38}$$

If $s'_{(sum)}$ is positive, its direction is toward the left, as previously defined, and if it is negative, it is toward the right. When $\theta = 0$, $s' = s$, and $p' = p$.

Numerous relationships are possible between ultrasonic waves incident on an interface between two media and the reflected and refracted waves that result. These are dependent upon the directions of the stresses and strains associated with the various waves and the acoustic properties of the materials. Only a brief discussion of these possibilities is presented in this chapter. For a detailed analysis the reader should consult Redwood (2).

IV.C. GENERAL CONSIDERATIONS OF INCIDENT WAVES

Figures 2.5 and 2.6 show the possible directions of strain in waves incident on a plane boundary between two media and the resulting reflected and refracted waves. Figure 2.5 illustrates a longitudinal wave impinging on the interface, and Fig. 2.6 illustrates an incident shear wave. The two conditions are shown separately for clarity.

A single equation may be derived by which the amplitude of each wave reflected or refracted may be related to that of the incident wave by applying the four boundary conditions of continuity of stress and strain across the interface to appropriate solutions to the wave equations (Sec. IV.A). This equation may be used to obtain a specific relationship by applying further conditions relative to the media and to the particular incident wave mode. For instance, a vacuum can propagate no ultrasonic energy. Therefore, waves incident on an interface between a vacuum and any other medium are reflected completely. In this case, both S_t and L_t are zero.

Generally, liquids can propagate no shear waves. Therefore, all shear wave components are zero in such a liquid medium.

Both shear and longitudinal waves can propagate through

isotropic solids. The elastic constants in an isotropic solid are equal in all directions. Therefore, only one shear-wave velocity and one longitudinal-wave velocity are possible in an unbounded isotropic medium. The ratio of the velocity of propagation of the longitudinal wave to that of the shear wave is

$$\frac{c_L}{c_s} = \sqrt{\frac{2(1 + \sigma)}{1 - 2\sigma}} \tag{2.39}$$

where σ is Poisson's ratio.

Both shear and longitudinal waves can propagate through anisotropic solids. The velocities of propagation through such media depend upon the orientations of the strains with respect to the crystallographic orientations of the media. Macroscopic properties of the media depend upon the grain orientation (random or oriented). Propagation characteristics, attenuation, and velocity are affected by the relationship between the wavelength and the grain size.

IV.D. DEVELOPMENT OF GENERAL EQUATIONS FOR REFLECTION AND REFRACTION WHERE MODE CONVERSION IS POSSIBLE

A wave equation may be written for each of the types of waves which may be propagated in the media. Redwood (2) uses the scalar potential ϕ, corresponding to longitudinal waves, and the vector potential ψ, corresponding to shear waves. The corresponding wave equations are

$$\nabla^2\phi = \frac{1}{c_L^{\,2}} \frac{\partial^2\phi}{\partial t^2} \tag{2.40}$$

$$\nabla^2\psi = \frac{1}{c_s^{\,2}} \frac{\partial^2\psi}{\partial t^2} \tag{2.41}$$

where $c_L^{\,2} = \frac{\lambda' + 2\mu}{\rho}$, $c_s^{\,2} = \frac{\mu}{\rho}$ and λ' and μ are Lame's elastic constants, μ being the elastic shear modulus mentioned in Sec. III. A. of this chapter.

To obtain the general solutions to the wave equations, using rectangular coordinates, we recall that all of the waves are related to the phase velocity c_p. Referring to Eq. (2.16) and Figs. 2.5 and 2.6, the phase velocity is

$$c_p = \frac{c_{1L}}{\sin \alpha_1} = \frac{c_{1s}}{\sin \alpha_{s1}} = \frac{c_{2L}}{\sin \alpha_{L2}} = \frac{c_{2s}}{\sin \alpha_{s2}} \tag{2.42}$$

The most general solutions for the potential functions are:

$$\phi_1 = \left\{ \zeta_{Li} e^{-j(\frac{\omega}{c_{1L}} \cos \alpha_1) x} + \zeta_{Lr} e^{j(\frac{\omega}{c_{1L}} \cos \alpha_1) x} \right\} e^{-j(\frac{\omega}{c_{1L}} \sin \alpha_1) z} e^{j\omega t} \tag{2.43}$$

$$\psi_1 = \left\{ \zeta_{si} e^{-j(\frac{\omega}{c_{1s}} \cos \alpha_{s1}) x} + \zeta_{sr} e^{j(\frac{\omega}{c_{1s}} \cos \alpha_{s1}) x} \right\} e^{-j(\frac{\omega}{c_{1s}} \sin \alpha_{s1}) z} e^{j\omega t} \tag{2.44}$$

$$\phi_2 = \zeta_{LT} e^{-j(\frac{\omega}{c_{2L}} \cos \alpha_{L2}) x} e^{-j(\frac{\omega}{c_{2L}} \sin \alpha_{L2}) z} e^{j\omega t} \tag{2.45}$$

$$\psi_2 = \zeta_{sT} e^{-j(\frac{\omega}{c_{2s}} \cos \alpha_{s2}) x} e^{-j(\frac{\omega}{c_{2s}} \sin \alpha_{s2}) z} e^{j\omega t} \tag{2.46}$$

in which the normals to the wave fronts are in the xz plane, and the interface lies in the yz plane; and where

ζ_{Li} is the displacement amplitude of the incident longitudinal wave in the direction of wave propagation,

ζ_{Lr} is the displacement amplitude of the reflected longitudinal wave in the direction of wave propagation,

ζ_{si} is the displacement amplitude of the incident shear wave in a direction normal to the direction of wave propagation,

ζ_{sr} is the displacement amplitude of the reflected shear wave in a direction normal to the direction of wave propagation,

ζ_{LT} is the displacement amplitude of the transmitted longitudinal wave in the direction of wave propagation, and

ζ_{sT} is the displacement amplitude of the transmitted shear wave normal to the direction of wave propagation.

Referring to Fig. 2.7, the displacements are assumed to be parallel to the xz plane and the interface is in the yz plane. According to the boundary conditions (at x = 0)

$$(\zeta_{Li} + \zeta_{Lr}) \sin \alpha_1 + (\zeta_{si} - \zeta_{sr}) \cos \alpha_{s1} = \zeta_{sT} \cos \alpha_{s2} + \zeta_{LT} \sin \alpha_{L2} \tag{2.47}$$

(displacements parallel to interface)

$$(\zeta_{Li} - \zeta_{Lr}) \cos \alpha_1 - (\zeta_{si} + \zeta_{sr}) \sin \alpha_{s1} = \zeta_{LT} \cos \alpha_{L2} - \zeta_{sT} \sin \alpha_{s2} \tag{2.48}$$

(displacements normal to interface)

The shear stresses and the tensile stresses due to both shear and longitudinal waves are determined (1) by algebraically summing the stresses parallel to the interface and acting through the interface, i.e., those components of stress causing shear on the interface, and (2) by summing the stresses acting on and normal to the interface.

First, consider the conditions for which all displacements and the direction of propagation are parallel to the xz plane and for which the interface lies in the yz plane (Fig. 2.7).

The components of stress acting parallel to the interface are given by

$$s_s = -j \frac{\rho c_s^2}{\omega} \left(\frac{\partial^2 \psi}{\partial x \partial t} + \frac{\partial^2 \phi}{\partial x \partial t}\right)_z \tag{2.49}$$

Using the directions for displacements of Fig. 2.7, the shear displacements parallel to z caused by the longitudinal waves are $\zeta_L \sin \alpha_L$ and those caused by the shear waves are $\zeta_s \cos \alpha_s$. In the equations that follow, the positive direction is considered to be to the right and downward in both displacements and propagation.

From Eq. (2.43) and Fig. 2.7

$$\left.\frac{\partial^2\phi_1}{\partial x\partial t}\right|_z = \left(\frac{\omega^2}{c_{1L}}\cos\alpha_1\right)\left\{\zeta_{Li}e^{-j\left(\frac{\omega}{c_{1L}}\cos\alpha_1\right)x} - \zeta_{LR}e^{j\left(\frac{\omega}{c_{1L}}\cos\alpha_1\right)x}\right\}$$

$$\times e^{-j\left(\frac{\omega}{c_{1L}}\sin\alpha_1\right)z}\, e^{j\omega t}\,\sin\alpha_1 \tag{2.50a}$$

and at x = 0

$$\left.\frac{\partial^2\phi_1}{\partial x\partial t}\right|_z = \frac{\omega^2}{2c_{1L}}\left\{(\zeta_{Li}-\zeta_{Lr})\sin 2\alpha_1\right\}e^{-j\left(\frac{\omega}{c_p}\right)z}\, e^{j\omega t} \tag{2.50b}$$

Similarly, from Eq. (2.44) through (2.46) and Fig. 2.7 at x = 0

$$\left.\frac{\partial^2\psi_1}{\partial x\partial t}\right|_z = \frac{\omega^2}{2c_{1s}}\left\{(\zeta_{si}+\zeta_{sr})(1+\cos 2\alpha_{s1})\right\}e^{-j\left(\frac{\omega}{c_p}\right)z}\, e^{j\omega t} \tag{2.51}$$

$$\left.\frac{\partial^2\phi_2}{\partial x\partial t}\right|_z = \frac{\omega^2}{2c_{2L}}\left\{\zeta_{LT}\sin 2\alpha_{L2}\right\}e^{-j\left(\frac{\omega}{c_p}\right)z}\, e^{j\omega t} \tag{2.52}$$

and

$$\left.\frac{\partial^2\psi_2}{\partial x\partial t}\right|_z = \frac{\omega^2}{2c_{2s}}\left\{\zeta_{sT}(1+\cos 2\alpha_{sz})\right\}e^{-j\frac{\omega}{c_p}z}\, e^{j\omega t} \tag{2.53}$$

where c_p is phase velocity as defined previously, subscripts 1 and 2 refer to waves in Medium 1 and Medium 2, respectively, and subscripts L and s refer to longitudinal and shear waves, respectively. The displacements ζ are the actual amplitudes of the displacements.

Applying Eq. (2.49) to Eqs. (2.50) through (2.53) and setting the sum of the stresses due to the waves in Medium 1 equal to the sum of those in Medium 2, we obtain

$$\rho_1 c_{1s}^2\left\{(\zeta_{Li} - \zeta_{Lr})\sin 2\alpha_1 + \left(\frac{c_{1L}}{c_{1s}}\right)(\zeta_{si} + \zeta_{sr})(1 + \cos 2\alpha_{s1})\right\} =$$

$$\rho_2 c_{2s}^2\left\{\left(\frac{c_{1L}}{c_{2L}}\right)\zeta_{LT}\sin 2\alpha_{L2} + \left(\frac{c_{1L}}{c_{2s}}\right)\zeta_{sT}(1 + \cos 2\alpha_{s2})\right. \qquad (2.54)$$

for stresses parallel to the interface.

When $\alpha_1 = \alpha_{s1} = \alpha_{L2} = \alpha_{s2} = 0$

$$\rho_1 c_{1s} c_{1L}(\zeta_{si} + \zeta_{sr}) = \rho_2 c_{2s} c_{1L}$$

Therefore, Eq. (2.54) may be written

$$\rho_1 c_{1s}^2\left\{(\zeta_{Li} - \zeta_{Lr})\sin 2\alpha_1 + \left(\frac{c_{1L}}{c_{1s}}\right)(\zeta_{si} + \zeta_{sr})\cos 2\alpha_{s1}\right\} =$$

$$\rho_2 c_{2s}^2\left\{\left(\frac{c_{1L}}{c_{2L}}\right)\zeta_{LT}\sin 2\alpha_{L2} + \left(\frac{c_{1L}}{c_{2s}}\right)\zeta_{sT}\cos 2\alpha_{s2}\right\} \qquad (2.54a)$$

The components of stress acting in a direction normal to the interface are

$$p_L = -j\,\frac{\rho}{\omega}\left\{c_L^2\left(\frac{\partial^2\phi}{\partial x\partial t} + \frac{\partial^2\psi}{\partial x\partial t}\right)_x + (c_L^2 - 2c_s^2)\left(\frac{\partial^2\phi}{\partial z\partial t} + \frac{\partial^2\psi}{\partial z\partial t}\right)\right\} \qquad (2.55)$$

Following the procedures used in obtaining Eqs. (2.54) and (2.54a) and using Fig. 2.7, at x = 0

$$\left.\frac{\partial^2\phi_1}{\partial x\partial t}\right|_x = \left\{\left(\frac{\omega^2}{c_{1L}}\right)[\zeta_{Li} + \zeta_{Lr}]e^{-j\frac{\omega}{c_p}z}e^{j\omega t}\right\}\cos^2\alpha_1 \qquad (2.56)$$

$$\left.\frac{\partial^2\psi_1}{\partial x\partial t}\right|_x = \left\{-\left(\frac{\omega^2}{c_{1s}}\right)[\zeta_{si} - \zeta_{sr}]e^{-j\frac{\omega}{c_p}z}e^{j\omega t}\right\}\sin\alpha_{s1}\cos\alpha_{s1}$$

(2.57)

$$\left.\frac{\partial^2\phi_2}{\partial x\partial t}\right|_x = \left(\frac{\omega^2}{c_{2L}}\right)\zeta_{LT}e^{-j\frac{\omega}{c_p}z}e^{j\omega t}\cos^2\alpha_{L2} \qquad (2.58)$$

$$\left.\frac{\partial^2\psi_2}{\partial x\partial t}\right|_x = -\left(\frac{\omega^2}{c_{2s}}\right)\zeta_{sT} e^{-j\frac{\omega}{c_p}z} e^{j\omega t} \sin\alpha_{s2}\cos\alpha_{s2} \tag{2.59}$$

$$\left.\frac{\partial^2\phi_1}{\partial z\partial t}\right|_z = \left(\frac{\omega^2}{c_{1L}}\right)[\zeta_{Li} + \zeta_{Lr}] e^{-j\frac{\omega}{c_p}z} e^{j\omega t} \sin^2\alpha_1 \tag{2.60}$$

$$\left.\frac{\partial^2\psi_1}{\partial z\partial t}\right|_z = \left(\frac{\omega^2}{c_{1s}}\right)[\zeta_{si} - \zeta_{sr}] e^{-j\frac{\omega}{c_p}z} e^{j\omega t} \cos\alpha_{s1}\sin\alpha_{s1} \tag{2.61}$$

$$\left.\frac{\partial^2\phi_2}{\partial z\partial t}\right|_z = \left(\frac{\omega^2}{c_{2L}}\right)\zeta_{LT} e^{-j\frac{\omega}{c_p}z} e^{j\omega t} \sin^2\alpha_{L2} \tag{2.62}$$

and

$$\left.\frac{\partial^2\psi_2}{\partial z\partial t}\right|_z = \left(\frac{\omega^2}{c_{2s}}\right)\zeta_{sT} e^{-j\frac{\omega}{c_p}z} e^{j\omega t} \sin\alpha_{s2}\cos\alpha_{s2} \tag{2.63}$$

Equating the stresses acting in a direction normal to the interface in Medium 1 to those in Medium 2 determined according to Eq. (2.55) gives

$$\rho_1 c_{1L}[\zeta_{Li} + \zeta_{Lr}]\cos 2\alpha_{s1} - \rho_1 c_{1s}(\zeta_{si} - \zeta_{sr})\sin 2\alpha_{s1} =$$

$$\rho_2 c_{2L}\zeta_{sT}\cos 2\alpha_{s2} - \rho_2 c_{2s}\zeta_{sT}\sin 2\alpha_{s2} \tag{2.64}$$

Equations (2.42), (2.47), (2.48), (2.54), and (2.64) may be used to determine the directions and relative amplitudes of waves reflected or refracted at an interface between two isotropic media if the incidence angle and the relative velocities of sound are known. In using these equations, it is necessary to consider:

(1) which waves can be propagated in the media,

(2) whether the critical angle of any wave mode has been exceeded, in which case that particular mode is completely reflected,

(3) in which direction the incident wave is polarized, if it is a shear wave.

Other phenomena also occur under various conditions. For instance, boundary disturbances or surface waves may be generated which travel along a boundary or surface of a medium. The existence of these waves has been observed by schlieren techniques in steel and aluminum immersed in water even at normal incidence. There are several types of these boundary and surface waves. The two best known types are Love waves, in which the particle motion is parallel to the surface and normal to the direction of propagation, and Rayleigh waves, in which the particle motion is elliptical and parallel to a plane normal to the surface. Both of these types of waves were mentioned in this chapter (Sec. III.A.).

The effective depth of penetration of a surface wave is approximately one wavelength, the intensity dropping exponentially with depth.

Rayleigh waves are often used, as are longitudinal and shear waves, for nondestructive testing.

When the direction of polarization of a shear wave is parallel to the plane of the interface, the components of stress in a direction normal to the interface are zero and no longitudinal waves are generated.

When the direction of polarization of an incident shear wave is at an angle both to the plane of the interface and to a plane normal to the interface which contains the normal to the wave front, the solutions describing reflection and refraction involve displacements corresponding to all three directions in a rectangular coordinate system. The solutions may be obtained in the same manner that Eqs. (2.47), (2.48), (2.54), and (2.64) were derived. In isotropic media, these formulas may be used by applying an angular correction to compensate for the third dimension. Anisotropic media, however, present a much more difficult problem in that the elastic constants vary with the direction. This gives rise to birefringence, that is, the formation of two waves of the same mode and frequency but which travel at different velocities.

A shear wave traveling through an oriented anisotropic material and polarized at an angle to the orientation undergoes a change in polarization due to birefringence. This phenomenon has been investigated as a means of measuring residual stresses in materials, since stress induces anisotropy.

IV.E. WAVE INCIDENT ON A LIQUID/SOLID PLANE INTERFACE, SEMI-INFINITE MEDIA

To demonstrate the use of Eqs. (2.47), (2.48), (2.54), and (2.64) to calculate the relative amplitudes of reflected and transmitted waves, we use the example of an ultrasonic wave traveling through a liquid medium encountering a plane, solid surface at an angle α_1. The solid is isotropic. Under these conditions, the waves transmitted into the solid may be one shear wave, a longitudinal wave, and a boundary disturbance under certain conditions. Only longitudinal waves exist in the liquid. Therefore, for the case of the longitudinal wave in liquid incident on a liquid-solid interface, Equations (2.47), (2.48), and (2.64) become:

$$(\zeta_{Li} + \zeta_{Lr}) \sin \alpha_1 = \zeta_{sT} \cos \alpha_{s2} + \zeta_{LT} \sin \alpha_{L2} \tag{2.47a}$$

$$(\zeta_{Li} - \zeta_{Lr}) \cos \alpha_1 = \zeta_{LT} \cos \alpha_{L2} - \zeta_{sT} \sin \alpha_{s2} \tag{2.48a}$$

$$\rho_1 c_{1L}[\zeta_{Li} + \zeta_{Lr}] = \rho_2 c_{2L}\zeta_{LT} \cos 2\alpha_{s2} - \rho_2 c_{2s}\zeta_{sT} \sin 2\alpha_{s2} \tag{2.64a}$$

The ratios of the displacements in terms of the characteristic impedances (ρc), velocities (c), and the angles of incidence, reflection, and refraction are obtained by solving Eqs. (2.47a), (2.48a), and (2.64a) simultaneously.

When $\alpha_{s2} < \alpha_{L2} < 90°$, assuming negligible boundary disturbances, solving Eqs. (2.47a), (2.48a), and (2.64a) for the displacement amplitudes of the transmitted and reflected waves relative to the displacement amplitude of the incident wave gives

$$\frac{\zeta_{Lr}}{\zeta_{Li}} = \frac{\left\{ [2\rho_2 c_{2s} \sin \alpha_{s2} \sin \alpha_{L2} + \rho_2 c_{2L} \cos 2\alpha_{s2}] \cos (\alpha_1 - \alpha_{s2}) + [\rho_2 c_{2s} \sin 2\alpha_{s2} \sin \alpha_1 + \rho_1 c_{1L} \cos \alpha_{s2}] \cos (\alpha_{L2} - \alpha_{s2}) \right\}}{-\left\{ [2\rho_2 c_{2s} \sin \alpha_{s2} \sin \alpha_{L2} + \rho_2 c_{2L} \cos 2\alpha_{s2}] \cos (\alpha_1 + \alpha_{s2}) + [2\rho_2 c_{2s} \sin \alpha_{s2} \sin \alpha_1 + \rho_1 c_{1L}] \cos (\alpha_{L2} - \alpha_{s2}) \right\}} \tag{2.65}$$

$$\frac{\zeta_{LT}}{\zeta_{Li}} = \frac{\left\{ [2\rho_2 c_{2s} \sin \alpha_{s2} \sin \alpha_1 + \rho_1 c_{1L}] \cos (\alpha_1 - \alpha_{s2}) + [2\rho_2 c_{2s} \sin \alpha_{s2} \sin \alpha_1 + \rho_1 c_{1L}] \cos (\alpha_1 + \alpha_{s2}) \right\}}{-\left\{ [2\rho_2 c_{2s} \sin \alpha_{s2} \sin \alpha_{L2} + \rho_2 c_{2L} \cos 2\alpha_{s2}] \cos (\alpha_1 + \alpha_{s2}) + [2\rho_2 c_{2s} \sin \alpha_{s2} \sin \alpha_1 + \rho_1 c_{1L}] \cos (\alpha_{L2} - \alpha_{s2}) \right\}} \tag{2.66}$$

$$\frac{\zeta_{sT}}{\zeta_{Li}} = \frac{[\rho_2 c_{2L} \cos 2\alpha_{s2} \sin \alpha_1 - \rho_1 c_{1L} \sin \alpha_{L2}] \sin 2\alpha_1}{\left\{ [\rho_1 c_{1L} \cos \alpha_{s2} + \rho_2 c_{2s} \sin 2\alpha_{s2} \sin \alpha_1] \sin (\alpha_{L2} + \alpha_1) + [\rho_2 c_{2L} \cos 2\alpha_{s2} \sin \alpha_1 - \rho_1 c_{1L} \sin \alpha_{L2}] \cos (\alpha_{s2} + \alpha_1) \right\}} \tag{2.67}$$

At the critical angle for longitudinal waves in the solid, $\alpha_{L2} = 90°$ and

$$\sin \alpha_1 = \frac{c_{1L}}{c_{2L}}$$

where c_{1L} is the velocity of sound in the liquid and c_{2L} is the velocity of longitudinal-wave propagation in the solid. Neglecting the boundary disturbance, the energy is partitioned between a reflected longitudinal wave and a transmitted shear wave and the relative amplitudes are given by

$$\frac{\zeta_{Lr}}{\zeta_{Li}} = \frac{\left\{[\rho_2 c_{2L} \cos 2\alpha_{s2}] \cos(\alpha_1 - \alpha_{s2}) + [2\rho_2 c_{2s} \sin \alpha_{s2} \sin \alpha_1 + \rho_1 c_{1L}] \cos(-\alpha_{s2})\right\}}{-\left\{[\rho_2 c_{2L} \cos 2\alpha_{s2}] \cos(\alpha_1 + \alpha_{s2}) + [2\rho_2 c_{2s} \sin \alpha_{s2} \sin \alpha_1 + \rho_1 c_{1L}] \cos(-\alpha_{s2})\right\}} \tag{2.65a}$$

$$\frac{\zeta_{sT}}{\zeta_{Li}} = \frac{\left\{\rho_2 c_{2L} \cos 2\alpha_{s2} \sin 2\alpha_1\right\}}{\left\{[\rho_1 c_{1L} \cos \alpha_{s2} + \rho_2 c_{2s} \sin 2\alpha_{s2} \sin \alpha_1] + [\rho_2 c_{2L} \cos 2\alpha_{s2} \cos(\alpha_{s2} + \alpha_1)]\right\}} \tag{2.67a}$$

At normal incidence, the shear-wave components vanish from Eqs. (2.47), (2.48), and (2.64), and all other angles are zero. Three waves remain; an incident longitudinal wave, a reflected longitudinal wave, and a transmitted longitudinal wave. Equations (2.48) and (2.64) become

$$\zeta_{Li} - \zeta_{Lr} = \zeta_{LT} \tag{2.48b}$$

$$\rho_1 c_{1L}(\zeta_{Li} + \zeta_{Lr}) = \rho_2 c_{2L} \zeta_{LT} \tag{2.64b}$$

from which

$$\frac{\zeta_{LT}}{\zeta_{Li}} = \frac{2\rho_1 c_{1L}}{\rho_2 c_{2L} + \rho_1 c_{1L}} \tag{2.66a}$$

$$\frac{\zeta_{Lr}}{\zeta_{Li}} = \frac{\rho_2 c_{2L} - \rho_1 c_{1L}}{\rho_2 c_{2L} + \rho_1 c_{1L}} \tag{2.65b}$$

Equations (2.66a) and (2.65b) differ from Eqs. (2.26) and (2.27) in that the latter give the ratios of the displacements in terms of the potential functions while Eqs. (2.66a) and (2.65b) give the relationships between the displacement amplitudes.

At the critical angle for shear waves in the solid $\alpha_{s2} = 90°$ and $\sin \alpha_1 = c_{1L}/c_{2s}$ where c_{2s} is the velocity of shear-wave propagation in the solid. At this point, the energy is partitioned between a reflected longitudinal wave and a surface or boundary wave. Actually, these boundary waves may occur at lesser angles of incidence and often cause annoying interferences in certain types of measurements.

IV.F. SHEAR WAVE AT SOLID/SOLID INTERFACE POLARIZED PARALLEL TO THE PLANE OF THE INTERFACE

When a shear wave in which the displacement is parallel to the interface is incident on a plane interface between two solids, only shear waves are reflected and refracted. If the two media are isotropic, only one shear wave is transmitted across the boundary and one is reflected. In this case, all terms pertaining to longitudinal components in Eqs. (2.47), (2.48), (2.54), and (2.64) vanish, leaving

$$(\zeta_{si} - \zeta_{sr}) \cos \alpha_{s1} = \zeta_{sT} \cos \alpha_{s2} \tag{2.47a}$$

$$\rho_1 c_{1s}(\zeta_{si} + \zeta_{sr})(1 + \cos 2\alpha_{1s}) = \rho_2 c_{2s} \zeta_{sT} \ (1 + \cos 2\alpha_{s2}) \tag{2.47c}$$

from which

$$\frac{\zeta_{sT}}{\zeta_{si}} = \frac{2\rho_1 c_{1s}(1 + \cos 2\alpha_{1s}) \cos \alpha_{1s}}{\left\{\rho_2 c_{2s}(1 + \cos 2\alpha_{s2}) \cos \alpha_{s1} + \rho_1 c_{1s}(1 + \cos 2\alpha_{s1}) \cos \alpha_{s2}\right\}} \tag{2.68}$$

for the transmitted wave, and

$$\frac{\zeta_{sr}}{\zeta_{si}} = \frac{\left\{\rho_2 c_{2s}(1 + \cos 2\alpha_{s2}) \cos \alpha_{s1} - \rho_1 c_{1s}(1 + \cos 2\alpha_{s1}) \cos \alpha_{s2}\right\}}{\left\{\rho_2 c_{2s}(1 + \cos 2\alpha_{s2}) \cos \alpha_{s1} + \rho_1 c_{1s}(1 + \cos 2\alpha_{s1}) \cos \alpha_{s2}\right\}} \tag{2.69}$$

for the reflected wave.

At normal incidence, $\alpha_{1s} = \alpha_{2s} = 0$ and Eqs. (2.68) and (2.69) become

$$\frac{\zeta_{sT}}{\zeta_{si}} = \frac{2\rho_1 c_{1s}}{\rho_2 c_{2s} + \rho_1 c_{1s}} \tag{2.68a}$$

and

$$\frac{\zeta_{sr}}{\zeta_{si}} = \frac{\rho_2 c_{2s} - \rho_1 c_{1s}}{\rho_2 c_{2s} + \rho_1 c_{1s}} \tag{2.69a}$$

IV.G. REFLECTION, REFRACTION, AND MODE CONVERSION IN GENERAL APPLICATIONS OF ULTRASONIC ENERGY

The many equations presented in this chapter represent an attempt at presenting the reader with the basic mechanics of wave propagation and their influence on various conditions faced in the actual applications of ultrasonic energy. A wave propagating through common media may encounter interfaces of every conceivable geometrical configuration. A multiplicity of incidence angles exist for the common geometries in which the interface is neither plane nor infinite. In such cases, energy may be scattered, refracted, and converted to other wave forms (mode conversion), as a review of the equations presented in this chapter would reveal.

Equations pertaining to anisotropic conditions have not been presented. These can be derived following the techniques used in deriving the equations for isotropic conditions and considering the directional characteristics of the elastic constants. Both Kolsky (3) and Redwood (2) deal with propagation through anisotropic media and the reader may consult those works for more detailed information. There are 21 independent elastic constants relating stress to strain in anisotropic crystals. All 21 are necessary to define the stress-strain relations in the most complex crystals. Crystals with planes of symmetry have fewer independent elastic constants; for instance, a cubic crystal has three and an isotropic material has only two.

A multicrystalline anisotropic solid may be treated as an

isotropic solid if the wavelength is large compared to the size of the crystals and if the crystals are randomly oriented. The anisotropy will, however, affect the attenuation of waves passing through the material.

If the crystals in the material are oriented in the material so that the elastic constants in the direction of orientation differ from those in a direction normal to the orientation, the velocity of wave propagation also will differ in these two directions.

V. TRANSMISSION THROUGH THIN PLATES

When an acoustic plane wave of finite cross section traveling in a medium encounters a plate that is very thin relative to the wavelength, the plate flexes with the medium. The pressure on the two faces of the plate are approximately equal in magnitude. Only that energy associated with flexing the plate is lost and the intensity of the transmitted wave approximately equals that of the incident beam if the medium on either side of the plate is the same.

When the plate thickness is of the same order of magnitude as the wavelength in the plate, internal reflections from the interfaces affect the amplitude of the transmitted and reflected waves. For instance, reflections of a longitudinal wave from the back surface of a plate one-quarter wavelength thick, are 180° out of phase with the incident wave at normal incidence. No energy is transmitted through the plate under these conditions. If the plate is one-half wavelength thick, transmission is maximum and only those losses associated with attenuation within the material of the plate affect the amplitude of the transmitted wave. No reflections occur at boundaries between two media with equal acoustic impedances.

The extent of the interference of pulses of ultrasonic energy within plates in which the wavelength and the thickness are of the same order of magnitude depends upon the ratio of thickness to wavelength (d/λ), the coupling to the surrounding medium, the pulse length and shape, the spectrum of the pulse, and the internal-absorption losses. When d/λ equals $\frac{1}{4}$, $\frac{3}{4}$, etc. (to a limit depending upon pulse length), maximum destructive interference occurs within the plate. When d/λ is $\frac{1}{2}$, 1, etc., maximum constructive interference (resonance) occurs and the energy is transferred through the plate with minimum loss.

The relative transmission is given by

$$\frac{I_T}{I_I} = \frac{e^{-2\alpha d}}{\left[1 + \frac{1}{4}\left(\frac{Z_2}{Z_1} - \frac{Z_1}{Z_2}\right)^2 \sin^2 \frac{\omega d}{c_2}\right]} \tag{2.70}$$

If d is small and α is negligible

$$\frac{I_T}{I_I} = \frac{1}{\left[1 + \frac{1}{4}\left(\frac{Z_2}{Z_1} - \frac{Z_1}{Z_2}\right)^2 \sin^2 \frac{\omega d}{c_2}\right]} \tag{2.70a}$$

where Z_1 is the acoustic impedance of the surrounding medium and Z_2 is the acoustic impedance of the plate.

VI. DIFFRACTION

Diffraction refers to the modification of wave fronts in passing by edges of opaque bodies, through narrow slits, or in being reflected or transmitted from surfaces. In the diffraction of light at the edge of an opaque object, rays appear to be deflected, producing fringes of light and dark, or colored bands. A similar modification of other waves (as sound and electromagnetic waves) occur which cause the curvature of waves around objects in their paths. Diffraction prevents full utilization of a wave front or the bringing of the wave front to a focus.

Diffraction effects are of concern in many ultrasonic applications such as in the measurement of attenuation of sound in materials, nondestructive inspection of materials, underwater sound, and imaging. It accounts for the distribution of the wave amplitude in the sound field radiating from a source and in the spreading of the beam from the source.

VI.A. HUYGENS' PRINCIPLE

Equations for describing diffraction are derived by use of Huygens' principle, according to which, every vibrating point on a wave front is regarded as the center of a new disturbance. The new disturbances act as point sources, each emitting a spherical wave which is assumed to produce an effect only along its own wave front. Propagation from those sources is always assumed to be forward. In the use of Huygens' principle, it is also assumed that the wavelength is much smaller than the dimensions of the surfaces or obstacles encountered.

Huygens' principle is illustrated in Fig. 2.8. Spherical waves vibrating sinusoidally are shown emanating from points A and B on a wave front. The two waves are of equal amplitude and phase in the direction of the wave front. The arrival of the two new wave fronts at points P_1 and P_4 is such that they are in phase and thus reinforce each other (neglecting the fact that, at these points, the waves meet obliquely). At points P_2 and P_5, they are 180° out of phase, and, therefore, destructive interference occurs at these points. At point P_3, the wave from B leads that from A by 90° and the amplitude of the disturbance at that point is equal to the amplitude of either wave taken separately.

Huygens' principle is used to calculate radiation patterns from transmitters of various configurations. Consider the radiation from a circular piston source of radius a, Fig. 2.9. The pressure dp at the point P(0, x) due to the radiation from the incremental area dS is

$$dp = -j \frac{\rho f v_o (r dr d\phi)}{d} e^{j \frac{\omega}{c} (d-ct)} \tag{2.71}$$

where d is the distance from dS to $P_{(0,x)}$, r is the distance from the origin to dS, and v_o is the particle velocity at dS.

The total contribution to the radiation pressure at $P_{(0,x)}$ of all increments dS is, therefore

$$p = -j(\rho f v_o) \int_0^{2\pi} \int_0^a \frac{e^{j\frac{\omega}{c}(d-ct)}}{d} \, r dr d\phi \tag{2.72}$$

From Fig. 2.9, for small values of x/a

$$d = \sqrt{r^2 + x^2}$$

or

$$p = -j(\rho f v_o) \int_0^{2\pi} \int_0^a \frac{e^{j\frac{\omega}{c}\left[\sqrt{r^2 + x^2} - ct\right]} r dr d\phi}{\sqrt{r^2 + x^2}} \tag{2.72a}$$

Therefore

$$p = -j2\rho c v_o \left\{\cos \frac{1}{2}\frac{\omega}{c}\left[\sqrt{a^2 + x^2} - x\right] + j \sin \frac{1}{2}\frac{\omega}{c}\left[\sqrt{a^2 + x^2} + x\right]\right\} \times e^{-j\omega t} \sin \frac{1}{2}\frac{\omega}{c}\left[\sqrt{a^2 + x^2} - x\right] \tag{2.73}$$

From Eq. (2.73) it can be seen that p is a <u>maximum</u> when

$$\frac{1}{2}\frac{\omega}{c}\left[\sqrt{a^2 + x^2} - x\right] = \frac{\Pi}{2}, \frac{3\Pi}{2}, \frac{5\Pi}{2} \ldots \frac{2m + 1}{2}\Pi$$

or solving for x

$$x = \frac{4a^2 - (2m + 1)^2 \lambda^2}{4(2m + 1)\lambda} \qquad (m = 0,1,2,3,\ldots) \tag{2.74}$$

The maximum closest to the radiator occurs when $2a = (2m + 1)\lambda$. Likewise, p is a <u>minimum</u> when

$$x = \frac{a^2 - n^2\lambda^2}{2n\lambda} \qquad (n = 1,2,3,4\ldots) \tag{2.75}$$

The minimum closest to the radiator occurs when a = n. Eq. (2.73) thus describes the pressure distribution <u>along the axis</u> of the radiated beam in the <u>near field</u>. A similar approach is used in determining the pressure distribution at any other point, on or off the axis.

The value of x at m = 0 is the extent of the near field. When $\lambda \ll a$, the near field extends to a distance which is approximately equal to a^2/λ. The far field begins at this point. Pressure distribution at greater distances may be calculated on or off the axis by assuming that $x \approx d$.

Calculations of sound-field patterns become very difficult for other types of radiator surfaces.

Consider, now, diffraction around a disk-shaped obstacle at the center of a circular beam (Fig. 2.10).

According to Huygens' principle, the plane in which the obstruction (radius a_2) is located may be considered as a new source of sound. Following the same procedure as before, the pressure distribution along the axis is

$$p = -\rho v_o f e^{-j\omega t} \left(\frac{c}{\omega}\right) \int_0^{2\pi} \int_{a_2}^{a_1} \frac{e^{j\frac{\omega}{c}\sqrt{r^2 + x^2}}}{\sqrt{r^2 + x^2}} \left(j\frac{\omega}{c}\right) r\,dr\,d\phi$$

$$= -\rho c v_o \left\{ e^{j\frac{\omega}{c}\sqrt{a_1^2 + x^2}} - e^{j\frac{\omega}{c}\sqrt{a_2^2 + x^2}} \right\} e^{-j\omega t}$$

$$= -j2\rho c v_o \left\{ \cos \frac{1}{2}\left(\frac{\omega}{c}\right)\left(\sqrt{a_1^2 + x^2} - \sqrt{a_2^2 + x^2}\right) \right.$$

$$\left. + j \sin \frac{1}{2}\left(\frac{\omega}{c}\right)\left(\sqrt{a_1^2 + x^2} + \sqrt{a_2^2 + x^2}\right) \right\}$$

$$\times e^{-j\omega t} \sin \frac{1}{2} \left(\frac{\omega}{c}\right) \left[\sqrt{a_1^2 + x^2} - \sqrt{a_2^2 + x^2}\right] \tag{2.76}$$

The pressure maxima along the axis in the near field occur when

$$\frac{1}{2} \left(\frac{\omega}{c}\right) \left[\sqrt{a_1^2 + x^2} - \sqrt{a_2^2 + x^2}\right] = \frac{\Pi}{2}, \frac{3\Pi}{2}, \frac{5\Pi}{2} \ldots \frac{2m + 1}{2} \Pi$$

i.e., at

$$x = \frac{\sqrt{(a_1^2 - a_2^2)^2 - (a_1^2 + a_2^2) \frac{\lambda^2}{2} (2m+1)^2 + \frac{\lambda^4}{4} (2m+1)^4}}{\lambda(2m+1)}$$

$$(m = 0,1,2,3\ldots) \tag{2.77}$$

The <u>pressure minima</u> along the axis in the near field occur when

$$\frac{1}{2} \left(\frac{\omega}{c}\right) \left(\sqrt{a_1^2 + x^2} - \sqrt{a_2^2 + x^2}\right) = 0, \Pi, 2\Pi, \ldots\ldots n\Pi$$

from which

$$x = \frac{\sqrt{(a_1^2 - a_2^2)^2 - 2(a_1^2 + a_2^2)\lambda^2 n^2 + \lambda^4 n^4}}{2\lambda n}$$

$$(n = 1,2,3\ldots) \tag{2.78}$$

The start of the far field occurs at the final maximum, that is, when m = 0.

The maximum closest to the obstruction occurs when

$$(a_1^2 - a_2^2)^2 = (a_1^2 + a_2^2) \frac{\lambda^2}{2} (2m+1)^2 + \frac{\lambda^4}{4} (2m+1)^4 \tag{2.79}$$

The minimum closest to the obstruction occurs when

$$(a_1^2 - a_2^2)^2 = 2(a_1^2 + a_2^2)\lambda^2 n^2 + \lambda^4 n^4 \tag{2.80}$$

Equations (2.79) and (2.80) determine the distance of the "dark-shadow" zone due to the disk-shaped obstruction. The distance corresponding to the axial fluctuations between maxima and minima might be called a "gray-shadow" zone. These two zones would correspond to the umbra and penumbra observed in shadows in the path of a light beam. These conditions are of special interest to any application based on shadow techniques, such as through-transmission testing for delaminations in thin plates or imaging of a sound field.

VI.B. DIFFRACTION IN THREE-DIMENSIONAL SPACE

From the previous analysis and the premises upon which it was based, one may quickly discern that along any plane in the near field normal to the axis of the propagating beam, the pressure distribution is not uniform. In the case of a piston-type radiator which is several wavelengths in diameter, the pattern appears as a series of concentric rings, corresponding to pressure maxima and minima. The result is the production of a main beam in the direction of the axis and of side lobes in which energy propagates from the source at angles to the axis. The spread of the main beam from a disk-shaped radiator is determined by the relation

$$\sin(\theta/2) = 0.6\lambda/a \tag{2.81}$$

where θ is the angle describing the spreading of the effective part of the beam, a is the radius of the emitter, and λ is the wavelength of sound in the propagating medium.

VI.C. DIRECTIVITY PATTERN

Directivity pattern (or beam pattern) refers to the relative response of a receiver (or pressure or intensity of the radiated wave), plotted as a function of position with respect to the transmitter. The pattern in a specified plane is usually given, and may be referred to either polar or rectangular coordinates. The directivity patterns are functions of the wavelength in the medium into which the wave is radiated and of the dimensions of the radiation surface.

VI.D. FOCUSING

As is true of light, ultrasonic energy may be focused. A significant difference results from the longer lengths of sound waves. Because of the longer wavelengths, sound focuses in a region rather than at a point, and the smallest practical focal region is a sphere one wavelength in diameter. In any type of focusing device, the sharpness of the focus is proportional to the ratio of the aperture of the device to the wavelength. The best focus is obtained when the focal length is the distance to the final pressure peak in the near field.

Focusing devices are often used in nondestructive testing to improve resolution of small defects. Focusing has also been used to obtain very high intensities for processing applications and in neurosurgery. Two methods of focusing ultrasound have been used: (1) acoustic lenses and (2) curved radiation surfaces.

VII. STANDING WAVES

Standing waves are periodic waves having a fixed distribution in space which is the result of interference of progressive waves of the same frequency and kind. Such waves are characterized by the existence of nodes or partial nodes and antinodes that are fixed in space. Their use is common in ultrasonic applications.

The equation for the acoustic pressure of an infinite and progressive wave is

$$p_o = P_o e^{-\alpha x + j(\omega t - kx)} \tag{2.82}$$

where P_o is the pressure amplitude of the wave as it leaves the source, α is the attenuation of the sound in the medium, and k is ω/c. Upon encountering an infinite plane surface that is parallel to the wave front, part of the energy is reflected, depending upon the relationship between acoustic impedances of the two media. During the reflection process, the wave also may undergo a shift in phase. If the distance between the source and the reflecting surface is L, the pressure amplitude of the incident wave at L is

$$P_1 = P_o e^{-\alpha L} \tag{2.83}$$

The pressure amplitude of the reflected wave at L is

$$P_2 = P_o e^{-\alpha L - 2\alpha_o - 2j\beta} \tag{2.84}$$

and the pressure of the reflected wave is

$$P_2 = P_o e^{-2\alpha L - 2\alpha_o + \alpha x + j(\omega t - 2\beta - kL + kx)} \tag{2.85}$$

where α_o accounts for the loss of energy on reflection and β is the phase shift due to reflection. The pressure p at a given distance x from the source is the algebraic sum of all contributing waves. Considering only a single reflection, the pressure p is

$$p = P_o[e^{-\alpha x + j(\omega t - kx)} - e^{-2\alpha_o - 2\alpha L + \alpha x + j(\omega t - kL - 2\beta + kx)}] \qquad 0 < x \leq L \tag{2.86}$$

Since $e^{j\theta} + e^{-j\theta} = 2 \cos \theta$ and $e^{j\theta} - e^{-j\theta} = 2j \sin \theta$, Eq. (2.86) may be written

$$p = P_o\{e^{-\alpha x}[\cos(\omega t - kx) + j \sin(\omega t - kx) - e^{-2\alpha_o - 2\alpha L + \alpha x} \cos(\omega t - kL - 2\beta + kx) + j \sin(\omega t - kL - 2\beta + kx)]\} \tag{2.86a}$$

The pressure amplitude fluctuates along x owing to the reflected waves. If the reflecting surface were located an infinite distance from the source or if the impedance at the interface between the two media were perfectly matched, the pressure amplitude would decrease only as a function of α. The standing-wave ratio, SWR, is the sum of the two amplitudes of Eq. (2.86a) divided by the difference between these two amplitudes. Then

$$\text{SWR} = \frac{P_o[e^{-\alpha x} + e^{-2\alpha_o - 2\alpha L + \alpha x}]}{P_o[e^{-\alpha x} - e^{-2\alpha_o - 2\alpha L + \alpha x}]} = \frac{e^{-\alpha x} + e^{-2\alpha_o - 2\alpha L + \alpha x}}{e^{-\alpha x} - e^{-2\alpha_o - 2\alpha L + \alpha x}} \tag{2.87}$$

If α is negligible, Eq. (2.87) reduces to

$$\mathrm{SWR} = \frac{1 + e^{-2\alpha_o}}{1 - e^{-2\alpha_o}} = \coth \alpha_o \tag{2.87a}$$

(see Fig. 2.11.

$$\frac{d_{min}}{\lambda} = \frac{n}{2} - \frac{\beta}{2\Pi} \qquad (n = 1,2,3,4,5...) \tag{2.88}$$

VIII. DOPPLER EFFECT

The Doppler effect is a change in frequency from that of a transmitted wave detected by a receiver due to various conditions of relative motion between the receiver, the transmitter, the medium, and reflecting surfaces. Several ultrasonic applications involve the Doppler effect. The change in frequency is due to the effect of the motion on the rate at which a complete wave (successive pressure peaks) passes the receiver; i.e., the period of the wave relative to the receiver may be compressed or expanded, depending upon the nature of the motion. The various possibilities are (1) stationary transmitter and medium and moving receiver, (2) stationary receiver and medium and moving source, (3) stationary source and receiver and moving medium, (4) moving source and receiver in stationary medium, (5) moving source and medium and stationary receiver, (6) moving medium and receiver and stationary source, (7) moving source, medium, and receiver, and (8) any of the conditions (1) through (7) plus a moving reflector (target).

When the receiver approaches a transmitter at a vector velocity $\bar{v}_r$ in a stationary medium, the time between pressure peaks at the receiver is

$$T_D = \lambda/(c_o + |\bar{v}_r|) \tag{2.89}$$

where λ is the wavelength in the medium, T_D is the Doppler period of the sound wave, and c_o is the velocity of sound in the medium. Therefore, the frequency at the receiver is

$$f_r = 1/T_D = (c_o + |\bar{v}_r|/\lambda) = \frac{f_o}{c_o}(c_o + |\bar{v}_r|) \tag{2.90}$$

where f_o is the transmitted frequency. If the velocity of the receiver, $\bar{v}_r$, is in the opposite direction

$$f_r = 1/T_D = (c_o - |\bar{v}_r|/\lambda) = f_o\ (c_o - |\bar{v}_r|)/c_o \qquad (2.90a)$$

If the velocity of the receiver, $\bar{v}_r$, is at an angle θ_r to a line through the receiver and transmitter, the received frequency is

$$f_r = f_o\ (c_o + |\bar{v}_r| \cos \theta_r)/c_o \qquad (2.90b)$$

where, if $\theta_r < \frac{\Pi}{2}$, the sign of the cosine is positive and indicates that the motion brings the transmitter and receiver closer together, and, if $\theta_r > \frac{\Pi}{2}$, the sign of the cosine is negative and the source and receiver are moving farther apart.

If the medium and the receiver are stationary and the source is moving at a velocity $\bar{v}_s$, the distance between the source and the receiver changes by $(|\bar{v}_s| \cos \theta_s)T_o$ during one period of the source frequency, T_o. Therefore, the distance between successive peaks in the Doppler wave is

$$\lambda_D = (c_o - |\bar{v}_s| \cos \theta_s)T_o \qquad (2.91)$$

and the received frequency is

$$f_r = c_o/\lambda_D = c_o/(c_o - |\bar{v}_s| \cos \theta_s)T_o$$
$$= f_o\ c_o/(c_o - |\bar{v}_s| \cos \theta_s) \qquad (2.92)$$

where the sign of the cosine is opposite to the direction of the source relative to the receiver.

When both the receiver and transmitter are moving

$$f_r = (c_o + |\bar{v}_r| \cos \theta_r)f_o/(c_o - |\bar{v}_s| \cos \theta_s) \qquad (2.93)$$

In Eq. (2.93), if both θ_s and θ_r are less than 90°, relative motion between source and receiver is generally toward each other and the signs of the cosine terms are positive. For example, if both are moving on a line toward each other, $\cos \theta_r = \cos \theta_s = 1$ and

$$f_r = (c_o + |\bar{v}_r|)f_o/(c_o - |\bar{v}_s|) \qquad (2.93a)$$

If they are moving in opposite directions, the signs of the velocity terms in the numerator and denominator of Eq. (2.93a) are reversed.

If the transmitter and receiver are stationary, and the medium is moving at a constant velocity $\bar{v}_m$, the effect of $\bar{v}_m$ is to change the apparent velocity of sound in the medium. The period of the wave passing the receiver is the same as the period of the transmitter frequency and no Doppler shift occurs. If the velocity of the medium varies, the wavelength expands and contracts with the variation and $\frac{d|\bar{v}_m|}{d\lambda} \cos \theta_m = f_o$. However, the time that it takes a wave front to travel from the transmitter to the receiver depends upon the sum of the velocities $(c_o + |\bar{v}_m| \cos \theta_m)$, where θ_m is the angle between the direction of the velocity vector and a straight line connecting the transmitter and the receiver. The travel time t is

$$t = d/(c_o + |\bar{v}_m| \cos \theta_m) \tag{2.94}$$

When $\theta_m < 90°$, $\cos \theta_m$ is positive and $|\bar{v}_m| \cos \theta_m$ is directed toward the receiver. When $\theta_m > 90°$, $\cos \theta_m$ is negative and directed toward the transmitter.

Although there is no Doppler effect due to a moving medium between stationary transmitters and receivers, a moving medium does affect the Doppler frequency when either or both transmitter and receiver are moved relative to the other. The equation is

$$f_r = f_o \frac{[c_o + |\bar{v}_m| \cos \theta_m + |\bar{v}_r| \cos \theta_r]}{[c_o + |\bar{v}_m| \cos \theta_m - |\bar{v}_s| \cos \theta_s]} \tag{2.95}$$

When a Doppler technique is used to detect a moving object, the frequency of the wave received at the transmitter after the wave has reflected from the object is approximately

$$f_r = f_o\left\{1 + 2\,(|\bar{v}_s| \cos \theta_s - |\bar{v}_t| \cos \theta_t)/c_o\right\} \tag{2.96}$$

where $\bar{v}_t$ is the velocity of the target and θ_t and θ_s are the instantaneous angles between the directions which the corresponding velocity vectors $\bar{v}_t$ and $\bar{v}_s$ of the target and the source respectively make with a straight line through the source and target. Or if the fluid is moving at a constant velocity $\bar{v}_m$

$$f_r = f_o\left\{1 + \frac{2(|\bar{v}_s| \cos\theta_s - |\bar{v}_t| \cos\theta_t)}{c_o + |\bar{v}_m| \cos\theta_m}\right\} \tag{2.96a}$$

where θ_m is the instantaneous angle between the velocity vector of the fluid and the radius vector from the target to the source.

When a stationary transmitter and stationary receiver are used in a bistatic arrangement, i.e., the transmitter and receiver are separated by a distance, the Doppler frequency caused by a moving target in stationary medium is

$$f_D = \frac{f_o|\bar{v}_t|}{c_o} (\cos\theta_{s'} + \cos\theta_{r'}) \tag{2.96b}$$

where $\theta_{s'}$ is the angle between the radius vector from the source S to the target T and the velocity vector, and $\theta_{r'}$ is the angle between the radius vector from T to receiver R and the velocity vector. As a result, the frequency of the wave as detected by the receiver is

$$f_r = f_o[1 + |\bar{v}_t| (\cos\theta_{s'} + \cos\theta_{r'})/c_o] \tag{2.96c}$$

A medium moving at varying velocities and directions may cause variations in the total path length of an ultrasonic beam incident on a reflecting surface and cause an effect similar to the effects caused by a moving surface.

The effect of a constant fluid velocity on the Doppler frequency of a bistatic arrangement of transmitter and receiver is

$$f_D = f_o |\bar{v}_t| \left\{\frac{\cos\theta_{s'}}{c_o - |\bar{v}_m| \cos\theta_{m'}} + \frac{\cos\theta_{r'}}{c_o + |\bar{v}_m| \cos(\theta_{r'} + \theta_{s'} - \theta_{m'})}\right\} \tag{2.96d}$$

where $\theta_{m'}$ is the angle between the fluid velocity vector and the radius vector from the source to the target. The receiver frequency, f_r, is

$$f_r = f_o\left\{1 + f_o\,|\bar{v}_t\left[\frac{\cos\theta_{s'}}{c_o - |\bar{v}_m|\cos\theta_{m'}} + \frac{\cos\theta_{r'}}{c_o + |\bar{v}_m|\cos(\theta_{r'} + \theta_{s'} - \theta_{m'})}\right]\right\} \qquad (2.96e)$$

(see Fig. 2.12).

IX. BEATS

When two simple-harmonic sound waves of nearly the same intensity but of slightly different frequencies are combined linearly, the sum of the amplitudes of vibration shows maxima and minima at periodic intervals. The result is a third wave which is approximately simple harmonic at a frequency equal to the difference between the frequencies of the initial waves. This phenomenon, called beating, is sometimes used in ultrasonic applications.

X. ATTENUATION OF AN ULTRASONIC WAVE

Attenuation refers to the diminishing of intensity of a wavefront as it progresses through a medium. In practice, several factors contribute to attenuation. These factors include the spreading of the beam, scattering, absorption due to various mechanisms, and mode conversion resulting in partitioning of the energy among two or more wave modes each traveling at its own velocity.

The effect of attenuation on a periodic wave is demonstrated by the equation:

$$\zeta = \zeta_o e^{-\alpha x} e^{j\omega(t-x/c)} \qquad (2.97)$$

where α is the loss coefficient in nepers per centimeter.

The intensity I is proportional to the square of the displacement. Therefore, if I_o is the initial intensity, watts per square centimeter, of a progressive ultrasonic wave, the intensity at distance x is

$$I = I_o e^{-2\alpha x} \tag{2.98}$$

X.A. ATTENUATION DUE TO BEAM SPREADING

If a spherical wave front of uniform intensity (Fig. 2.13), suffers no loss in traveling from its source, the total energy (P_T) at each spherical surface of total area S at distance r from the center of the source is always

$$P_T = I_r S_r \tag{2.99}$$

where I_r is the intensity at distance r and S_r is the total area of a sphere or segment of the sphere through which the total energy W_T passes.

The area of a sphere is $4\pi r^2$. Let r_1 and r_2 correspond to the ratio of two concentric spheres containing a point source at their centers radiating a total energy P_T. If there are no losses, the total energy passing through S_1 corresponding to r_1 is equal to that through S_2 corresponding to r_2. From Eq. (2.99)

$$P_T = I_1 S_1 = I_2 S_2 = 4\pi r_1^2 I_1 = 4\pi r_2^2 I_2 \tag{2.99a}$$

Therefore

$$I_1 r_1^2 = I_2 r_2^2 \tag{2.100}$$

Equation (2.100) is true for all radiation patterns of spherical segments if the conditions of uniform intensity, no losses, and constant direction are met.

X.B. ATTENUATION DUE TO SCATTERING

Scattering is a common cause of attenuation of ultrasonic energy. The basic principles which cause this scattering of ultrasonic energy have been discussed previously in the sections on refraction and reflection. A precise formula for scattering including all conditions of wave modes, obstacle configurations, and material properties is impractical. However, analyses based upon scattering from basic geometrical forms such as spheres, cylinders, and cubes are usually sufficient.

Scattering from obstacles may affect the attenuation in two ways: (1) by single scattering or (2) by multiple scattering. In single scattering, energy is deflected once by an obstacle in the path of the beam and is thus lost to the main beam. In multiple scattering, a ray of energy may be scattered many times by successive scatterers, with the following results: (1) all of the energy may be lost to the main beam (2) part of the energy may be lost and part returned and (3) all of the energy may be returned to the main beam.

When energy in a beam has undergone multiple scattering, but remains in the beam, phase relationships between the components of the wave are affected. This is due to the differences in path length traveled by components of the energy, and the fact that some of the energy may be converted to slower wave modes.

Scattering of waves from an obstacle is a function of the scattering cross section S_{cs} of the obstacle. Scattering cross section is defined as the ratio of the total energy scattered per unit time to the energy per unit area carried per unit time by the incident wave, the unit area being normal to the direction of propagation of the plane incident wave. The ultrasonic attenuation is directly proportional to the scattering cross section. From the definition of S_{cs}, it follows that IS_{cs} is the total energy scattered per single scatterer per unit time at any point x where the intensity is I.

X.B.1. Scattering From a Cylindrical Obstruction in a Homogeneous Medium

The total ultrasonic power scattered by a hollow cylinder in a homogeneous medium in which the wave front is parallel to the axis of the cylinder is (4)

$$P_s = \begin{cases} (6\pi^5 a^4/\lambda) I_o & (\lambda >> 2\pi a) \\ (2aI_o) & (\lambda_1 \leq 2\pi a \leq \lambda_2) \\ (4aI_o) & (\lambda << 2\pi a) \end{cases} \qquad (2.101)$$

where I_0 is the intensity of the incident wave within the material, W/cm^2; P_s is the total power scattered per unit length, W; and a is the radius of the cylinder, cm. The limiting value of total scattered power for very short wavelengths is the power contained in a beam twice as wide as the cylinder.

X.B.2. Scattering From a Spherical Obstruction in a Homogeneous Medium

The total ultrasonic power scattered by a spherical cavity in an isotropic medium is approximately

$$P_s = 2\pi a^2 I_o \qquad \lambda << 2\pi a \tag{2.102}$$

where a is the radius of the sphere.

X.B.3. Scattering From a Disk-Shaped Cavity in the Path of an Ultrasonic Beam

If a disk-shaped cavity of radius a_1 obstructs a collimated beam of ultrasonic energy of radius a_2 and the axis of the disk coincides with the axis of the collimated beam, the approximate total energy remaining within the beam past the obstruction is

$$P_t = [a_2^2 - a_1^2]P_i/a_2^2 \qquad \lambda < a_1 \tag{2.103}$$

where P_i is the total energy of the incident beam.

If the disk lies in a plane that is other than normal to the direction of propagation, the radius a_1 in Eq. (2.103) is the radius of the normal projection on the axis of the beam, except as the angle between the two axes approaches $\pi/2$.

X.B.4. Scattering From an Elastic Isotropic Sphere in a Homogeneous Medium

From the previous discussions regarding reflections and refraction at an interface between media of different elastic properties, one may realize that scattering from elastic obstructions within the path of an ultrasonic beam may become quite complicated. Waves are generated inside the obstacle and these may include both longitudinal waves and shear waves.

The scattering cross section S_{cs} of an elastic sphere is (5)

$$S_{cs\ \text{elastic sphere}} = \frac{4\pi}{9} g_e k_1 (4a^6) \tag{2.104}$$

where

$$g_e = \left(\frac{3(\kappa_1/k_1)^2}{[3(\kappa_2/k_2)^2 - 4](\mu_2/\mu_1) + 4} - 1\right)^2$$
$$+ \frac{1}{3}\left(1 + 2(\kappa_1/k_1)^3\right)[\kappa_2/\kappa_1)(\mu_2/\mu_1) - 1]$$
$$+40[2 + 3(\kappa_1/k_1)^5] \frac{[(\mu_2/\mu_1) - 1]^2}{\left\{2[3(\kappa_1/k_1)^2 + 2](\mu_2/\mu_1)+[9(\kappa_1/k_1)^2 - 4]\right\}}$$

and where $k_m = \omega/c_{1m}$, c_{1m} = bulk longitudinal velocity, $\kappa_m = \omega/c_{sm}$, c_{sm} = shear-wave velocity, μ_m = shear modulus of elasticity, and Subscript 1 refers to the medium in which the scatterer is embedded and Subscript 2 refers to the material of the obstacle. g_e also may be given in terms of acoustic impedances, densities, and moduli, or of velocities and densities.

For very small spherical cavities, $\genfrac{}{}{0pt}{}{k_1}{\kappa_1}a \ll 1$, the scattering cross section is

$$S_{cs(\text{spherical cavity})} = (4\pi/9)g_c(k_1a)^6/k_1^{\ 6} \tag{2.105}$$

where

$$g_c = (4/3) + 40\left\{\frac{2 + 3(\kappa_1/k_1)^5}{[4 - 9(\kappa_1/k_1)^2]^2}\right\} - \frac{3}{2}(\kappa_1/k_1)^2$$
$$+(2/3)(\kappa_1/k_1)^3 + (9/16)(\kappa_1/k_1)^4$$

Similar equations may be derived for cylinders and other configurations.

XVI.B.5. <u>Scattering in Practice</u>

In practice, an ultrasonic wave propagating in a scattering medium encounters many scatterers. Typical examples are grain boundaries in metals, porosity in any media, and bubbles in liquids. The total effect on the attenuation is a summation of individual effects.

According to Mason and McSkimin (6), the attenuation in a metal is

$$\alpha = \beta_1 f + \beta_2 f^4 \qquad \lambda \geq 3D \tag{2.106}$$

where

D is the average diameter of the grains, β_1 is a constant corresponding to hysteresis, and β_2 is a constant corresponding to scattering.

Scattering loss from grain boundaries in metals results from the random orientation of the grains with respect to the directions of wave propagation. Since, in an anisotropic material, the elastic constant varies with orientation, scattering occurs at intersections between grain boundaries. Materials at the grain boundaries which differ in density and elasticity from the body of the grains also contribute to scattering.

According to Rayleigh (7), scattering of energy from a single particle is given by the formula

$$\frac{\text{S.A.}}{\text{I.A.}} = \frac{\pi V}{R\lambda^2}\left[\frac{\Delta\kappa}{\kappa} + \frac{\Delta\rho}{\rho}\cos\theta\right] \tag{2.107}$$

where $\frac{\text{S.A.}}{\text{I.A.}}$ is the ratio between the amplitude of the scattered wave to that of the incident, V is the volume of the scatterer, R is the radius of the scatterer, κ is the elasticity of the medium, $\Delta\kappa$ is the difference in elasticity between the particle and the medium, ρ is the density of the medium, and $\Delta\rho$ is the difference in density between the particle and the medium.

According to Mason and McSkimin (6), the Rayleigh scattering law leads to the following scattering term for longitudinal waves in metals

$$\alpha = \frac{\pi^4 D^3 f^4}{3c_L^4}\left\langle\left[\frac{c_{11}' - \langle c_{11}'\rangle_{av}}{\langle c_{11}'\rangle_{av}}\right]^2\right\rangle_{av} \tag{2.108}$$

and for shear waves

$$\alpha = \frac{4\pi^3 D^3 f^4}{9\, c_s^4} \left\langle \left[\frac{c_{44}' - \langle c_{44}' \rangle_{av}}{\langle c_{44}' \rangle_{av}} \right] \right\rangle_{av} \qquad (2.109)$$

where c_L is the longitudinal-wave velocity in the metal, c_s is the shear-wave velocity in the metal, and α is attenuation in Np/cm.

$$c_{11}' = c_{11}(\ell_1^2 + m_1^2)^2 + c_{33}m_1^4 + (2c_{13} + 4c_{44})n_1^2(\ell_1^2 + m_1^2)$$

(for hexagonal crystals, any orientation)

$$c_{44}' = c_{11}\left[(\ell_1\ell_2 + m_1m_2)^2 + \left(\frac{\ell_1m_2 - m_1\ell_2}{2}\right)^2\right]$$

$$- \frac{c_{12}}{2}[\ell_1m_2 - m_1\ell_2]^2 + 2c_{13}n_1n_2(\ell_1\ell_2 + m_1m_2)$$

$$+ c_{33}n_1^2n_2^2 + c_{44}[m_1n_2 + n_1m_2)^2 + (n_1\ell_2 + \ell_1n_2)^2]$$

(for hexagonal crystals, any orientation)

$$c_{11}' = c_{11}(\ell_1^4 + m_1^4 + n_1^4) + (2c_{12} + 4c_{44})(\ell_1^2m_1^2 + \ell_1^2n_1^2 + m_1^2n_1^2)$$

(for cubic crystals, any orientation)

$$c_{44}' = c_{11}(\ell_1^2\ell_2^2 + m_1^2m_2^2 + n_1^2n_2^2$$

$$+ 2c_{12}[\ell_1\ell_2m_1m_2 + n_1n_2(\ell_1\ell_2 + m_1m_2)]$$

$$+ c_{44}[(\ell_1\ell_2 + m_1m_2)^2 + (\ell_1\ell_2 + n_1n_2)^2 + (m_1m_2 + n_1n_2)^2]$$

(for cubic crystals, any orientation)

where ℓ_1, ℓ_2, ℓ_3, $m_1, \ldots n_3$ are the direction cosines between the new set of axes and the crystallographic axes x, y, and z according to the relation

	x	y	z
x'	ℓ_1	m_1	n_1
y'	ℓ_2	m_2	n_2
z'	ℓ_3	m_3	n_3

c_{11}, c_{12}, ... c_{44} are the elastic constants corresponding to the various crystallographic orientations. (1 Np/cm = 264 db/ft). Mason and McSkimin (6) give the scattering factors for cubic metals as

$$\left\langle\left[\frac{c_{11}'-\langle c_{11}'\rangle_{av}}{\langle c_{11}'\rangle_{av}}\right]^2\right\rangle_{av} = \frac{4}{21}\left[\frac{2(c_{12}-c_{11})+4c_{44}}{5c_{11}+2(c_{12}-c_{11})+4c_{44}}\right]^2 \tag{2.110}$$

$$\left\langle\left[\frac{c_{44}'-\langle c_{44}'\rangle_{av}}{\langle c_{44}'\rangle_{av}}\right]^2\right\rangle_{av} = \frac{17}{5760}\left[\frac{2(c_{11}-c_{12})-4c_{44}}{\langle c_{44}'\rangle_{av}}\right]^2 \tag{2.111}$$

and for hexagonal metals

$$\left\langle\left[\frac{c_{11}'-\langle c_{11}'\rangle_{av}}{\langle c_{11}'\rangle_{av}}\right]^2\right\rangle_{av} = \frac{4}{1575}\Big[48c_{11}^2-64c_{11}c_{33}+28c_{33}^2 -16c_{11}(c_{13}+c_{44})+4c_{33}(2c_{13}+4c_{44}) +3(2c_{13}+4c_{44})^2\Big]/\langle c_{11}'\rangle_{av}^2 \tag{2.112}$$

and

$$\left\langle\left[\frac{c_{44}'-\langle c_{44}'\rangle_{av}}{\langle c_{44}'\rangle_{av}}\right]^2\right\rangle_{av} = \frac{1}{45}\left[\frac{c_{11}-c_{12}-2c_{44}}{\langle c_{44}'\rangle_{av}}\right]^2 \tag{2.113}$$

For cubic metals

$$\langle c_{11}'\rangle_{av} = c_{11} + \frac{[2(c_{12}-c_{11})+4c_{44}]}{5} \tag{2.114}$$

and

$$\langle c_{44}'\rangle_{av} = c_{44} + \frac{[c_{11}-c_{12}]-2c_{44}}{12} \tag{2.115}$$

and for hexagonal metals

$$\langle c_{11}' \rangle_{av} = \frac{8}{15} c_{11} + \frac{3}{15} c_{33} + \frac{2}{15} (2c_{13} + 4c_{44}) \qquad (2.116)$$

and

$$\langle c_{44}' \rangle_{av} = \frac{1}{3} \left(\frac{c_{11} - c_{12}}{2}\right) + \frac{2}{2} c_{44} \qquad (2.117)$$

Later studies have shown Eq. (2.106) to be applicable to a limited range of ratios D/λ. Discrepancies between scattering factors and α given by Eqs. (2.108) through (2.117) and later studies also are common. Papadakis (8) gives a more thorough discussion of such scattering. These discrepancies show the extreme difficulty of accounting for all practical conditions in a simple theory. However, the ultrasonics specialist should be aware of the factors which influence attenuation, and Eqs. (2.106) through (2.117) serve this purpose for hysteresis and scattering from grain boundaries in metals. Scattering is also caused by inclusions, microcracks, and other forms of discontinuities.

X.C. ATTENUATION DUE TO HYSTERESIS

Hysteresis in an ultrasonic wave refers to a lag between the imposed stress and the resulting strain. Consequently, a plot of strain with increasing stress does not coincide with that for decreasing stress. As the stress is varied over a complete cycle, the stress-strain curve forms a characteristic hysteresis loop. The area inside this loop corresponds to energy lost by hysteresis during the cycle. Therefore, this type of absorption is proportional to the frequency (corresponding to the term $\beta_1 f$ in Eq. (2.106).

X.D. ATTENUATION DUE TO OTHER MECHANISMS

Attenuation is caused by many factors other than scattering. Any mechanism which results in the removal of energy or conversion of energy from the original state to a new form is a cause of attenuation. Some of these causes are (1) frictional losses due to relative motion between adjacent surfaces, for instance adjacent surfaces in laminated structures and powder metal compactions of less than 100 percent theoretical density; (2) conduction of heat from high stress regions to low stress regions and to regions adjacent to the ultrasonic beam; (3) micro eddy currents; (4) motions of atoms in a lattice caused by stresses which effect a divergence from a preferential distribution of the atoms; (5) viscosity (gases and liquids), and (6) dislocations in solids.

Mason and others (9) discuss many of these mechanisms in some detail.

XI. RELAXATION

When a constant, dc voltage is applied across an RC series circuit or an RL series circuit, Fig. 2.14, the final values of charge and current are not obtained instantaneously, but they are arrived at after a finite period of time. For instance, for the circuit of Fig. 2.14(a), the charge buildup on the condenser after closing the switch is

$$q = CE[1 - e^{-t/RC}] \qquad (2.118)$$

where q is the quantity of charge on the condenser at time t after closing the switch, C is the capacitance of the condenser, E is the voltage across the circuit, and R is the resistance. Similarly, the current buildup in the RL circuit of Fig. 2.14(b) is

$$i = (E/R)[1 - e^{-Rt/L}] \qquad (2.119)$$

where i is the current through the circuit at time t and L is the circuit inductance.

From Eqs. (2.118) and (2.119) it is obvious that the quantities RC and L/R govern the time required for the charge or current, respectively, to reach a certain percentage of its final value. Because of their influence on the time required to reach a percentage of their final value, these quantities, RC and L/R, are called time constants. The time constants also are equal to the time that it would take the charge or current to reach its final value if these quantities continued to change at their initial rates.

When the time t exactly equals the time constant, the quantities in the brackets of Eqs. (2.118) and (2.119) are

$$1 - e^{-1} = 0.632 \qquad (2.120)$$

In acoustics, various phenomena cause a lag between an applied strain and the resulting stress, or vice versa. These phenomena are called relaxation phenomena. In comparison with the electrical analogues of Eqs. (2.118) and (2.119), we can define a relaxation time as the time required for the dependent variable to increase from zero to $(1 - e^{-1})$ or 0.632 of its final value. The acoustical equivalent of Eqs. (2.118) and (2.119) is

$$S = (p/\rho c^2)(1 - e^{-t/\tau})$$

where S is the applied strain; τ is the relaxation time; and p is the steady-state value of pressure.

Relaxation causes absorption and this type of absorption is the one most commonly observed at high ultrasonic frequencies (megahertz range). At low frequencies, absorption due to relaxation is negligible. But, as the frequency increases, the absorption also increases, rising to a peak at a relaxation frequency f, and then decreases. Absorption due to relaxation goes to zero at a higher frequency corresponding to a condition where energy transfer from the sound wave to the particular effect to which the relaxation is related (for instancc vibration motion of molecules) has no time to occur. At this same frequency, the elastic modulus of the medium increases, causing a corresponding increase in the velocity of sound. This increase in velocity of sound is called velocity dispersion.

Relaxational types of absorption include (1) those due to conduction of heat, (2) those due to viscosity, or (3) those due to any other process which transfers acoustical energy to other modes, such as rotation or vibration of atoms within a molecule, or which transfers energy from the acoustical wave to another energy form where it is temporarily stored and returned. Some materials exhibit more than one relaxation frequency.

XII. CAVITATION

Many of the useful effects of ultrasonic energy are associated with cavitation, a term used to describe the formation of cavities, or bubbles, in a liquid medium. During the rarefaction portion of the cycle, when the pressure in the wave is below ambient, gas pockets may form and expand with the impressed field. These gases may be of two types: (1) those that have been dissolved or trapped in minute bubbles in the liquid or on surfaces in contact with the liquid and (2) vapors of the liquid itself. The first of these produces gaseous cavitation, and this form is of relatively low intensity. The second type called vaporous cavitation is of fairly high intensity.

Not all phenomena associated with cavitation appear to be explained completely by either vaporous or gaseous cavitation. If the pressure within the cavity is lower than the vapor pressure of the liquid during the expansion phase, the bubble is a result of fragmentation due to the tensile stress imposed by the ultrasonic wave being equal to the tensile strength of the liquid. This type of cavitation is very intense. The tensile strength of the liquid imposes an upper limit on the amplitude of the stress of the ultrasonic wave used to produce cavitation.

It is unlikely that the true tensile strength of the liquid is ever reached in practice because most liquids contain nuclei about which cavitation bubbles originate. These nuclei may consist of dispersed dust particles, prominences on immersed surfaces, or minute gas bubbles.

High localized stresses are developed during the formation and subsequent collapse of cavitation bubbles. Free chemical ions are produced within the vicinity of the bubble walls. High localized temperatures also may be present. Sometimes weak flashes of light are produced--an effect called sonoluminescence. The peak values of either the temperature, sometimes estimated to be as high as 13,000°F or the pressures (possibly as high as 75,000 psi) can be determined only theoretically because if these temperatures and pressures are obtained, their duration is extremely brief and they are minutely localized.

Some effects produced in the presence of cavitation include increased chemical activity (including reactions which would not occur in the absence of cavitation), erosion of surfaces, rupture or fragmentation of suspended particles, emulsification of liquid mixtures, and dispersion of small particles in the liquid. Applications based upon these phenomena will be discussed in later chapters.

The importance of cavitation to ultrasonic processing has prompted a considerable amount of research and many publications with respect to the physics and associated effects of this phenomenon (10-12).

The onset of cavitation occurs at intensities, or cavitation thresholds, that depend upon such factors as the sizes of nuclei, ambient pressure, amount of dissolved gases, vapor pressure, viscosity, surface tension, and the frequency and duration of the ultrasonic energy.

When cavitation occurs, it not only dissipates the ultrasonic energy, but it impedes transmission of the sound past it. Each bubble is a scattering site, and the scattering cross section includes both the bubble and the surrounding medium that is most directly affected by the bubble oscillations.

To be effective, that is, to cause the effects associated with the expansion and violent collapse of cavitation bubbles, the bubble must be capable of expanding with the rarefaction part of the cycle of the impressed field and of collapsing before the total pressure reaches its minimum value. That is, the bubble must reach the size where it will collapse catastrophically in less than one-quarter cycle of the impressed wave. Therefore, generation of intense cavitation depends upon the relationship between the dimensions of the nuclei, the wavelength of the sound field, and the intensity of the sound field. Bubbles larger than a critical radius, R_c, will not expand to an unstable size for catastrophic collapse before the pressure in the wave starts to increase. Frederick (11) gives the following relationships for R_c:

$$R_c = \frac{1}{\omega}\sqrt{3\gamma P_o/\rho} \qquad P_o >> 2\sigma/R_c \qquad (2.122a)$$

$$R_c = \frac{1}{\omega}\sqrt{6\gamma\sigma/\rho R_c} \qquad P_o << 2\sigma/R_c \qquad (2.122b)$$

where R_c is the critical bubble radius (cm), γ is the ratio of specific heats of the gas in the bubble, σ is the surface tension of the liquid (dyn/cm), P_o is the hydrostatic pressure (atm), and ρ is the density of the liquid (g/cm^3).

The intensity, or violence, of the cavitation bubble is a function of the ratio of the bubble radius at the instant before collapse to the minimum radius before expansion.

The effect of surface tension on bubble size is indicated by the equation

$$P_i = P_o + \frac{2\sigma}{R_o} \qquad (2.123)$$

where P_i is the vapor pressure inside the bubble and R_o is the radius of a bubble which is stable under these conditions.

Viscosity affects the rate of growth and collapse of cavitation bubbles. For this reason, very high viscosities may preclude the generation of cavitation.

In some applications, high ambient pressures have been used to suppress cavitation in coupling fluids used to transmit focused energy into specimens located in the focal region. This technique eliminates losses due to scattering from the cavities.

Figure 2.15 is a photograph of ultrasonic horns designed to produce high intensities for processing in liquids (a) after very little service, (b) after exposure to ultrasonic cavitation for approximately 4 hr in detergent and water, and (c) after exposure to ultrasonic cavitation for only a few minutes in an acidic environment. Figure 2.15(b) is an example of cavitation-induced erosion. Figure 2.15(c) is an example of cavitation-induced corosion. The frequency was 20 kHz.

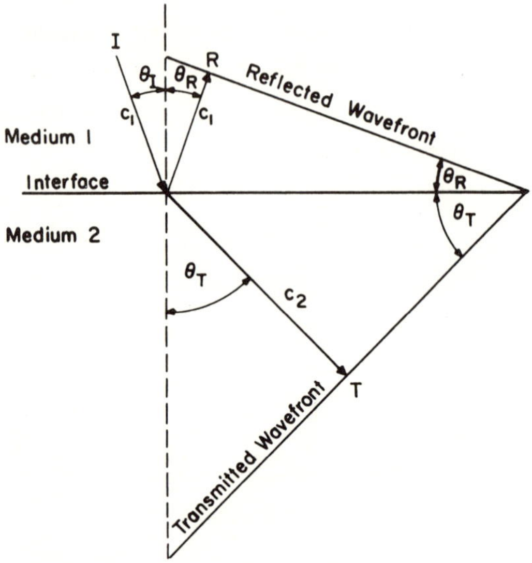

FIG. 2.1. Development of Snell's Law.

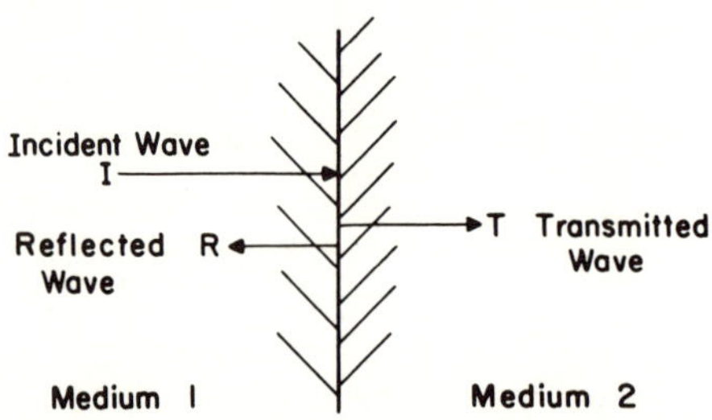

FIG. 2.2. Reflections and transmission at an interface between two media of different acoustic impedances.

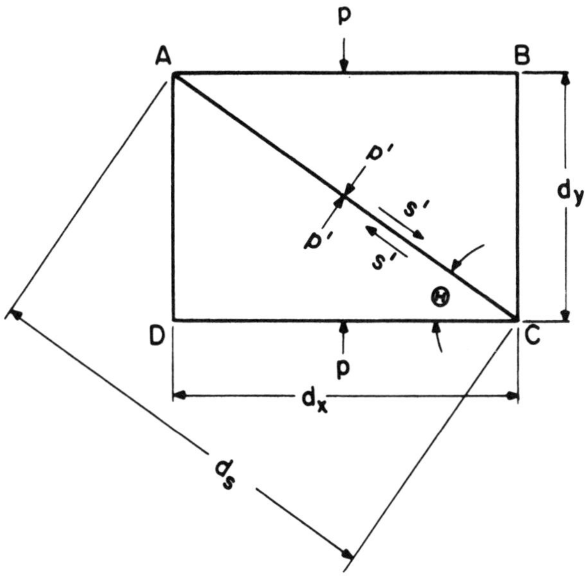

FIG. 2.3. Stresses on a plane through a block subjected to compression between two parallel faces.

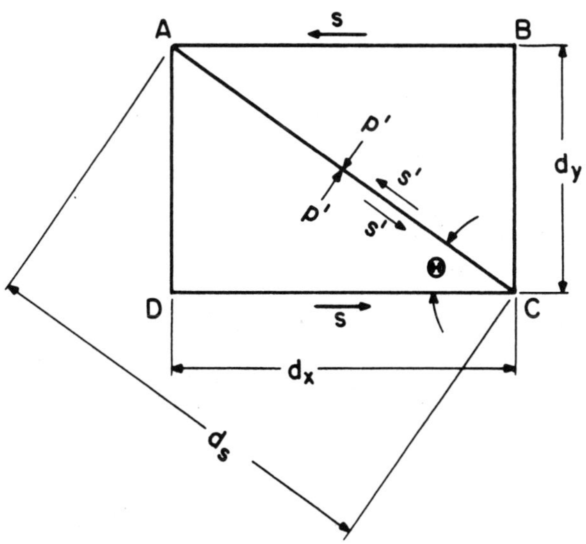

FIG. 2.4. Stresses on a plane through a block subjected to shear between two parallel faces.

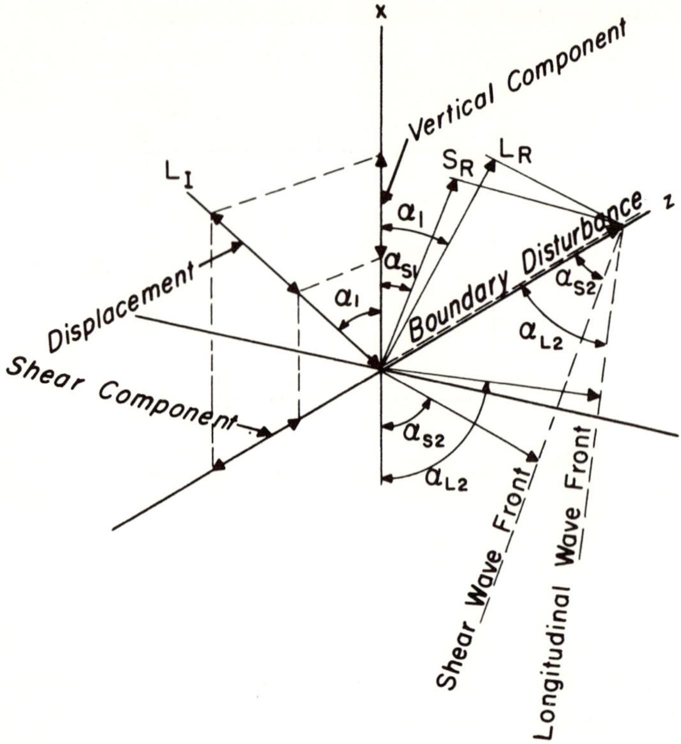

FIG. 2.5. Longitudinal wave incident on a boundary between two media.

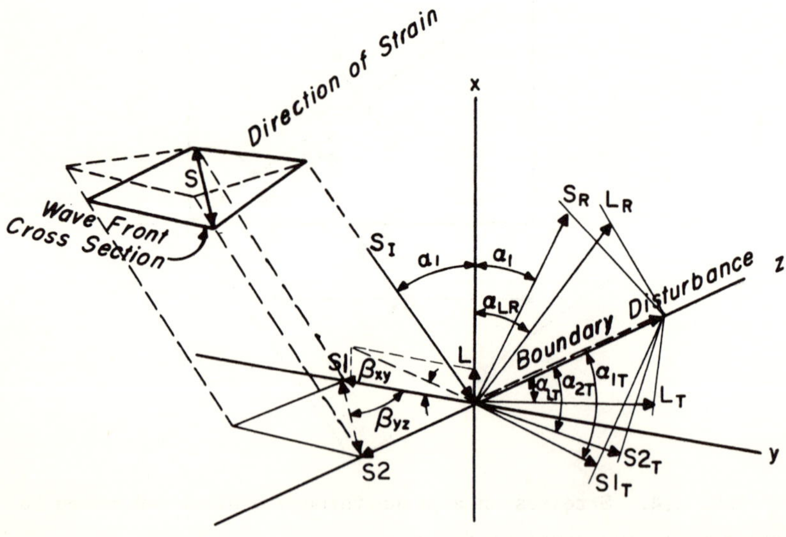

FIG. 2.6. Shear wave incident on a boundary between two media.

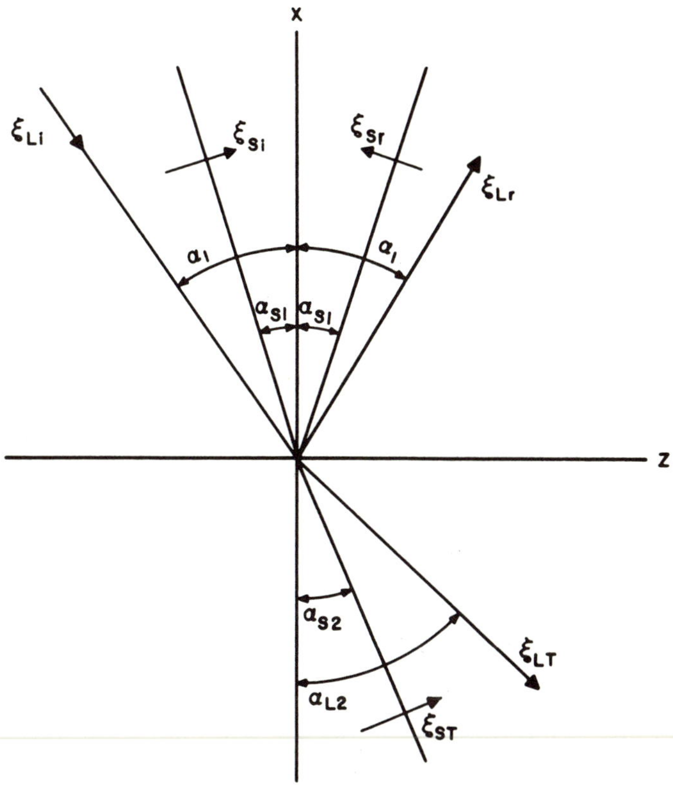

FIG. 2.7. Displacements in incident, reflected, and refracted waves at an interface between two solid media.

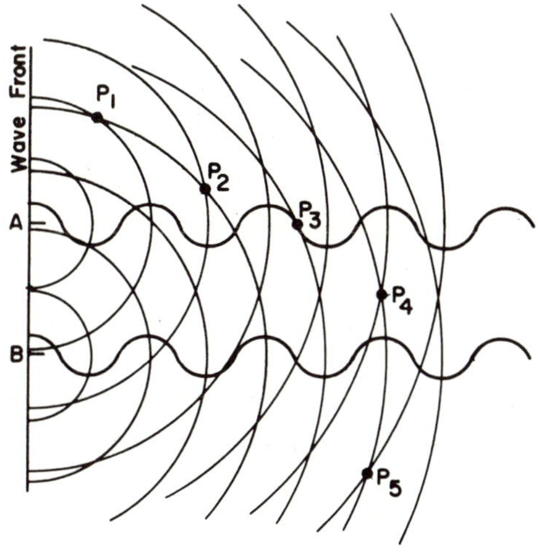

FIG. 2.8. Demonstration of Huygens' principle.

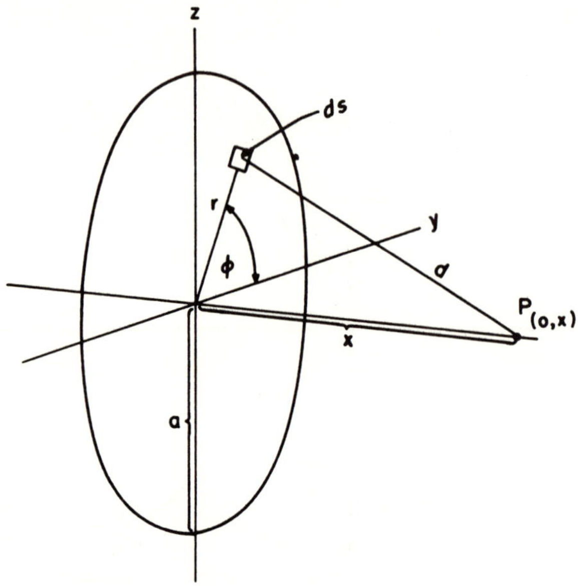

FIG. 2.9. Radiation from a circular piston.

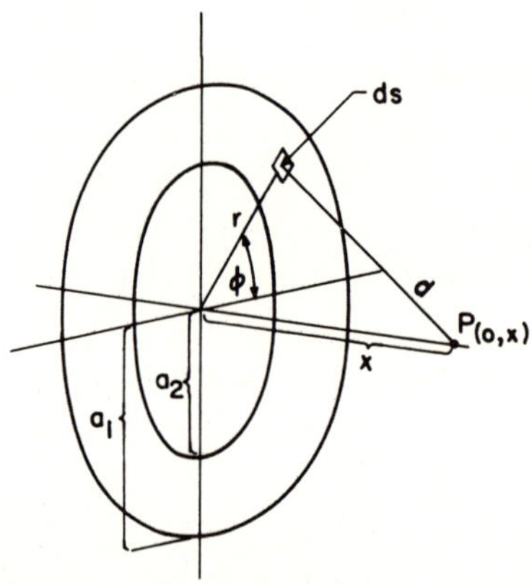

FIG. 2.10. Pressure distribution on the axis of a circular sound beam obstructed by a disk-shaped obstacle.

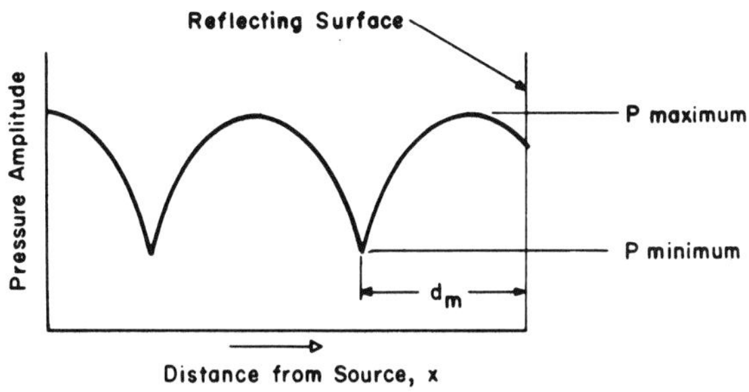

FIG. 2.11. Dimensions for measuring standing wave ratio.

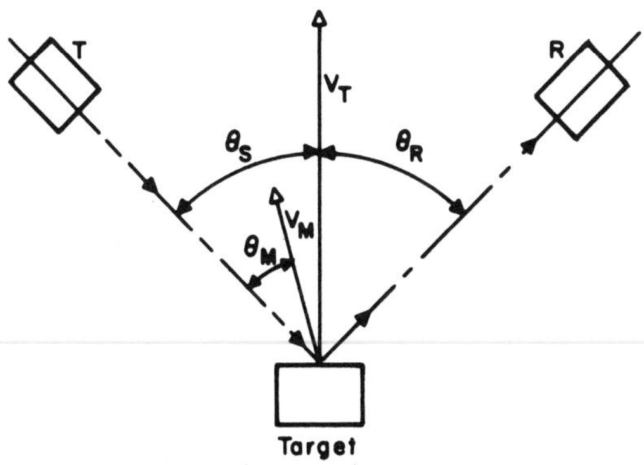

FIG. 2.12. Conditions for producing Doppler effect in a bistatic arrangement of transducers in a flowing medium.

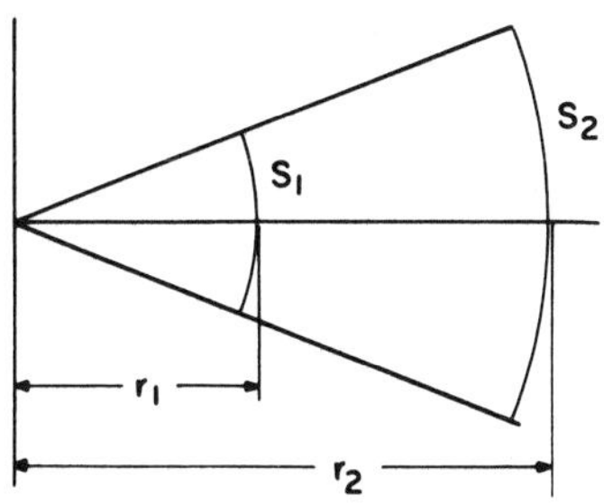

FIG. 2.13. Spreading of spherical wave front.

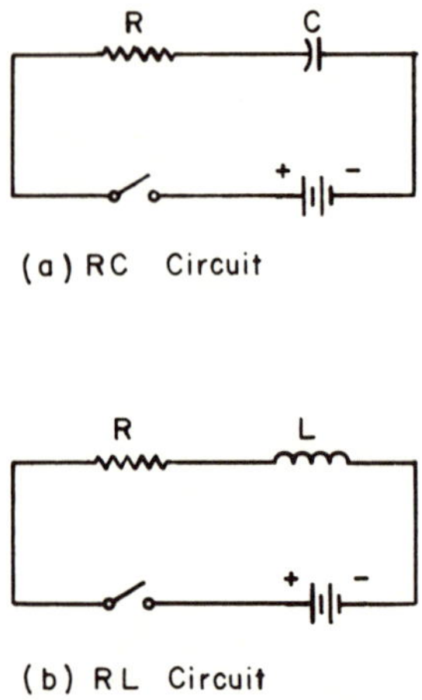

FIG. 2.14. RC and RL circuits.

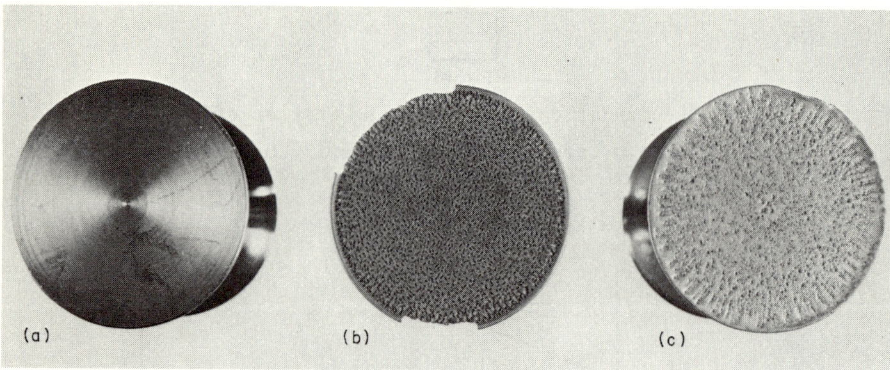

FIG. 2.15. Effects of cavitation on the radiating surfaces of Monel horns vibrating at 20 kHz: (a) after little use; (b) after approximately 4 hr exposure to intense cavitation in detergent and water; (c) after a few seconds in a dilute acid.

REFERENCES

1. J. K. Roberts and A. R. Miller, Heat and Thermodynamics, Interscience, New York, 1960.

2. M. Redwood, Mechanical Waveguides, Pergammon, New York, Oxford London, and Paris, 1960.

3. H. Kolsky, Stress Waves in Solids, Oxford at the Clarendon Press (Amen House, London), E. C. 4, 1953.

4. P. M. Morse, Vibration and Sound, McGraw-Hill, New York, 1948.

5. C. F. Ying and R. Truell, J. Appl. Phys., V. 27, 1086-1097 (1956).

6. W. P. Mason and H. J. McSkimin, J. Acoust. Soc. Am., V. 19, 464-473 (1947).

7. Lord Rayleigh, The Theory of Sound, Vol. 2, 1st. Am. Ed., Dover, New York, 1945.

8. E. P. Papadakis, in Physical Acoustics Principles and Methods (W. P. Mason, ed.), Vol. 4B, Academic, New York, 1968.

9. W. P. Mason, Physical Acoustics and the Properties of Solids, Van Nostrand, Princeton, N. J., 1958.

10. H. G. Flynn, in Physical Acoustics Principles and Methods (W. P. Mason, ed.), Vol. 1B, Academic, New York, 1964.

11. J. R. Frederick, Ultrasonic Engineering, Wiley, New York, 1965.

12. T. F. Hueter and R. H. Bolt, Sonics, Wiley, New York and Chapman and Hall, London, 1955.

Chapter 3

FUNDAMENTAL EQUATIONS EMPLOYED IN ULTRASONIC DESIGN AND APPLICATIONS

I. INTRODUCTION

To cover the wide range of usefulness of ultrasonic energy, the design of ultrasonic systems and of their applications involves the use of many equations. The most significant of these in ultrasonics is the general wave equation, to which most of the others are related. The derivation of the general wave equation and specific solutions are presented in this chapter. These derivations and solutions are intended to help the reader in understanding the principles involved in the propagation of ultrasonic waves. They also serve as a guide in deriving other equations that are not readily available to the reader. In addition to its tutorial aspects, this chapter provides a large number of the types of equations for which the designer of ultrasonic equipment and applications often has a need (as do Chaps. 2 and 4). Hopefully, these collections of equations will prove to be time savers to those who have such needs.

The oscillating mass-loaded spring is a common consideration in engineering. In some systems the combination is used to isolate vibrations of one mass from another, as in quieting machinery or household appliances, and resonance in these cases is to be avoided. In other systems, the combination is used to produce high-intensity vibrations, as in electrodynamic vibrators, and in these cases, the system is driven at resonance. The mass-loaded spring, or simple spring-mass oscillator, has engineering significance. It also serves as a fitting introduction to the derivation of the wave equation.

As mentioned in Chap. 2, an ultrasonic wave may be considered as consisting of a series of infinitesimally small mass elements elastically connected to its nearest neighbors. The spring constants are the elastic moduli. The wave equations are developed according to this concept.

By applying appropriate restrictions (boundary conditions) to the wave equation, solutions are obtained for designing various components of ultrasonic systems, for analyzing waves or causes of changes in wave structure during propagation, or for making measurements of engineering quantities.

Acoustic impedance must be considered in the design of every ultrasonic system. It is the primary factor controlling the efficiency and operation of a system. Therefore, the concepts of mechanical and acoustical impedances are introduced early in this chapter and are defined in terms that are related to the solutions of the wave equations. Acoustic impedances were used in Chap. 2 with respect to reflection and transmission of ultrasonic energy at a boundary between two media. They are used extensively in Chap. 4, which is devoted to the design of elements of an ultrasonic system, and in Chap. 5, which is devoted to the design of ultrasonic transducers.

This chapter includes the derivation of the equation for the simple spring-mass oscillator and applies the same principles to the derivation of the general wave equation. The general solution to the plane-wave equation is given and applied to longitudinally vibrating uniform bars.

The transverse-wave equation is introduced in order to present the important functions of vibrations of bars vibrating in transverse modes (flexure). There is considerable interest in the measurement of flexural vibrations in bars as a means of obtaining useful engineering data, particularly elastic moduli. To obtain these data requires an understanding of the role of these quantities in the vibrations of the bars. Also, flexural modes are sometimes used in ultrasonic transducers or systems.

Consideration of the general wave equation and of the particular form applied to transverse waves provides a background for investigating plate waves (or so-called Lamb waves). These waves are used widely in the evaluation of materials (nondestructive testing). Lamb waves are of two types: symmetrical and asymmetrical. The symmetrical waves correspond to longitudinal waves and asymmetrical waves are flexural waves. Hence, discussion of these waves is reserved until after the discussions relative to the general wave equation and its derivative, the transverse-wave equation, and concludes the chapter.

The design formulas presented in Chap. 4 are specific solutions to the plane-wave equation.

II. SIMPLE SPRING/MASS OSCILLATOR

II.A. THE IDEAL CONDITION

An acoustic wave can exist in a medium only because the medium has mass and elastic properties. Sound cannot travel through a vacuum. The incremental mass affected by an ultrasonic wave and its elastic couplings are analogous to an oscillating mass attached to springs. Therefore, as a preliminary to developing the wave equations, consider the action of such a mass attached to two springs as illustrated in Fig. 3.1. For simplicity, the system of Fig. 3.1 is assumed to be ideal; i.e., no energy is dissipated when the mass is set in motion. In actual practice, the ideal conditions can be approached closely enough that equations derived in this section can be applied to engineering designs in which energy losses are very small. Where these losses are significantly large, the equations of the next section apply.

To fulfill the hypothetical conditions, it must be assumed that the mass rests on a frictionless surface. Since the surface is horizontal, the gravitational pull on the mass is constant, regardless of the position of the mass. In addition, assume that the system is located in a vacuum so that when the mass moves there is no loss of energy due to wind resistance. Under such conditions, when the mass is displaced to the right by a distance, x, the spring to the left is subjected to tension and the spring on the right to compression, the force tending to restore the mass to its original position. If the total force exerted by the springs on mass M is F_s pounds and the combined spring constant [defined as the force required to displace the springs over unit distance (lb/in.)] is k, the magnitude of the total force F_s exerted on the mass M is

$$F_s = -k\,x \tag{3.1}$$

where x is the distance in inches that the mass is displaced from rest position. The negative sign is used because the force is in a direction opposite to the displacement x. If the mass M is released, the force F_s moves the mass back toward the rest position. Since mass has inertia, the restoring force causes an acceleration given by

$$F_M = Ma = M\frac{dv}{dt} = M\frac{d^2x}{dt^2} \tag{3.2}$$

where a is the acceleration of the mass (in./sec^2), M is w/g, v is the instantaneous velocity of the mass (in./sec), w is the weight of mass (lb), x is the displacement of the mass (in.), g is gravity (386 in./sec^2), and t is time (sec).

The force F_M is directly opposed to the force F_s and equal to F_s in magnitude, i.e.,

$$F_M - F_s = 0 \tag{3.3}$$

or

$$M\frac{d^2x}{dt^2} + kx = 0 \tag{3.3a}$$

The solution for Eq. (3.3a) is

$$x = A\cos\sqrt{\frac{k}{M}}\,t + B\sin\sqrt{\frac{k}{M}}\,t \tag{3.4}$$

Therefore, under ideal conditions (vacuum, frictionless horizontal surface) the simple oscillator of Fig. 3.1 will vibrate sinusoidally with an angular frequency $\omega = 2\pi f = \sqrt{k/M}$ where f is frequency of oscillation. Since the system is lossless, the total potential energy stored in the spring at the maximum displacement is converted to kinetic energy when the mass returns to rest position. This stored kinetic energy of the mass causes a displacement in the opposite direction which is equal in magnitude to the initial displacement. Without energy losses the operation would be repeated indefinitely. Often the mass of the spring is not negligible. In this case, one-third of the mass of the spring is added to the M of Eq. (3.4). This equation also applies to an oscillating mass suspended by a spring.

In the real situation, in which energy is lost from the system, for instance through friction, the oscillations become damped. The amplitude of vibrations decreases exponentially if the damping factor remains constant.

II.B. THE REAL CONDITION

To consider the real condition, assume that the mass is suspended on a spring. In this case the resistance to motion imposed by the air tends to remove energy from the system that is otherwise stored to promote oscillation. Slight losses in the spring which tend to heat the spring also contribute to the removal of such energy. The obvious result is that, unless energy lost is replaced by a driving force, the vibratory system will eventually stop oscillating.

The losses thus described are dependent upon the velocity of the body in vibration. Usually, the resistive force F_r is directly proportional to the velocity. Therefore, it can be expressed mathematically as

$$F_r = - R_m \sqrt{\frac{d\zeta}{dt}} \tag{3.5}$$

where R_m is called the mechanical resistance of the system and ζ is the displacement of the system.

Introducing the resistance term into Eq. (3.3a) produces the following relationship describing the motion of a simple oscillator

$$M \frac{d^2\zeta}{dt^2} + R_m \frac{d\zeta}{dt} + k\zeta = 0 \tag{3.6}$$

The general solution to Eq. (3.6) is

$$\begin{aligned} \zeta &= e^{-\frac{R_m t}{2M}} \left[K_1 \cos\sqrt{\frac{k}{M'}}\, t + K_2 \sin\sqrt{\frac{k}{M'}}\, t\right] \\ &= e^{-\alpha t} \left[K_1 \cos\sqrt{\frac{k}{M'}}\, t + K_2 \sin\sqrt{\frac{k}{M'}}\, t\right] \end{aligned} \tag{3.7}$$

where

$$M' = \frac{4M^2k}{4kM - R_m^{\ 2}}$$

$$\alpha = \frac{R_m}{2M} \quad \text{(the damping factor)}$$

Therefore, the damped angular frequency, ω_d, is

$$\omega_d = \sqrt{\frac{k}{M'}} = \sqrt{\omega_o^{\ 2} - \alpha^2} = 2\pi f_d \tag{3.8}$$

where

ω_o is $\sqrt{k/M}$ (the undamped angular frequency)

When $t = 0$, $K_1 = \zeta_o$, the displacement from the equilibrium position at time zero. The maximum amplitude depends upon the initial conditions and the manner of initiating the oscillations.

The solution to Eq. (3.6) may also be written

$$\zeta = Ke^{-\alpha t} \cos\,(\omega_d t + \phi) \tag{3.9}$$

where K and ϕ are amplitude and phase constants, respectively, and depend upon the initial conditions. Therefore, by measuring the rate of decay of such a vibratory system, it is possible to determine α and, hence, R_m.

Equation (3.8) shows that the actual resonant frequency of a mass on a spring is less than the resonant frequency of a theoretically undamped system by a factor which depends upon the damping resistance. When this damping resistance is very small, $\omega_o \approx \omega_d$.

II.C. EFFECT OF DAMPING ON PHASE RELATIONSHIPS--FORCED VIBRATIONS IN A SIMPLE OSCILLATOR

The simple oscillator (mass-loaded spring) may serve to illustrate effects to be expected in more complicated systems. For instance, assume that the simple damped oscillator, whose motion is described by Eq. (3.6), is driven by a force $Fe^{j\omega t}$. The equation describing this condition is

$$M \frac{d^2\zeta^*}{dt^2} + R_m \frac{d\zeta^*}{dt} + k\zeta^* = Fe^{j\omega t} \tag{3.10}$$

where ζ^* is a complex displacement. The solution of Eq. (3.10) is the sum of two parts: a steady-state term which depends upon F and a transient term containing two arbitrary constants. The transient term is obtained by setting F = 0, which gives Eq. (3.6) for which the general solution is Eq. (3.7). After a period of time, the damped portion of the solution becomes negligible, leaving only the steady-state solution.

If the steady-state solution is assumed to be of the form

$$\zeta^* = K^* e^{j\omega t}$$

where K^* is a complex constant, substituting in Eq. (3.10) gives

$$K^* (-\omega^2 M + j\omega R_m + k) = F \tag{3.11}$$

For Eq. (3.11) to be true for all times, t

$$K^* = \frac{F}{(k - \omega^2 M) + j\omega R_m}$$

and, therefore, the complex displacement ζ^* may be written

$$\zeta^{*} = \frac{-jFe^{(\omega t - \phi)}}{\omega R_m - j(k - \omega^2 M)} = \frac{-jFe^{j(\omega t - \phi)}}{\omega\{R_m + j(\omega M - \frac{k}{\omega})\}} \tag{3.12}$$

where ϕ is the phase angle between the driving force and the velocity and is given by

$$\phi = \tan^{-1} \frac{\omega M - \frac{k}{\omega}}{R_m}$$

By definition, the complex velocity $v^{*} = \frac{d\zeta^{*}}{dt}$, or

$$v^{*} = \frac{d\zeta^{*}}{dt} = \frac{Fe^{j(\omega t - \phi)}}{R_m + j\ (\omega M - \frac{k}{\omega})} \tag{3.13}$$

Also, the mechanical impedance Z_m is the ratio between the driving force and the velocity. From Eq. (3.13), the impedance is

$$Z_m = R_m + j(\omega M - \frac{k}{\omega}) \tag{3.14}$$

and Eq. (3.13) may be rewritten

$$v^{*} = \frac{Fe^{j(\omega t - \phi)}}{Z_m} \tag{3.15}$$

The real term, R_m, of Eq. (3.14) corresponds to a resistance and the imaginary term corresponds to reactive impedances. The electrical analogue is a series RLC circuit where R_m corresponds to resistance R, M corresponds to inductance L, and 1/k corresponds to capacitance C.

These principles may be applied to more complicated ultrasonic systems which are not covered in this book by methods illustrated previously.

III. WAVE EQUATIONS

"The wave equation" is a familiar term in the realm of physics. It is used in one form or another in the study of electromagnetic radiation, in the study of solid-state physics, and, of course, in the study of acoustics. The following section is devoted to the derivation of the plane-wave equation in terms applicable to ultrasonics. The general equation is, then, inferred from the derivation of the plane-wave equation.

III.A. PLANE-WAVE EQUATION

An elastic medium may be considered as consisting of a series of homogeneous, incremental elements or masses, each of density ρ, thickness dx, and cross-sectional area S. When a net longitudinal force is applied to the element, the element accelerates according to Eq. (3.2). Therefore, a finite period of time is required for the element to acquire a velocity other than its initial velocity.

As soon as the element moves from its initial position, it applies a force to succeeding elements which, in turn, undergo acceleration. At point x, the force on the face of the element located between x and x + dx is

$$F_x = -YS \frac{d\zeta}{dx} \qquad (3.16)$$

where Y is the modulus of elasticity of the medium in pounds per square inch and ζ is the displacement in inches. At distance x + dx, the force on the opposite face of the element is

$$-\left(F_x + \frac{\partial F_x}{\partial x}\, dx\right)$$

Therefore, the net force across the element is

$$F_x - \left(F_x + \frac{\partial F_x}{\partial x}\, dx\right) = -\frac{\partial F_x}{\partial x}\, dx$$

$$= YS \frac{\partial^2 \zeta}{\partial x^2}\, dx \qquad (3.17)$$

By replacing M with ρSdx and a with $\partial^2\zeta/\partial t^2$ in Eq. (3.2), the inertial force of the element is

$$F_m = \rho S \frac{\partial^2\zeta}{\partial t^2} dx \tag{3.18}$$

where ρ is the density of the medium.

For equilibrium to exist, the elastic forces on the element must equal the inertial forces imposed by the element. Therefore, from Eqs. (3.17) and (3.18)

$$\rho S \frac{\partial^2\zeta}{\partial t^2} dx = YS \frac{\partial^2\zeta}{\partial x^2} dx \tag{3.19}$$

or

$$\frac{\partial^2\zeta}{\partial t^2} = \frac{Y}{\rho}\frac{\partial^2\zeta}{\partial x^2} = c^2 \frac{\partial^2\zeta}{\partial x^2} \tag{3.20}$$

which is the longitudinal- or plane-wave equation with no losses. Because c is the velocity at which the stress moves from element to element in the system, it is the speed of sound in the medium. Here, Y is Young's modulus of elasticity if the medium is a solid, slender bar, small in cross section compared to a wavelength.

Equation (3.20) is the plane-wave equation. Transverse waves on a string obey the same wave equation; however, the terms are different.

A method similar to that used in deriving Eq. (3.20) can be used to derive equations governing acoustic behavior in configurations in which S is variable with x. For example, in Chap. 4 a horn equation

$$\frac{1}{c^2}\frac{\partial^2\zeta}{\partial t^2} - \frac{1}{S}\frac{\partial S}{\partial x}\frac{\partial\zeta}{\partial x} - \frac{\partial^2\zeta}{\partial x^2} = 0$$

or (3.21)

$$\frac{\partial^2 v}{\partial x^2} + \frac{1}{S}\frac{\partial S}{\partial x}\frac{\partial v}{\partial x} + \frac{\omega^2}{c^2} v = 0$$

is used to derive formulas for designing various solid horn configurations when the cross-sectional dimensions at any point x are small compared to a wavelength.

III.B. THE GENERAL WAVE EQUATION

Both Eq. (3.20) and Eq. (3.21) describe wave motion in only one direction (here defined as the x direction). The more general case considers all possible directions of propagation. The general wave equation, then, is

$$\frac{\partial^2 \zeta}{\partial t^2} = c^2\left(\frac{\partial^2}{\partial x^2} + \frac{\partial^2}{\partial y^2} + \frac{\partial^2}{\partial z^2}\right)\zeta = c^2\nabla^2\zeta \tag{3.22}$$

The term $\nabla^2\zeta$ may be expressed in any system of coordinates which best suits the configuration of the boundaries, i.e., Cartesian coordinates for a rectangular configuration, spherical coordinates for a spherical system, etc.

In a nonhomogeneous medium, velocity c is a function of space coordinates. Many different solutions to the wave equation for nonhomogeneous media are possible. The velocity c as used in this book, in general, will be an integrated average within certain coordinates. However, it often is necessary for various applications to consider nonhomogeneous systems.

IV. SOLUTION OF THE PLANE-WAVE EQUATION. LINEAR SYSTEM

IV.A. GENERAL SOLUTION

The general solution for Eq. (3.20) is

$$\zeta = f_1(ct - x) + f_2(ct + x) \tag{3.23}$$

which mathematically may or may not be periodic. The solution, f_1 (ct - x) represents a wave moving in the positive x direction and f_2 (ct + x) represents a wave moving in the negative x direction. The functions f_1 and f_2 are determined by the boundary conditions determined by a specific problem.

As a practical example, let Eqs. (3.20) and (3.23) describe longitudinal-wave motion in a thin bar. Assuming that the solution of Eq. (3.20) can be rewritten as a product of two functions that individually depend on one variable only,

$$\zeta = X(x)\,T(t) \tag{3.24}$$

Differentiating Eq. (3.24) gives

$$\frac{\partial^2 \zeta}{\partial t^2} = X \frac{\partial^2 T}{\partial t^2}$$

and (3.25)

$$\frac{\partial^2 \zeta}{\partial x^2} = T \frac{\partial^2 X}{\partial x^2}$$

Substituting Eq. (3.25) in Eq. (3.20) gives

$$X \frac{\partial^2 T}{\partial t^2} = c^2 T \frac{\partial^2 X}{\partial x^2} \tag{3.26}$$

The variables may be separated by dividing both sides of Eq. (3.26) by the product XT; therefore

$$\frac{1}{T} \frac{\partial^2 T}{\partial t^2} = \frac{c^2}{X} \frac{\partial^2 X}{\partial x^2} \tag{3.27}$$

To be valid for all possible values of x and t, it follows that both sides of Eq. (3.27) must be equal to a common constant. For this example, let

$$\frac{1}{T} \frac{\partial^2 T}{\partial t^2} = \frac{c^2}{X} \frac{\partial^2 X}{\partial x^2} = -p^2 \tag{3.27a}$$

Then the following ordinary differential equations can be written

$$\frac{d^2 T}{dt^2} + p^2 T = 0$$

(3.28)

and

$$\frac{d^2 X}{dx^2} + \frac{p^2}{c^2} X = 0$$

The solutions of Eq. (3.28) are

$$T = A \cos pt + B \sin pt$$

(3.29)

and

$$X = C \cos \frac{p}{c} x + D \sin \frac{p}{c} x$$

where p has the dimensions of radians per second and is identical with $\omega = 2\pi f$ defined previously.

Therefore, the solution of Eq. (3.20) may be written as follows:

$$\begin{aligned} \zeta &= X(x)\ T(t) \\ &= [C \cos \frac{\omega x}{c} + D \sin \frac{\omega x}{c}][A \cos \omega t + B \sin \omega t] \\ &= \Big\{\frac{1}{2}\ [AC + BD] \cos [\frac{\omega t}{c} - \omega t] + [AC - BD] \cos [\frac{\omega x}{c} + \omega t] \\ &\quad + [BC + AD] \sin [\frac{\omega x}{c} + \omega t] + [BC - AD] \sin (\omega t - \frac{\omega x}{c})\Big\} \end{aligned}$$

(3.30)

IV.B. FREE-FREE LONGITUDINALLY VIBRATING UNIFORM BAR

The simplest example of the use of Eq. (3.30) is in describing the vibratory characteristics of a slender bar, free at each end, and in longitudinal half-wave resonance (Fig. 3.2). In this example, the boundary conditions are:

(1) At $x = 0$ and at $x = \ell$, $\partial\zeta/\partial x = 0$

(2) At $x = 0$, $\zeta = \zeta_o$ and $v = v_o$

Differentiating Eq. (3.30) with respect to x gives

$$\begin{aligned} \frac{\partial\zeta}{\partial x} = \frac{\omega}{2c}\Big\{&-[AC + BD] \sin (\frac{\omega x}{c} - \omega t) - [AC - BD] \sin (\frac{\omega x}{c} + \omega t) \\ &+ [BC + AD] \cos (\frac{\omega x}{c} + \omega t) + [BC - AD] \cos (\omega t - \frac{\omega x}{c}) \end{aligned}$$

(3.31)

Applying the boundary condition that at $x = 0$, $\partial\zeta/\partial x = 0$ reduces Eq. (3.31) to

$$\frac{\partial \zeta}{\partial x} = \frac{\omega}{c} [BD \sin \omega t + BC \cos \omega t] = 0 \tag{3.32}$$

Since Eq. (3.32) must be true for all values of t, BD = 0 and BC = 0. Substituting these values into Eq. (3.30) gives

$$\zeta = [AC] \cos (\frac{\omega x}{c}) \cos (\omega t) + [AD] \sin (\frac{\omega x}{c}) \sin (\omega t) \tag{3.33}$$

When x = 0

$$\zeta = \zeta_o = [AC] \cos \omega t = \zeta_m \cos \omega t \tag{3.34}$$

where ζ_m is the maximum displacement of the end of the bar where x = 0. Therefore

$$\zeta = \zeta_m \cos (\frac{\omega x}{c}) \cos \omega t + [AD] \sin (\frac{\omega x}{c}) \sin \omega t \tag{3.35}$$

$$v = \frac{\partial \zeta}{\partial t} = -\omega \zeta_m \cos (\frac{\omega x}{c}) \sin (\omega t) + [AD]\omega \sin (\frac{\omega x}{c}) \cos (\omega t) \tag{3.36}$$

According to the boundary conditions, at x = 0

$$v = v_o = -\omega \zeta_m \sin (\omega t) \tag{3.37}$$

or

$$v_m = \omega \zeta_m$$

The remaining unknown quantity is [AD]. This quantity is determined by applying the boundary condition that at x = ℓ

$$\frac{\partial \zeta}{\partial t} = 0$$

which gives

$$\frac{\partial \zeta}{\partial x} = -\frac{\omega}{c} \zeta_m \sin (\frac{\omega x}{c}) \cos (\omega t) + [AD] \frac{\omega}{c} \zeta_m \cos (\frac{\omega x}{c}) \sin (\omega t) \tag{3.38}$$

$$= 0 + [AD] \frac{\omega}{c} (-1) \sin \omega t$$

Therefore, [AD] = 0

and

$$\zeta = \zeta_m \cos\left(\frac{\omega x}{c}\right) \cos \omega t \tag{3.39}$$

describes the displacement along a slender bar in longitudinal-half-wave resonance. The particle velocity at any point x along the bar is

$$v = -\omega\zeta_m \cos\left(\frac{\omega x}{c}\right) \sin(\omega t) = j\omega\zeta \tag{3.40}$$

The particle acceleration at any point x along the bar is

$$a = \frac{\partial v}{\partial t} = -\omega^2\zeta_m \cos\left(\frac{\omega x}{c}\right) \cos(\omega t) \tag{3.41}$$

$$= -\omega^2\zeta$$

IV.C. STRESS IN A VIBRATING UNIFORM BAR

Formulas were derived in the previous sections on the assumption that stresses were developed in proportion to the strain coincident with the acoustic wave. Equation (3.16) defined the assumed condition mathematically and may be rewritten

$$\frac{F_x}{S} = -Y\frac{\partial\zeta}{\partial x} = s = \text{stress}$$
$$= j\frac{Y}{\omega}\frac{\partial v}{\partial x} \tag{3.42}$$

Equations (3.16) and (3.42) assume that Y is constant regardless of whether the material is in compression or in tension. It also must be assumed that the material is isotropic. If Y is defined as Young's modulus, in the dynamic case the bar must be slender compared to the wavelength (which, for the half-wave resonant bar, is twice the length of the bar). Other moduli will be considered in later sections as the need arises.

From Eqs. (3.39), (3.40), and (3.42), it follows that

$$s = -Y\frac{\partial\zeta}{\partial x} = Y\frac{\omega\zeta_m}{c} \sin\left(\frac{\omega x}{c}\right) \cos(\omega t) \tag{3.42a}$$

or

$$s = j\,\frac{Y}{\omega}\,\frac{\partial v}{\partial x} = j\,Y\,\frac{\omega\zeta_m}{c}\,\sin\left(\frac{\omega x}{c}\right)\,\sin(\omega t) \qquad (3.42b)$$

IV.D. MECHANICAL IMPEDANCE

Impedance is the primary factor controlling the efficiency and operation of a system, whether the system is electrical or acoustical.

Impedances in an electrical circuit limit the flow of electrical current in accordance with the magnitude and type of impedances. The analogue for an acoustical system relates force (or stress) to velocity.

A force F impressed on an increment of an acoustical system imparts a velocity of motion, v, to the increment in proportion to F. Expressed mathematically

$$F = Zv \qquad (3.43)$$

Equation (3.43) is analogous to the electrical case

$$E = Z_e I \qquad (3.44)$$

The force F corresponds to the voltage E, velocity potential v to current I, and proportionality constant Z to electrical impedance $Z_e = R_e + jX_e$. Therefore, $Z = R + jX$ may be defined as mechanical impedance.

A long, slender bar is the acoustical equivalent to an electrical transmission line with distributed constants. The impedance at any point x is determined by dividing the force at x by the velocity potential at x. For the free-free bar without losses

$$Z = F/v = sS/v = -j\,\frac{SY}{c}\,\tan(\omega x/c)$$

$$= -jS\rho c\,\tan(\omega x/c) \qquad (3.45)$$

where $c = \sqrt{Y/\rho}$. The product, ρc, of Eq. (3.45) is called the characteristic impedance of the medium. Standing alone, it is a pure resistance.

Equation (3.45) shows a purely reactive impedance. The reason for this is that losses have been ignored in deriving the wave equation and the equations of vibration leading to Eq. (3.45). Equation (3.45) is a close approximation to the actual impedance if the loss factors are very small compared to the reactive components of the system. However, if the losses are significant, the wave equation must take them into account. The plane-wave equation can be corrected for losses by the addition of the term R_m/ρ as follows:

$$\frac{\partial^2\zeta}{\partial t^2} + \frac{R_m}{\rho}\frac{\partial\zeta}{\partial t} = c^2\frac{\partial^2\zeta}{\partial x^2} \tag{3.46}$$

or

$$(\omega^2 - j\omega R_m)v + c^2\frac{\partial^2 v}{\partial x^2} = 0 \tag{3.46a}$$

where ω is angular frequency that would occur if there were no losses in the system.

The solution to Eq. (3.46a) is

$$v = A\cos\frac{\omega_d}{c}x + B\sin\frac{\omega_d}{c}x \tag{3.47}$$

where

$$\omega_d = \pm\sqrt{\omega^2 - j\omega R_m/\rho} = \pm\omega\sqrt{1 - jR_m/\rho\omega} \tag{3.47a}$$

IV.E. QUALITY FACTOR (Q)

If it were possible to have a vibratory system in which no loss of energy occurs, the system would vibrate indefinitely. Energy introduced into such a system would be stored and increase the intensity of the oscillations. As soon as energy is extracted from the system, whether to do work or by absorption, the vibratory characteristics are affected. Energy losses are always present in a real system. These losses are functions of velocities in the vibratory system so that, when energy is supplied to the system, the amplitude rises to an equilibrium level where the energy supplied exactly equals the energy dissipated. Thus, it can be shown that when the driving force is suddenly removed from an oscillatory system, the amplitude of vibration decreases exponentially, since the energy losses decrease with decreasing amplitude of vibration.

As has been shown, the ratio of energy stored to energy dissipated affects the amplitude of vibration and the decay rate of the oscillations. In addition, it controls the bandwidth of the driven system, defined as the total width of the steady-state response curve measured in hertz between the frequency below resonance and that above resonance at which the power is one-half the power at the resonance peak. This ratio of energy stored to energy dissipated is represented by the ratio of the reactive component of the impedance of the system to the resistive component. This ratio is called the Q of the system. If the resistive component is very small compared with the reactance, the bandwidth is approximately the product of the angular frequency at resonance and the inverse of the Q.

V. THE TRANSVERSE-WAVE EQUATION

The form of the wave equation that applies to bars vibrating in flexure (see Fig. 3.3), i.e., the transverse-wave equation, is

$$\frac{\partial^2 \zeta}{\partial t^2} = -\kappa^2 c^2 \frac{\partial^4 \zeta}{\partial x^4} \tag{3.48}$$

where ζ is the transverse displacement of the bar from its normal configuration, in., κ is the radius of gyration of the cross section of the bar (in.), and c is $\sqrt{Y/\rho}$ (in./sec). Equation (3.48) is derived in much the same manner as was Eq. (3.20) except that bending moments (M) and shear forces (F_y) are considered rather than the simple longitudinal forces considered previously. Even by limiting the displacement and slope of the bar to small values so that variations in angular momentum are negligible, Eq. (3.48) is still an approximation in which second-order terms involving $(dx)^2$ are ignored. The procedure is as follows (referring to Fig. 3.3).

At static equilibrium, moments about the left end of the segment of Fig. 3.3 are

$$M_x - M_{(x + dx)} - (F_y)_{(x + dx)} dx = 0 \tag{3.49}$$

For small dx

$$M_{(x + dx)} = M_x + \frac{\partial M}{\partial x} dx$$

and

$$(F_y)_{(x + dx)} = (F_y)_x + (\frac{\partial F_y}{\partial x} dx$$

These expressions introduced into Eq. (3.49) give

$$M_x - M_x - \frac{\partial M}{\partial x} dx - (F_y)_x dx - (\frac{\partial F_y}{\partial x} dx) dx = 0$$

from which

$$F_y = -\frac{\partial M}{\partial x} = YS\kappa^2 \frac{\partial^3 \zeta}{\partial x^3} \tag{3.50}$$

if the second-order term involving $(dx)^2$ is neglected. Equation (3.50) approximates the true relationship between F_y and y only if the displacement and slope of the bar are small enough that variations in angular momentum may be neglected.

The net upward force dF_y acting on the segment dx is

$$dF_y = (F_y)_x - (F_y)_{(x + dx)} = -(\frac{\partial F_y}{\partial x}) dx = -YS\kappa^2 \frac{\partial^4 \zeta}{\partial x^4} dx \tag{3.51}$$

which will give the segment an upward acceleration, or

$$dF_y = (\rho S dx) \frac{\partial^2 \zeta}{\partial t^2} = -YS\kappa^2 \frac{\partial^4 \zeta}{\partial x^4} dx \tag{3.52}$$

or

$$\frac{\partial^2 \zeta}{\partial t^2} = -\kappa^2 c^2 \frac{\partial^4 \zeta}{\partial x^4} \tag{3.48}$$

VI. SOLUTION OF THE TRANSVERSE-WAVE EQUATION

The general solution of the transverse-wave equation [Eq. (3.48)] is

$$y = e^{j\omega t}\left(\bar{A}e^{\frac{\omega x}{v}} + \bar{B}e^{\frac{-\omega x}{v}} + \bar{C}e^{j\frac{\omega x}{v}} + \bar{D}e^{-j\frac{\omega x}{v}}\right) \tag{3.53}$$

where $\bar{A}$, $\bar{B}$, $\bar{C}$, and $\bar{D}$ are complex amplitude constants,

$v = \sqrt{\omega c \kappa}$ (phase velocity),

$c = \sqrt{Y/\rho}$ (wave velocity), and

κ is radius of gyration of the cross section (1).

It can be seen that phase velocity v is a function of frequency as well as geometry. Since this is true, when a wave of transverse vibrations containing many frequency components is impressed on a bar, the wave shape changes as the wave progresses along the bar because the higher frequency components travel faster than the lower frequency components.

Only the real part of Eq. (3.53) is of any consequence in practical applications. Equation (3.53) may, therefore, be rewritten in the form

$$y = (\cos \omega t + \bar{j} \sin \omega t)[\bar{A}(\cosh \frac{\omega x}{v} + \sinh \frac{\omega x}{v} + \bar{B}(\cosh \frac{\omega x}{v} - \sinh \frac{\omega x}{v})$$

$$+ \bar{C}(\cos \frac{\omega x}{v} + j \sin \frac{\omega x}{v}) + \bar{D}(\cos \frac{\omega x}{v} - j \sin \frac{\omega x}{v})]$$

$$= (\cos \omega t + \phi)[(\bar{A}+\bar{B}) \cosh \frac{\omega x}{v} + (\bar{A}-\bar{B}) \sinh \frac{\omega x}{v} + (\bar{C}+\bar{D}) \cos \frac{\omega x}{v}$$

$$+ \bar{j}(\bar{C}-\bar{D}) \sin \frac{\omega x}{v}] \tag{3.53a}$$

Letting A equal the real component of $(\bar{A} + \bar{B})$,
B equal the real component of $(\bar{A} - \bar{B})$,
C equal the real component of $(\bar{C} - \bar{D})$,
D equal the real component of $\bar{j}\ (\bar{C} - \bar{D})$;

we can write

$$\zeta = \cos(\omega t + \phi)[A \cosh \frac{\omega x}{v} + B \sinh \frac{\omega x}{v} + C \cos \frac{\omega x}{v} + D \sin \frac{\omega x}{v}] \quad (3.54)$$

which contains only real amplitude terms A, B, C, and D. This simplification is justified by the harmonic form of ζ. In Eq. (3.54), ϕ is the initial phase angle of the motion.

VI.A. CLAMPED-FREE UNIFORM BAR

A clamped-free bar, such as a cantilever, is rigidly mounted at one end so that it is perfectly restrained from moving at that end but unrestrained and free to vibrate at the other end. Under these conditions, at the clamped end, $(x = 0)$, the displacement ζ is 0 and the slope $\partial\zeta/\partial x$ is 0 for all values of t. At the free end $(x = \ell)$, no external force is applied and both the bending moment and the shearing force are zero in a plane through the end face of the bar. Therefore, at the free end $(x = \ell)$, $\partial^2\zeta/\partial x^2 = 0$ and $\partial^3\zeta/\partial x^3 = 0$.

Applying the boundary conditions at $x = 0$ to Eq. (3.54) we obtain

$$\zeta = 0 = A + C, \text{ or } A = -C$$

and

$$\frac{\partial\zeta}{\partial x} = 0 = B + D, \text{ or } B = -D$$

from which

$$\zeta = \cos(\omega t + \phi)[A(\cosh \frac{\omega x}{v} - \cos \frac{\omega x}{v} + B(\sinh \frac{\omega x}{v} - \sin \frac{\omega x}{v})] \quad (3.55)$$

Applying the boundary conditions at $x = \ell$ to Eq. (3.55) gives

$$\frac{\partial^2\zeta}{\partial x^2} = 0 = \cos(\omega t + \phi)\left(\frac{\omega}{v}\right)^2 [A(\cosh \frac{\omega\ell}{v} + \cos \frac{\omega\ell}{v}) + B(\sinh \frac{\omega\ell}{v} + \sin \frac{\omega\ell}{v})] \quad (3.56)$$

and

$$\frac{\partial^3\zeta}{\partial x^3} = 0 = \cos(\omega t + \phi)\left(\frac{\omega}{v}\right)^3 [A(\sinh \frac{\omega\ell}{v} - \sin \frac{\omega\ell}{v}) + B(\cosh \frac{\omega\ell}{v} + \cos \frac{\omega\ell}{v})] \quad (3.57)$$

Therefore

$$A(\cosh \frac{\omega \ell}{v} + \cos \frac{\omega \ell}{v}) = -B(\sinh \frac{\omega \ell}{v} + \sin \frac{\omega \ell}{v}) \tag{3.56a}$$

and

$$A(\sinh \frac{\omega \ell}{v} - \sin \frac{\omega \ell}{v}) = -B(\cosh \frac{\omega \ell}{v} + \cos \frac{\omega \ell}{v}) \tag{3.57a}$$

which can be true only at discrete frequencies. These discrete, or allowed, frequencies are determined by dividing Eq. (3.56a) by Eq. (3.57a), as follows:

$$\frac{\cosh \frac{\omega \ell}{v} + \cos \frac{\omega \ell}{v}}{\sinh \frac{\omega \ell}{v} - \sin \frac{\omega \ell}{v}} = \frac{-(\sinh \frac{\omega \ell}{v} + \sin \frac{\omega \ell}{v})}{-(\cosh \frac{\omega \ell}{v} + \cos \frac{\omega \ell}{v})}$$

from which

$$\cosh^2 \frac{\omega \ell}{v} + 2 \cosh \frac{\omega \ell}{v} \cos \frac{\omega \ell}{v} + \cos^2 \frac{\omega \ell}{v} = \sinh^2 \frac{\omega \ell}{v} - \sin^2 \frac{\omega \ell}{v} \tag{3.58}$$

or

$$\cosh \frac{\omega \ell}{v} \cos \frac{\omega \ell}{v} = 1 \tag{3.58a}$$

Thus the resonance frequencies of a fixed-free bar resonating in flexure are determined by Eq. (3.58a).

From the identities

$$\cosh 2\theta = \cosh^2 \theta + \sinh^2 \theta$$

$$\cos 2\theta = \cos^2 \theta - \sin^2 \theta$$

Equation (3.58a) may be rewritten

$$(\cosh^2 \frac{\omega \ell}{2v} + \sinh^2 \frac{\omega \ell}{2v})(\cos^2 \frac{\omega \ell}{2v} - \sin^2 \frac{\omega \ell}{2v}) = -1 \tag{3.59}$$

Dividing both sides of Eq. (3.59) by

$\cosh^2 \frac{\omega \ell}{2v} \cos^2 \frac{\omega \ell}{2v}$, gives

$$(1 + \tanh^2 \frac{\omega \ell}{2v})(1 - \tan^2 \frac{\omega \ell}{2v}) = -\frac{1}{\cosh^2 \frac{\omega \ell}{2v} \cos^2 \frac{\omega \ell}{2v}}$$

$$= -(1 - \tanh^2 \frac{\omega \ell}{2v})(1 + \tan^2 \frac{\omega \ell}{2v}) \tag{3.60}$$

or

$$\tan^2 \frac{\omega \ell}{2v} \tanh^2 \frac{\omega \ell}{2v} = 1$$

or

$$\cot \frac{\omega \ell}{2v} = \pm \tanh \frac{\omega \ell}{2v}$$

The intersections of the curves corresponding to cot $\omega\ell/2v$ and $\pm$ tanh $\omega\ell/2v$ plotted against $(\omega\ell/2v)$ give the frequencies corresponding to the allowed modes of vibration of the clamped-free uniform bar. These curves are plotted in Fig. 3.4.

The intersections occur at

$$\frac{\omega \ell}{2v} = \frac{\pi}{4} (1.194, 2.988, 5, 7\ldots) \tag{3.61}$$

The numerals 5, 7, ... etc., follow from the fact that the hyperbolic tangent is nearly unity for all angles greater than π.

Since

$$v = \sqrt{\omega c \kappa}$$

or

$$v^2 = \omega c \kappa = 2\pi f c \kappa$$

it follows from Eq. (3.61) that

$$f = \frac{\pi c \kappa}{8\ell^2} (1.194^2, 2.988^2, 5^2, 7^2\ldots) \tag{3.62}$$

are allowed resonant frequencies of a fixed-free bar vibrating in flexure.

The nodes of the fixed-free bar (corresponding to $\zeta = 0$) are not evenly distributed along the bar. Since the hypothesis states that the bar is clamped at one end ($x = 0$), the clamped end is a nodal position. Here, both $\zeta = 0$ and $\partial\zeta/\partial x = 0$. This is the only node occurring at the fundamental frequency. One additional node occurs near the free end at the first overtone. This node is characterized by $\zeta = 0$ but $\partial\zeta/\partial x \neq 0$ and $\partial^2\zeta/\partial x^2 \neq 0$. Therefore, two different types of nodes exist for this mode. The second overtone includes nodes of these two types, one at the clamped end and the other near the free end, plus a third node characterized by $\zeta = 0$ and $\partial^2\zeta/\partial x^2 = 0$. Each successive overtone adds an additional node characterized by the conditions $\zeta = 0$ and $\partial^2\zeta/\partial x^2 = 0$, while also retaining the two types of nodes near the extremities of the lower modes.

These intermediate nodes associated with the second overtone and higher modes (f_3, f_4, $f_5 \ldots f_n$) can be determined by the equations

$$\zeta_n = \cos\,(\omega t + \phi_n)[A(\cosh \frac{\omega x}{v} - \cos \frac{\omega x}{v}) + B(\sinh \frac{\omega x}{v} - \sin \frac{\omega x}{v})]$$

$$= 0 \qquad (n = 3, 4, 5, \ldots) \qquad (3.55a)$$

and

$$\frac{\partial^2 \zeta_n}{\partial x^2} = \cos\,(\omega t + \phi_n)(\frac{\omega}{v})^2[A(\cosh \frac{\omega x}{v} + \cos \frac{\omega x}{v}) + B(\sinh \frac{\omega x}{v} + \sin \frac{\omega x}{v})] = 0 \qquad (3.56a)$$

Therefore

$$A(\cosh \frac{\omega x}{v} - \cos \frac{\omega x}{v}) = -B(\sinh \frac{\omega x}{v} - \sin \frac{\omega x}{v} \qquad (3.55b)$$

$$B(\cosh \frac{\omega x}{v} + \cos \frac{\omega x}{v}) = -B(\sinh \frac{\omega x}{v} + \sin \frac{\omega x}{v}) \qquad (3.56b)$$

Dividing Eq. (3.55b) by Eq. (3.56b) and simplifying gives

$$2 \cosh \frac{\omega x}{v} \sin \frac{\omega x}{v} = 2 \sinh \frac{\omega x}{v} \cos \frac{\omega x}{v} \qquad (3.63)$$

or

$$\tan \frac{\omega x}{v} = \tanh \frac{\omega x}{v}$$

Equation (3.63) is solved graphically as was Eq. (3.60) for discrete values of x corresponding to y = 0. The quantity $\omega x/v$ is greater than π and lies only in the quadrants where $\tan \omega x/v$ is positive, i.e. where

$$\frac{\omega x}{v} = \frac{5\pi}{4}, \frac{9\pi}{4}, \frac{13\pi}{4}, \ldots \qquad (3.64)$$

and

$$\frac{\omega}{v} = \frac{\pi}{2\ell}(5, 7, 9, \ldots) \qquad (3.61a)$$

corresponding to the second, third, fourth, etc., overtones. For example, the nodal positions for the second overtone are as follows:

First nodal position: $x = 0$

Second nodal position: $\frac{\omega x}{v} = \frac{5\pi}{4}$

and

$$\frac{\omega}{v} = \frac{5\pi}{2\ell} \quad \text{[from Eq. (3.61)]}$$

and

$$\frac{5\pi}{2\ell}\,x = \frac{5\pi}{4}$$

$$x = 0.5\ell \text{ (corresponding to } f_3\text{)}$$

The third nodal position of the second overtone is not accurately determined by Eq. (3.63) because the conditions that $\partial^2\zeta/\partial x^2 = 0$, on which Eq. (3.63) is based, does not exist at the node nearest the free end. Rather $\partial^2\zeta/\partial x^2 = 0$ occurs at the free end. Neither is $\partial\zeta/\partial x = 0$. However, ζ must be zero for a node to exist. Equations (3.61a) and (3.64) may be used to obtain a first approximation of the location of the third node of the second overtone, and this approximate value is $x = 0.9\ell$ which would be the true value if $\partial^2\zeta/\partial x^2 = 0$ at that position. The actual location of the third node is at $x = 0.868\ell$.

Actual nodal positions, corresponding frequencies, and phase velocities are given in Table 3.1.

TABLE 3.1

Locations of Nodes and Corresponding Phase Velocities of a Fixed-Free Bar in Transverse Vibration

Frequency	Phase velocity	Nodal positions
f_1	v_1	$x = 0$
f_2 (=$6.267f_1$)	$2.50v_1$	$x = 0, 0.774\ell$
f_3 (=$17.55f_1$)	$4.18v_1$	$x = 0, 0.5\ell, 0.868\ell$
f_4 (=$34.39f_1$)	$5.87v_1$	$x = 0, 0.356\ell, 0.644\ell, 0.905\ell$

The modes of transverse vibrations of a clamped-free bar are shown in Fig. 3.5.

VI.B. FREE-FREE BAR (BAR FREE AT BOTH ENDS)

Again referring to the general solution of the wave equation for a bar in transverse vibration [Eq. (3.54)], the boundary conditions for a free-free bar are $\partial^2\zeta/\partial x^2 = 0$ and $\partial^3\zeta/\partial x^3 = 0$ at both $x = 0$ and $x = \ell$.

Applying these conditions at $x = 0$

$$\frac{\partial^2\zeta}{\partial x^2} = \cos(\omega t + \phi)\frac{\omega^2}{v^2}[A \cosh 0 + B \sinh 0 - C \cos 0 - D \sin 0] = 0$$

and

$$\frac{\partial^3\zeta}{\partial x^3} = \cos(\omega t + \phi)\frac{\omega^3}{v^3}[A \sinh 0 + B \cosh 0 + C \sin 0 - D \cos 0] = 0 \tag{3.65}$$

from which $A - C = 0$ and $B - D = 0$.
Therefore

$$\zeta = \cos(\omega t + \phi)[A(\cosh \frac{\omega x}{v} + \cos \frac{\omega x}{v}) + B(\sinh \frac{\omega x}{v} + \sin \frac{\omega x}{v})] \tag{3.66}$$

is the equation of motion of a free-free bar in flexure.

At $x = \ell$

$$\frac{\partial^2\zeta}{\partial x^2} = \cos(\omega t + \phi)\frac{\omega^2}{v^2}[A \cosh \frac{\omega\ell}{v} + B \sinh \frac{\omega\ell}{v} - A \cos \frac{\omega\ell}{v} - B \sin \frac{\omega\ell}{v}] = 0$$

and

$$\frac{\partial^3\zeta}{\partial x^3} = \cos(\omega t + \phi)\frac{\omega^3}{v^3}[A \sinh \frac{\omega\ell}{v} + B \cosh \frac{\omega\ell}{v} + A \sin \frac{\omega\ell}{v} - B \cos \frac{\omega\ell}{v}] = 0 \tag{3.67}$$

Using the procedure of the previous section for the clamped-free bar, we obtain

$$\cosh\left(\frac{\omega\ell}{v}\right) \cos\left(\frac{\omega\ell}{v}\right) = 1$$

$$\tan\left(\frac{\omega\ell}{2v}\right) = \pm \tanh\left(\frac{\omega\ell}{2v}\right) \tag{3.68}$$

from which the allowed frequencies are obtained. The allowed frequencies corresponding to the intersection of tan $(\omega\ell/2v)$ with $\pm$ tanh $(\omega\ell/2v)$ plotted against $\omega\ell/2v$ are given by

$$\frac{\omega\ell}{2v} = \frac{\pi}{4}\,(3.0112,\ 5,\ 7,\ 9\ldots) \qquad (3.69)$$

and, since

$$v = \sqrt{\omega c\kappa}$$

$$f = \frac{\pi c\kappa}{8\ell^2}\,(3.0112^2,\ 5^2,\ 7^2,\ 9^2\ldots) \qquad (3.70)$$

As with the clamped-free bar, the overtones are not harmonics. The nodes are of two types. Those nearest the ends are characterized by $\zeta = 0$, but $\partial\zeta/\partial x \neq 0$ and $\partial^2\zeta/\partial x^2 \neq 0$. The nodes between those nearest the ends are characterized by $\zeta = 0$ and $\partial^2\zeta/\partial x^2 = 0$, and the same procedure as that used to determine the locations of the nodes of a clamped-free bar is used here. In comparison with the clamped-free bar, two nodes instead of one are associated with the fundamental frequency. The first overtone produces three and so on. Table 3.2 gives the relative frequencies, phase velocities, and positions of nodes of the fundamental and the first three overtones.

TABLE 3.2

Vibration Characteristics of a Free-Free Bar in Flexure

Frequency	Phase velocity	Nodal positions
f_1	v_1	$x = 0.224\ell,\ 0.776\ell$
f_2 $(2.756f_1)$	$1.66v_1$	$x = 0.132\ell,\ 0.5\ell,\ 0.868\ell$
f_3 $(5.404f_1)$	$2.32v_1$	$x = 0.094\ell,\ 0.356\ell,\ 0.644\ell,\ 0.906\ell$
f_4 $(8.933f_1)$	$2.99v_1$	$x = 0.073\ell,\ 0.277\ell,\ 0.5\ell,\ 0.723\ell,\ 0.927\ell$

The corresponding modes are plotted in Fig. 3.6. As one would expect, the positions of the nodes are symmetrical about the center of the bar for all modes. The shapes of the displacement curves for all modes corresponding to odd-numbered frequencies, including the fundamental f_1, are symmetrical about the center. However, the modes corresponding to even-numbered frequencies are asymmetrical, i.e., not symmetrical about the center. The latter modes exhibit true nodes at the center.

VI.C. CLAMPED-CLAMPED BAR (BAR CLAMPED AT BOTH ENDS)

A clamped-clamped bar, i.e., a bar which is rigidly clamped at each end, will vibrate at the same frequencies as the free-free bar of equal dimensions. Here, again, two types of nodes can occur. The nodes at the ends are characterized by $\zeta = 0$ and $\partial\zeta/\partial x = 0$. Those in between are characterized by $\zeta = 0$ and $\partial^2\zeta/\partial x^2 = 0$. The nodal positions do not coincide with those of the free-free bar but are as shown in Table 3.3.

TABLE 3.3

Locations of Nodes and Corresponding Phase Velocities of a Fixed-Fixed Bar Resonant in Transverse Vibration

Frequency	Phase velocity	Nodal positions
f_1	v_1	0, ℓ
f_2 (= 2.756 f_1)	1.66 v_1	0, 0.5ℓ, ℓ
f_3 (= 5.404 f_1)	2.32 v_1	0, 0.356ℓ, 0.643ℓ, ℓ
f_4 (=8.933 f_1)	2.99 v_1	0, 0.277ℓ, 0.5ℓ, 0.723ℓ, ℓ
f_5 (= 13.3446 f_1)	3.653 v_1	0, 0.227ℓ, 0.409ℓ, 0.591ℓ, 0.773ℓ, ℓ

VI.D. EFFECT OF GEOMETRY ON TRANSVERSE VIBRATIONS OF BARS

The effect of geometry on the vibratory characteristics of a bar depends upon the corresponding distribution of the mass of the bar.

The fundamental resonant frequencies of various tapered bars which are rigidly mounted at their large end are listed in Table 3.4.

TABLE 3.4

Resonant Frequencies of Tapered Bars Rigidly Mounted at the Large End

Geometry	Direction of vibration	Frequency, Hz
Wedge-shaped bar	Normal to parallel sides	$\frac{1.14}{\ell^2}\sqrt{\frac{Y\ b^2}{12\rho}}$
Wedge-shaped bar	Parallel to parallel sides	$\frac{0.85}{\ell^2}\sqrt{\frac{Y\ b^2}{12\rho}}$
Conical bar		$\frac{1.39}{\ell^2}\sqrt{\frac{Y\ a^2}{4\rho}}$

In Table 3.4, a is the radius of the cone at the base, in centimeters, ℓ is the length of the bar, in centimeters, Y is Young's modulus, in dynes per square centimeter, ρ is the density in grams per cubic centimeter, and b is the thickness of the bar at the base in the direction of vibration, in centimeters.

VII. PLATE WAVES

VII.A. GENERAL

The form of the wave equation for plates is

$$\nabla^4\zeta + \frac{3\rho(1 - \sigma^2)}{Yh^2}\ \frac{\partial^2\zeta}{\partial t^2} = 0 \tag{3.71}$$

where

ζ is the displacement amplitude,

ρ is the density of the material of the plate,

Y is the modulus of elasticity,

σ is Poisson's ratio,

h is the half-thickness of the plate, and

t is time.

The equation for the flexural bar, Eq. (3.48) (Sec. V) may be derived from Eq. (3.71) by applying the conditions of a thin bar in flexure. The thin bar in flexure is not constrained laterally by adjacent material. The stresses are assumed to be directed in planes that are parallel to the direction of bending. The directions of stress are (1) parallel to the length of the bar and (2) parallel to the direction of bending.

The stresses developed in plates in flexure are more complicated than those for the flexural bar because each incremental element in a plate under stress may be subjected to constraints in all directions. Hence, the term involving Poisson's ratio, Eq. (3.71).

Solutions to the plate equation allow an infinite number of vibrational conditions. For instance, the simple harmonic solution given by Morse (2) is

$$\zeta(r,\Phi) = \frac{\cos}{\sin}(m\Phi)\,\{AJ_m(\gamma r) + BI_m(\gamma r)\} \tag{3.72}$$

where J_m is cylindrical Bessel function,

I_m is hyberbolic Bessel function, and

γ is the wavelength constant for a plate.

For the circular plate of radius a clamped at the edge, the boundary conditions are that both ζ and $\partial\zeta/\partial r$ are zero at r = a. To satisfy these conditions, from Eq. (3.72)

$$\zeta(a,\Phi) = \frac{\cos}{\sin}(m\Phi)\{AJ_m(\gamma a) + BI_m(\gamma r)\} = 0$$

or

$$B = -A\left[\frac{J_m(\gamma a)}{I_m(\gamma a)}\right] \tag{3.73}$$

and

$$I_m(\gamma a)\frac{d}{dr}J_m(\gamma r) = J_m(\gamma a)\frac{d}{dr}I_m(\gamma r) \tag{3.74}$$

As with the flexural bar, these conditions can be satisfied only at discrete frequencies. Again referring to Morse (2), these frequencies are fixed by the values

$$\begin{array}{lll} \beta_{01} = 1.015, & \beta_{02} = 2.007, & \beta_{03} = 3.000 \\ \beta_{11} = 1.468 & \beta_{12} = 2.483, & \beta_{13} = 3.490 \\ \beta_{21} = 1.879 & \beta_{22} = 2.992 & \beta_{23} = 4.000 \end{array} \tag{3.75}$$

$$\beta_{mn} \xrightarrow[n\to\infty]{} n + \frac{m}{z}$$

where $\gamma_{mn} = (\pi/a)\beta_{mn}$. Therefore, the allowed frequencies for the normal plate clamped at its circumference are

$$f_{mn} = \frac{\pi\kappa}{2a^2}\sqrt{\frac{Y}{3\rho(1-\sigma^2)}}\,(\beta_{mn})^2 \tag{3.76}$$

From the values for β_{mn} given in Eq. (3.75), the allowed frequencies are

$$f_{01} = 0.9342\left(\frac{h}{a^2}\right)\sqrt{\frac{Y}{\rho(1-\sigma^2)}}$$

and

$$f_{mn} = \left(\frac{\beta_{mn}}{\beta_{01}}\right)^2 f_{01} = 0.9707\,(\beta_{mn})^2\,f_{01} \tag{3.77}$$

From β_{mn} values above in Eq. (3.75), and from Eq. (3.77), one may easily see that the overtones of the clamped plate are not harmonics. It can be shown similarly that this is generally true for all other boundary conditions for vibrating plates. Therefore, the velocity of propagation of plate waves is dependent not only upon the elastic properties and the density, but also upon the frequency of the oscillations.

VII.B. LAMB WAVES

Lamb waves are plate waves. As may be deduced from the previous discussions, there are an infinite number of Lamb waves each propagating at a velocity dictated by its frequency, thickness, density, and elasticity. In fact, two types are recognized--symmetrical and asymmetrical--and there also exists an infinite number of each of these types. Figure 3.7 illustrates the particle motion in each of the two types.

Lamb-wave equations may be derived either from Eq. (3.71) or from the general wave equation [Eq. (3.22) in Sec. III). Equations for shear waves and for longitudinal (or compressional) waves again may be written in the form of Eqs. (3.78) and (3.79), i.e.,

$$\nabla^2\phi = \frac{1}{c_L^2}\frac{\partial^2\phi}{\partial t^2} \quad \text{(compressional waves)} \tag{3.78}$$

$$\nabla^2\psi = \frac{1}{c_s^2}\frac{\partial^2\psi}{\partial t^2} \quad \text{(shear waves)} \tag{3.79}$$

Solutions to Eqs. (3.78) and (3.79) for infinite plates bounded by two planes parallel to and equidistant from the yz axis are (3):

$$\phi_1 = [A \cosh \frac{2\pi}{\lambda}\sqrt{1 - c_p^2/c_L^2}\; x]e^{j(\omega t - 2\pi z/\lambda)}$$

$$\psi_1 = [B \sinh \frac{2\pi}{\lambda}\sqrt{1 - c_p^2/c_s^2}\; x]e^{j(\omega t - 2\pi z/\lambda)}$$

$$\text{(symmetrical waves)} \tag{3.80}$$

$$\phi_2 = [A \sinh \frac{2\pi}{\lambda}\sqrt{1 - c_p^2/c_L^2}\; x]e^{j(\omega t - 2\pi z/\lambda)}$$

$$\psi_2 = [B \cosh \frac{2\pi}{\lambda}\sqrt{1 - c_p^2/c_s^2}\; x]e^{j(\omega t - 2\pi z/\lambda)}$$

$$\text{(asymmetrical waves)} \tag{3.81}$$

where c_L is the velocity of a longitudinal wave in the material of the plate, c_s is the velocity of shear waves, and c_p is the phase velocity.

The boundary conditions are that the stresses at the surfaces of the plate ($x = \pm h$) are zero. As used in Chap. 2 in the derivations of the relative amplitudes of reflected and refracted waves from a boundary, Hooke's law for isotropic media may be written

$$\begin{aligned} p_{xx} &= (\lambda' + 2\mu)\frac{\partial v}{\partial x} + \lambda\frac{\partial u}{\partial z} \\ &= \rho\left[c_L^2\frac{\partial v}{\partial x} + (c_L^2 - 2c_s^2)\frac{\partial u}{\partial z}\right] \\ p_{zx} &= \rho c_s^2\left(\frac{\partial u}{\partial x} + \frac{\partial v}{\partial z}\right) \end{aligned} \tag{3.82}$$

where λ' is Lame's constant,

μ is the shear modulus,

$u = \dfrac{\partial\phi}{\partial z} + \dfrac{\partial\psi}{\partial x}$, and

$v = \dfrac{\partial\phi}{\partial x} - \dfrac{\partial\psi}{\partial z}$.

Substitution of Eqs. (3.80) in Eqs. (3.82) and applying the boundary condition that the stress at $x = \pm h$ is zero gives the frequency equation for symmetrical Lamb waves which is

$$\frac{\tanh \omega h\sqrt{\dfrac{c_s^2 - c_p^2}{c_L^2 \quad c_p^2}}}{\tanh \omega h\sqrt{\dfrac{c_L^2 - c_p^2}{c_L^2 \quad c_p^2}}} = 4\left\{\frac{\sqrt{\left(1 - \dfrac{c_p^2}{c_L^2}\right)\left(1 - \dfrac{c_p^2}{c_s^2}\right)}}{\left(2 - \dfrac{c_p^2}{c_s^2}\right)}\right\}$$

$$c_p < c_s < c_L \tag{3.83}$$

When c_p is greater than c_s or both c_s and c_L, Eq. (3.83) becomes imaginary on both sides. The solutions then may be written

$$\frac{\tan \omega h \sqrt{\dfrac{c_p^2 - c_s^2}{c_s^2 \; c_p^2}}}{\tanh \omega h \sqrt{\dfrac{c_L^2 - c_p^2}{c_L^2 \; c_p^2}}} = 4 \left\{ \frac{\sqrt{\left(1 - \dfrac{c_p^2}{c_L^2}\right)\left(\dfrac{c_L^2}{c_s^2} - 1\right)}}{\left(2 - \dfrac{c_p^2}{c_s^2}\right)^2} \right\}$$

$$c_s < c_p < c_L \qquad (3.83a)$$

and

$$\frac{\tan \omega h \sqrt{\dfrac{c_p^2 - c_s^2}{c_s^2 \; c_p^2}}}{\tan \omega h \sqrt{\dfrac{c_p^2 - c_L^2}{c_L^2 \; c_p^2}}} = -4 \left\{ \frac{\sqrt{\left(\dfrac{c_p^2}{c_L^2} - 1\right)\left(\dfrac{c_p^2}{c_s^2} - 1\right)}}{\left(2 - \dfrac{c_p^2}{c_s^2}\right)^2} \right\}$$

$$c_s < c_L < c_p \qquad (3.83b)$$

Frequency equations for asymmetrical Lamb waves are obtained by substituting Eqs. (3.81) in Eqs. (3.82) and applying the boundary conditions of zero stress at $x = \pm h$. For the <u>asymmetrical Lamb waves</u>, the frequency equations are

$$\frac{\tanh \omega h \sqrt{\dfrac{c_s^2 - c_p^2}{c_s^2 \; c_p^2}}}{\tanh \omega h \sqrt{\dfrac{c_L^2 - c_p^2}{c_L^2 \; c_p^2}}} = \frac{1}{4} \left\{ \frac{\left[2 - \left(\dfrac{c_p^2}{c_s^2}\right)\right]^2}{\sqrt{\dfrac{(c_s^2 - c_p^2)(c_L^2 - c_p^2)}{c_s^2 \; c_p^2}}} \right\}$$

$$c_p < c_s < c_L \qquad (3.84a)$$

$$\frac{\tan \omega h \sqrt{\dfrac{c_p^2 - c_s^2}{c_s^2 \; c_p^2}}}{\tanh \omega \sqrt{\dfrac{c_L^2 - c_p^2}{c_L^2 \; c_p^2}}} = -\frac{1}{4}\left\{\frac{\left[2 - \dfrac{c_p^2}{c_s^2}\right]^2}{\sqrt{\dfrac{(c_p^2 - c_s^2)(c_L^2 - c_p^2)}{c_s^2 \; c_L^2}}}\right\}$$

$$c_s < c_p < c_L \qquad (3.84b)$$

and

$$\frac{\tan \omega h \sqrt{\dfrac{c_p^2 - c_s^2}{c_s^2 \; c_p^2}}}{\tan \omega h \sqrt{\dfrac{c_p^2 - c_L^2}{c_L^2 \; c_p^2}}} = -\frac{1}{4}\left\{\frac{\left[2 - \dfrac{c_p^2}{c_s^2}\right]^2}{\sqrt{\dfrac{(c_p^2 - c_s^2)(c_p^2 - c_L^2)}{c_s^2 \; c_L^2}}}\right\}$$

$$c_s < c_L < c_p \qquad (3.84c)$$

In Eqs. (3.83) and (3.84), given the plate thickness 2h and the material of the plate, all quantities become constant except c_p and the frequency f, and these are interrelated. The allowed frequencies (f) and velocities (c_p), which also are the velocities of the Lamb waves, are those combinations of f and c_p which satisfy Eqs. (3.83) and (3.84).

When a symmetrical or asymmetrical Lamb wave propagates in a plate with a phase velocity c_p equal to either the shear velocity c_s or the longitudinal velocity c_L, the displacement normal to the plate surfaces is zero and the surface vibrates in the horizontal direction only.

When a symmetrical or asymmetrical Lamb wave propagates through a plate with a phase velocity equal to $\sqrt{2}$ times the shear velocity, the displacement of the surfaces is in the vertical direction only, i.e., normal to the planes of the surfaces.

The phase velocity c_p of both the first symmetrical mode and the asymmetrical mode approaches the velocity of Rayleigh waves as a lower limit.

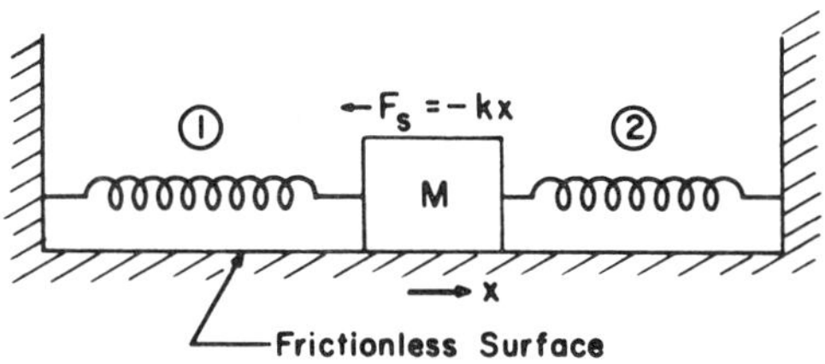

FIG. 3.1. Simple spring/mass oscillator.

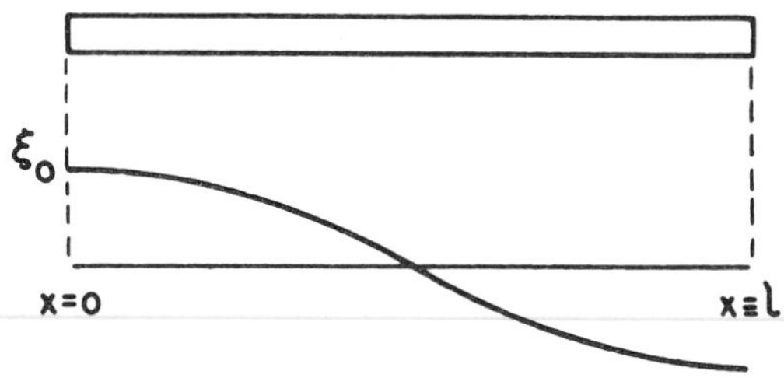

FIG. 3.2. Slender uniform bar in longitudinal half-wave resonance (free-free bar).

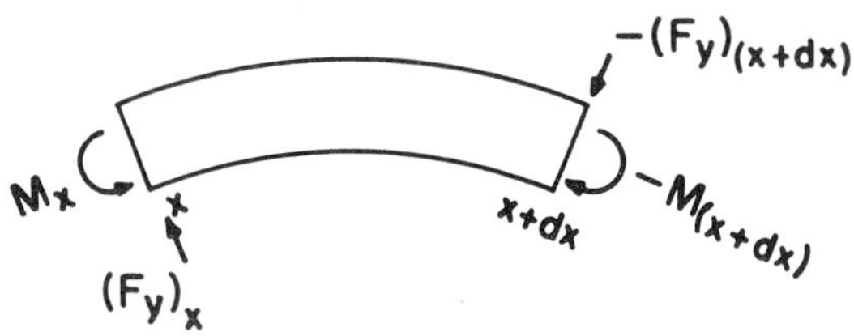

FIG. 3.3. Bending moments and shear forces in a bar in flexure.

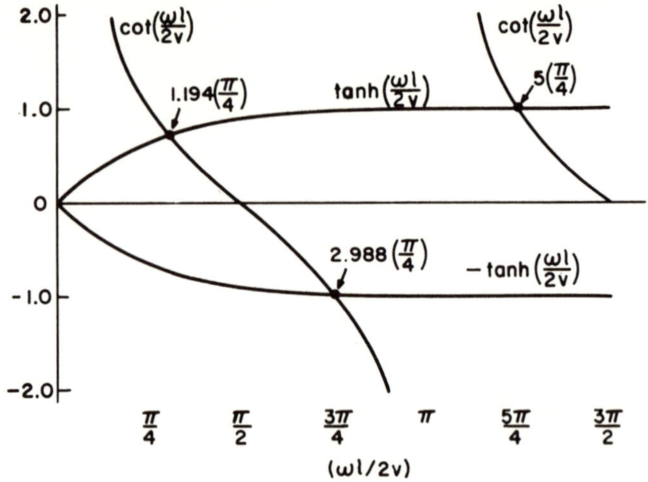

FIG. 3.4. Graphical determination of allowed frequencies for flexural resonance in fixed-free bars.

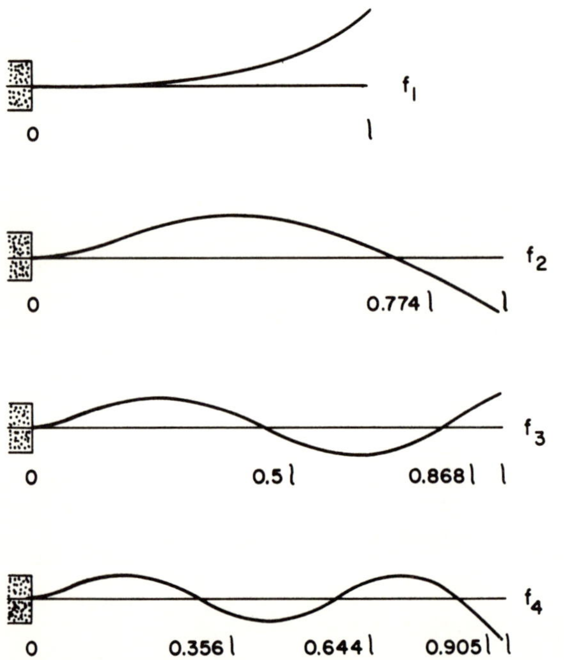

FIG. 3.5. The modes of transverse vibrations of a clamped-free bar.

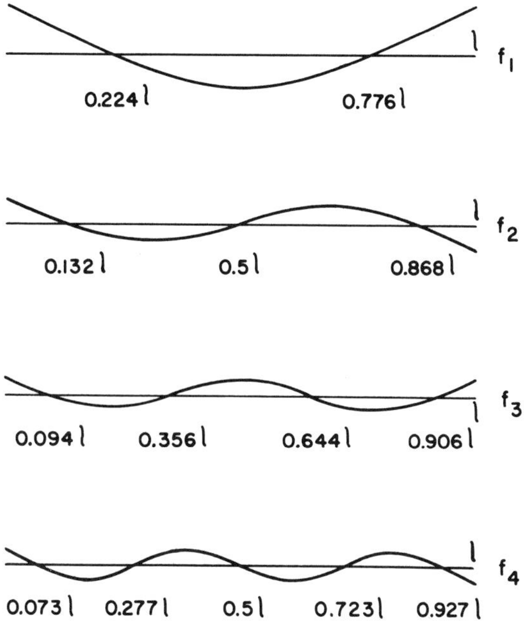

FIG. 3.6. The modes of transverse vibrations of a free-free bar.

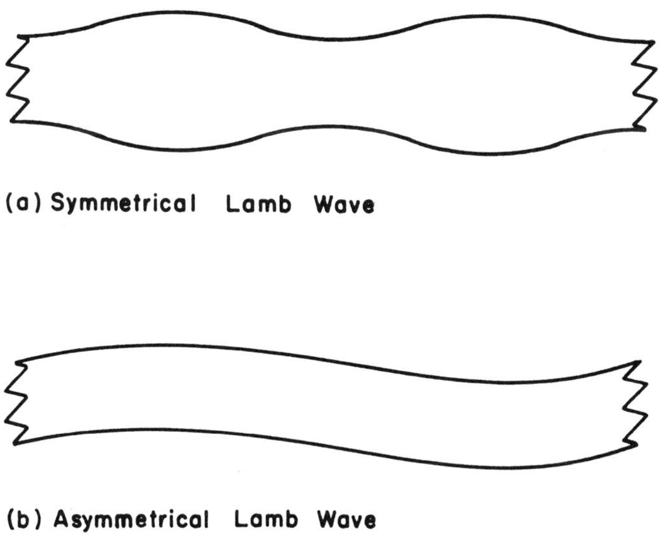

FIG. 3.7. Displacement characteristics of Lamb waves.

REFERENCES

1. L. E. Kinsler and R. R. Frey, Fundamentals of Acoustics, Wiley, New York, 1950.

2. P. M. Morse, Vibration and Sound, McGraw-Hill, New York, 1948.

3. D. C. Worlton, "Lamb Waves at Ultrasonic Frequencies," Hanford Atomic Products Operation, Report HW-60662, Richland, Washington: published by the U. S. Atomic Energy Commission, June 9, 1969.

Chapter 4

DESIGN OF ULTRASONIC HORNS FOR PROCESSING APPLICATIONS

I. INTRODUCTION

In machining and in various high-intensity processing applications, ultrasonic energy is transferred from the transducer into the medium subjected to treatment through transmission lines of various configurations. The amplitude of vibration at the radiating surface depends upon the geometry of the transmission line, its energy losses, and the amplitude of vibration at the driven end. A tapered transmission line produces a displacement amplification; i.e., the amplitude of vibration at resonance at the small end is greater than that at the larger end. These tapered elements are called mechanical amplifiers or horns.

The design of transducers for high-intensity use often involves combining segments of transmission lines that are composed of different materials; therefore, the same equations used in the design of ultrasonic horns, or mechanical amplifiers, are useful in designing transducers and other types of transmission lines. The so-called Langevin sandwich is an example of such a combination. The Langevin-sandwich type of transducer consists of piezoelectric elements sandwiched between two masses so that the fundamental resonance frequency is lower than the resonance frequency of any of the elements taken separately. These transducers are often designed by assuming that the masses are lumped and the piezoelectric plates are stiff springs. If the piezoelectric element is thin and the masses are comparatively long, it is more accurate to assume that the mass is distributed and this means that the equations for transmission lines must be used.

Because of the importance of horns in the application of high-intensity ultrasonic energy, this chapter is devoted to the development of equations for the design of these elements.

II. THE HORN EQUATION

In Chap. 3, the horn equation was given in the following forms:

$$\frac{1}{c^2}\frac{\partial^2 \zeta}{\partial x^2} - \frac{1}{S}\frac{\partial S}{\partial x}\frac{\partial \zeta}{\partial x} - \frac{\partial^2 \zeta}{\partial x^2} = 0 \qquad (3.21)$$

and

$$\frac{\partial^2 v}{\partial x^2} + \frac{1}{S}\frac{\partial S}{\partial x}\frac{\partial v}{\partial x} + \frac{\omega^2}{c^2} v = 0$$

Equation (3.21) is a special form of the wave equation and reduces to the plane-wave equation, Eq. (3.20), if the taper is zero, i.e., $\partial S/\partial x$ is zero. Thus, the uniform bar may be considered to be a horn with zero taper and with an amplification factor of unity.

III. TYPES OF HORNS

III.A. CYLINDER OR UNIFORM BAR AS AN ULTRASONIC HORN

The simplest horn is a cylinder, or uniform bar. Its characteristics are described by Eqs. (3.39), (3.40), (3.41), (3.42), and (3.45) in Chap. 3. The length of the horn is determined by Eq. (4.1)

$$c = \lambda f \tag{4.1}$$

where λ is the wavelength of the ultrasound. The length ℓ of a half-wave cylinder is

$$\ell = \frac{\lambda}{2} = \frac{c}{2f} \tag{4.2}$$

Since the cylinder is a uniform bar, the displacements at the two ends are equal in magnitude when the bar is in half-wave resonance, i.e., $\zeta_o = \zeta_\ell$.

III.B. STEPPED HORN (DOUBLE CYLINDER)

The stepped horn is a modification of the uniform bar (Fig. 4.1). The diameter at one end of the bar differs from that at the other end. Referring to Fig. 4.1, $D_1 > D_2$. Conservation of momentum would indicate that when the horn is resonant, v_2 must be greater than v_1 for equilibrium to be maintained. Since the displacement is proportional to the velocity potential v, $\zeta_1 < \zeta_2$. These statements may be proved by referring to the formulas for uniform bars (free-free bars) which are also applicable for the stepped horn when the different areas are taken into account.

Referring to Fig. 4.1 and assuming that the horn is homogeneous, the velocity potential v_1 at $x = \lambda/4 - dx$, is $v_1 = F_x/Z_1$. At $x = \lambda/4 + dx$, $v_2 = F_x/Z_2$. Therefore

$$\frac{v_1}{v_2} \simeq \frac{Z_2}{Z_1} = \frac{S_2\ Y_2\ c_1 \tan \omega \left(\frac{\ell - \ell/2}{c_2}\right)}{S_1\ Y_1\ c_2 \tan \left(\frac{\omega\ell}{2c_1}\right)} = \frac{S_2}{S_1} \tag{4.3}$$

since for this case $Y_2 = Y_1$ and $c_2 = c_1$.

Equation (4.3) shows that the velocities, and hence the displacements at the ends of the stepped horn of Fig. 4.1 are inversely proportional to the areas of the ends. Therefore, for cylindrical stepped horns, if $D_1/D_2 = 2$, $v_2/v_1 = 4$. In other words, the amplification factor is the ratio of the areas of the ends of the horn, with the higher amplitude being at the smaller end.

The plane of maximum stress in a half-wave uniform bar and of a half-wave stepped horn where the step occurs at the center (quarter-wave position) is located at the node. A horn fabricated without a fillet at the junction of the two cylinders, as in Fig. 4.1, would experience such high concentrations of stress at the node where the dimensions change abruptly that, at high intensities, it might soon fail in fatigue.

Common practice in designing horns for developing high sound intensities calls for sufficient filleting at the junction to prevent fatigue failure. An arbitrary rule is to make the radius of the fillet equal to the difference between the radii of the two cylinders. This may be impractical if the ratio of diameters is much greater than 2. Scratches or machine marks on the fillet may also cause failure for the same reason.

III.C. EXPONENTIALLY TAPERED HORN

The exponential horn (Fig. 4.2) is described mathematically by

$$S = S_o e^{-\gamma x} \tag{4.4}$$

where γ is a taper factor (1).

Differentiating S with respect to x and substituting in Eq. (3.21) gives the differential equation for the exponential horn:

$$\frac{1}{c^2}\frac{\partial^2 \zeta}{\partial t^2} + \gamma \frac{\partial \zeta}{\partial x} - \frac{\partial^2 \zeta}{\partial x^2} = 0 \tag{4.5}$$

If the motion is simple harmonic

$$\frac{\partial \zeta}{\partial t} = v = j\zeta\omega$$

or

$$\zeta = -j \frac{v}{\omega} \tag{4.6}$$

Substituting Eq. (4.6) in Eq. (4.5) gives

$$\frac{\partial^2 v}{\partial x^2} - \gamma \frac{\partial v}{\partial x} + \frac{\omega^2}{c^2} v = 0 \tag{4.7}$$

Equation (4.7) may be solved by assuming a solution of the form

$$v = K_o e^{\gamma x/2} U(x) \tag{4.8}$$

where K_o is a constant. Then

$$\frac{\partial v}{\partial x} = \frac{K_o}{2} \gamma e^{\gamma x/2} U + K_o e^{\gamma x/2} \frac{\partial U}{\partial x}$$

and

$$\frac{\partial^2 v}{\partial x^2} = \frac{K_o}{4} \gamma^2 e^{\gamma x/2} U + K_o \gamma e^{\gamma x/2} \frac{\partial U}{\partial x} + K_o e^{\gamma x/2} \frac{\partial^2 U}{\partial x^2}$$

Substituting these quantities for $\frac{\partial v}{\partial x}$ and $\frac{\partial^2 v}{\partial x^2}$ in Eq. (4.7) gives the ordinary differential equation

$$\frac{\partial^2 U}{\partial x^2} + \left(\frac{\omega^2}{c^2} - \frac{\gamma^2}{4}\right) U = 0 \tag{4.9}$$

for which the solution is

$$\begin{aligned} U &= k_1 \cos \left(\frac{\omega^2}{c^2} - \frac{\gamma^2}{4}\right)^{1/2} x + k_2 \sin \left(\frac{\omega^2}{c^2} - \frac{\gamma^2}{4}\right)^{1/2} x \\ &= k_1 \cos \frac{\omega x}{c'} + k_2 \sin \frac{\omega x}{c'} \end{aligned} \tag{4.10}$$

where $c' = \dfrac{c}{\sqrt{1 - \dfrac{\gamma^2 c^2}{4\omega^2}}}$

Substituting Eq. (4.10) in Eq. (4.8) gives

$$v = e^{\gamma x/2} \left[K_{01} \cos \frac{\omega x}{c'} + K_{02} \sin \frac{\omega x}{c'}\right] \tag{4.11}$$

where

$$K_{01} = K_0 k_1 \text{ and } K_{02} = K_0 k_2$$

The coefficients K_{01} and K_{02} may be determined by applying the appropriate boundary conditions:

At $x = 0$

$$v = V_o = K_{01}$$

and

$$\frac{\partial v}{\partial x} = 0$$

$$= e^{\gamma x/2} \left\{[-V_o \frac{\omega}{c'} \sin \frac{\omega x}{c'} + K_{02} \frac{\omega}{c'} \cos \frac{\omega x}{c'}] + \frac{\gamma}{2} [V_o \cos \frac{\omega x}{c'} + K_{02} \sin \frac{\omega x}{c'}\right\} = K_{02} \frac{\omega}{c'} + \frac{\gamma}{2} V_o \tag{4.12}$$

Therefore

$$K_{02} = -V_o \frac{\gamma c'}{2\omega} \tag{4.13}$$

giving

$$v = e^{\gamma x/2} V_o[\cos \frac{\omega x}{c'} - \gamma \frac{c'}{2\omega} \sin \frac{\omega x}{c'}] \tag{4.14}$$

as the equation for the velocity potential at distance x from the large end of an exponentially tapered horn.

The particle velocity of the large end of the horn V_o is described in simple harmonic motion, as previously stated: at $x = \ell$, the velocity potential $V_\ell = -e^{\gamma\ell/2} V_o$. Since according to Eq. (4.4) $S/S_o = \ell^{-\gamma x}$, then

$$V_\ell = -V_o \left(\frac{S_o}{S_\ell}\right)^{\frac{1}{2}} = -V_o \left(\frac{D_o}{D_\ell}\right)$$

and

$$\zeta_\ell = -\zeta_o \left(\frac{D_o}{D_\ell}\right) \tag{4.15}$$

where D_o and D_ℓ are diameters at $x = 0$ and $x = \ell$, respectively.

Equation (4.15) indicates that the displacements at the ends of an exponentially tapered horn are inversely proportional to the diameters at the ends.

The velocity node (v = 0) occurs where

$$\tan \frac{\omega x}{c'} = \frac{2\omega}{\gamma c'} \tag{4.16}$$

The stress of a horn is

$$s = j \frac{Y}{\omega} \frac{\partial v}{\partial x} \tag{4.17}$$

which, for exponential horn, is

$$s = -j \frac{Y}{\omega} V_o \left[\frac{\omega}{c'} + \frac{\gamma^2 c'}{4\omega}\right] e^{\gamma x/2} \sin \frac{\omega x}{c'} \tag{4.18}$$

Maximum stress occurs when $\frac{\partial s}{\partial x} = 0$; i.e., when

$$\tan \frac{\omega x}{c'} = - \frac{2\omega}{\gamma c'} \tag{4.19}$$

Comparison of Eqs. (4.16) and (4.19) shows that the velocity node occurs near the large-diameter end of the horn, while the point of maximum stress lies closer to the small-diameter end of the horn.

The length ℓ of the half-wave, exponential horn is given by the formula

$$\ell = \pi \frac{c'}{\omega} = \frac{c'}{2f} \tag{4.20}$$

In this case, $c' > c$ and, therefore, ℓ is longer than a uniform bar vibrating in half-wave resonance at the same frequency as the exponential horn.

The mechanical impedance at x of the exponential horn is

$$Z = \frac{sS}{v} = \frac{-j\ SY \left[\frac{\omega}{c'} + \frac{\gamma^2 c'}{4\omega}\right]}{\omega\left[\cot \frac{\omega x}{c'} - \frac{\gamma c'}{2\omega}\right]}$$

$$= \frac{-j\ SY \left[\frac{\omega}{c'} + \frac{\gamma^2 c'}{4\omega}\right] \tan \frac{\omega x}{c'}}{\omega\left[1 - \frac{\gamma c'}{2\omega} \tan \frac{\omega x}{c'}\right]} \tag{4.21}$$

where $S = S_o e^{-\gamma x}$ and S_o is the area at the large end of the horn.

III.D. WEDGE-SHAPED HORNS

The equation describing the taper of a wedge is (see Fig. 4.3)

$$S = S_1 + k_w x \tag{4.22}$$

where $k_w = (S_2 - S_1)/\ell$

Assuming again that, at resonance, the displacement is in simple harmonic motion and performing the essential differentiations, i.e.,

$$\frac{\partial S}{\partial x} = k_w$$

and

$$\frac{1}{S}\frac{\partial S}{\partial x} = \frac{k_w}{S_1 + k_w x}$$

and substituting these quantities in Eq. (3.21) gives the wave equation for the wedge in the form

$$\frac{\partial^2 v}{\partial x^2} + \frac{k_w}{S_1 + k_w x}\frac{\partial v}{\partial x} + \frac{\omega^2}{c^2} v = 0 \tag{4.23}$$

Equation (4.22) may be simplified by letting $u = S_1 + k_w x$. Therefore

$$u = f(x);\ f'(x) = k_w;\ f''(x) = 0$$

$$v = \phi(x);\ \phi'(x) = \frac{dv}{dx};\ \phi''(x) = \frac{d^2v}{dx^2}$$

so that

$$\frac{\partial v}{\partial u} = \frac{1}{k_w}\frac{\partial v}{\partial x}$$

$$\frac{\partial^2 v}{\partial u^2} = \frac{1}{k_w^2}\frac{\partial^2 v}{\partial x^2} \tag{4.24}$$

Substituting Eqs. (4.24) in Eq. (4.23) and letting $p^2 = \frac{\omega^2}{k_w^2 c^2}$, reduces the wedge-shaped horn equation to

$$\frac{\partial^2 v}{\partial u^2} + \frac{1}{u}\frac{\partial v}{\partial u} + p^2 v = 0 \tag{4.25}$$

The solution for Eq. (4.25) is

$$v = KJ_o\ (pu)$$

$$= KJ_o \left\{ \frac{\omega}{ck_w} [S_1 + k_w x] \right\} = KJ_o \left\{ \frac{\omega}{c} \left[\frac{S_1 \ell}{S_2 - S_1} + x \right] \right\} \tag{4.26}$$

where K is a constant of integration and J_o (pu) is a Bessel function of the first kind of order zero.

At $x = 0$, $v = V_1$, the velocity at the small end. Therefore

$$K = \frac{V_1}{J_o \left\{ \frac{\omega}{c} \left[\frac{\ell S_1}{S_2 - S_1} \right] \right\}}$$

and

$$v = \frac{V_1 \left\{ J_o \left[\frac{\omega}{c} \left(\frac{\ell S_1}{S_2 - S_1} \right) + x) \right] \right\}}{J_o \left[\frac{\omega}{c} \left(\frac{\ell S_1}{S_2 - S_1} \right) \right]} = j\omega\zeta \tag{4.27}$$

where V_1 represents simple harmonic motion at $x = 0$.

The stress is

$$s = j\,\frac{E}{\omega}\frac{\partial v}{\partial x}$$

$$= -j\rho c V_1 \frac{J_1 \left\{ \frac{\omega}{c} \left[\frac{\ell S_1}{S_2 - S_1} + x \right] \right\}}{J_o \left\{ \frac{\omega}{c} \left[\frac{\ell S_1}{S_2 - S_1} \right] \right\}} \tag{4.28}$$

where J_1 is a Bessel function of the first kind of order 1.

The impedance Z at x of the wedge is

$$Z_x = \frac{sS_x}{v}$$

$$= -j\rho c \left[S_1 + \left(\frac{S_2 - S_1}{\ell} \right) x \right] \left\{ \frac{J_1 \left\{ \frac{\omega}{c} \left[\frac{\ell S_1}{S_2 - S_1} + x \right] \right\}}{J_o \left\{ \frac{\omega}{c} \left[\frac{\ell S_1}{S_2 - S_1} + x \right] \right\}} \right\} \tag{4.29}$$

The velocity node is at a value of x where

$$J_o \left\{\frac{\omega}{c} \left[\frac{\ell S_1}{S_2 - S_1} + x\right]\right\} = 0 \qquad (4.30)$$

and the maximum stress occurs when

$$J_1 \left\{\frac{\omega}{c} \left[\frac{\ell S_1}{S_2 - S_1} + x\right]\right\} = \text{a maximum or a minimum} \qquad (4.31)$$

When $S_1 = 0$, Eqs. (4.27) through (4.31) are

$$v = V_1 J_o \left(\frac{\omega x}{c}\right) \qquad (4.27a)$$

$$s = -j\rho c V_1 \left\{J_1 \left(\frac{\omega x}{c}\right)\right\} \qquad (4.28a)$$

$$Z = -j\rho c \left(\frac{S_2 x}{\ell}\right) \left\{\frac{J_1 \left(\frac{\omega x}{c}\right)}{J_o \left(\frac{\omega x}{c}\right)}\right\} \qquad (4.29a)$$

$$J_o\left(\frac{\omega x}{c}\right) = 0 \quad \text{(velocity node)} \qquad (4.30a)$$

$$J_1\left[\frac{\omega x}{c}\right] = \text{a maximum or minimum} \quad \text{(maximum stress)} \qquad (4.31a)$$

The length ℓ is determined by the boundary conditions. If the wedge is a resonant, half-wave vibrator, at x = 0, ℓ, the stress is zero and $\frac{\partial v}{\partial x}$ is zero. Therefore, the quantitity

$$\frac{\omega}{c} \left[\frac{\ell S_1}{S_2 - S_1} + x\right]$$

is a zero of J_1 at x = 0 and at x = ℓ; i.e.,

$$\frac{\omega}{c} \left[\frac{\ell S_1}{S_2 - S_1}\right] = 0,\ 3.8317,\ 7.0156,\ 10.1735\ldots\ 13.3237\ldots \text{when } x = 0 \qquad (4.32a)$$

and correspondingly

$$\frac{\omega}{c} \left[\frac{\ell S_2}{S_2 - S_1}\right] = 3.8317,\ 7.0156,\ 10.1735,\ 13.3237,\ 16.4706\ldots \text{when } x = \ell \qquad (4.32b)$$

To determine the length ℓ for given ratios of area, divide Eq. (4.32b) by Eq. (4.32a) to obtain the ratio S_2/S_1; then substitute for S_2 in terms of S_1 in Eq. (4.32b) and calculate ℓ. The one exception occurs when $S_1 = 0$. In this case, $\omega\ell/c = 3.8317$ determines ℓ directly. Area ratios corresponding to consecutive zeros of Eqs. (4.32a) and (4.32b) and calculated lengths for half-wave resonances are given in Table 4.1.

TABLE 4.1

Calculated Lengths for Half-Wave Resonant Wedge-Shaped Horns

S_2/S_1 =	∞	1.832	1.515	1.310	1.236	1.000
ℓ =	3.83 c/ω	3.20 c/ω	3.16 c/ω	3.15 c/ω	3.14 c/ω	1.0 c/ω

The following example illustrates the method of calculating the values given in Table 4.1:

When $$\frac{\omega}{c}\left[\frac{\ell S_2}{S_2 - S_1}\right] = 7.0156 \qquad (4.32b)$$

$$\frac{\omega}{c}\left[\frac{\ell S_1}{S_2 - S_1}\right] = 3.8317 \qquad (4.32a)$$

Dividing Eq. (4.32b) by Eq. (4.32a) gives

$$S_2/S_1 = 7.0156/3.8317 = 1.832$$

Substituting this value in Eq. (4.32b) gives

$$\ell = 3.2\ c/\omega$$

Figure 4.4 is a curve for which wedge length is plotted in terms of c/ω as a function of area ratio S_1/S_2. Reasonably accurate lengths of wedge-shaped horns may be calculated by using this curve.

The ratio of velocities at the ends of the horn, and therefore the ratio of the corresponding displacement amplitudes, is determined by dividing the velocity at the small end ($x = 0$) by the velocity V_2 at the large end ($x = \ell$) of the horn. Then

$$\frac{V_1}{V_2} = \frac{J_o\left\{\frac{\omega}{c}\left[\frac{\ell S_1}{S_2 - S_1}\right]\right\}}{J_o\left\{\frac{\omega}{c}\left[\frac{\ell S_2}{S_2 - S_1}\right]\right\}} \tag{4.33}$$

when $S_1 = 0$

$$\frac{V_1}{V_2} = \frac{1}{J_o(\omega\ell/c)} = 2.48 \qquad \text{(for half-wave)} \tag{4.34}$$

where $J_o(\omega\ell/c)$ is the minimum of $J_o(X_n)$ and is equal to -0.4028.

III.E. CONICAL HORNS

The taper of a conical horn, Fig. 4.5, is given by (2)

$$\frac{d_1}{d} = \frac{\ell d_1}{\ell d_1 + x(d_2 - d_1)} \tag{4.35}$$

where d_1 is the diameter at the small end and d is the diameter at distance x from the small end. Then the area S at x is

$$S = \pi d^2/4 = \frac{\pi d_1^{\,2}}{4}\,\frac{(x_1 + x)^2}{x_1^{\,2}} \tag{4.36}$$

where $x_1 = \ell d_1/(d_2 - d_1)$.

Assuming simple harmonic motion for displacement at resonance and substituting for S, according to Eq. (4.36), in the horn equation [Eq. (3.21)] results in the following equation for the conical horn:

$$\frac{\partial^2 v}{\partial x^2} \quad \frac{2}{x_1 + x} \quad \frac{\partial v}{\partial x} \quad \frac{\omega^2}{c^2}\, v = 0 \tag{4.37}$$

The solution for Eq. (4.37) is

$$v = (x_1 + x)^{-1}\left[A \cos \frac{\omega(x_1 + x)}{c} + B \sin \frac{\omega(x_1 + x)}{c}\right] \tag{4.38}$$

At $x = 0$, $v = V_1$ and at $x = \ell$, $v = V_2$, both of which are maxima. Therefore, at $x = 0, \ell$, $\partial v/\partial x = 0$. Multiplying both sides of Eq. (4.38) by $(x_1 + x)$ and differentiating, yields

$$(x_1 + x)\frac{\partial v}{\partial x} = -v - A\frac{\omega}{c}\sin\left[\omega\frac{(x_1 + x)}{c}\right] + B\left(\frac{\omega}{c}\right)\cos\left[\omega\frac{(x_1 + x)}{c}\right] \tag{4.39}$$

By applying the boundary conditions to Eqs. (4.38) and (4.39), it can be shown that

$$A = V_2\left[(x_1 + \ell)\cos\frac{\omega(x_1 + \ell)}{c} - \frac{c}{\omega}\sin\frac{\omega(x_1 + \ell)}{c}\right] \tag{4.40}$$

and

$$B = V_2\left[\frac{c}{\omega}\cos\frac{\omega(x_1 + \ell)}{c} + (x_1 + \ell)\sin\frac{\omega(x_1 + \ell)}{c}\right] \tag{4.41}$$

Substituting these values for A and B in Eq. (4.38) yields

$$v = (x_1 + x)^{-1}V_2\left[(x_1 + \ell)\cos\frac{\omega}{c}(x - \ell) + \frac{c}{\omega}\sin\frac{\omega}{c}(x - \ell)\right] \tag{4.42}$$

where

$$x_1 = \frac{d_1}{d_2}\left(\frac{x_1}{x_1 + \ell}\right)$$

Since at $x = 0$, $v = V_1$, the ratio of velocities and, therefore, the displacement amplitude at the ends of the horn is

$$\frac{V_1}{V_2} = \frac{\zeta_1}{\zeta_2} = \frac{d_2}{d_1}\cos\frac{\omega\ell}{c} - \frac{c(d_2 - d_1)}{\omega\ell d_1}\sin\frac{\omega\ell}{c}$$

$$= \frac{1}{\cos\frac{\omega\ell}{c} - \left[\frac{\omega\ell d_1}{c(d_2 - d_1)}\right]\sin\left(\frac{\omega\ell}{c}\right)} \tag{4.43}$$

Thus, when $d_1 = 0$, $\frac{V_1}{V_2} = \sec\frac{\omega\ell}{c}$ (4.43a)

and

$$v = \frac{V_2}{x}\left[\ell\cos\frac{\omega}{c}(x - \ell) + \frac{c}{\omega}\sin\frac{\omega}{c}(x - \ell)\right] \tag{4.44}$$

Since at the velocity node, v = 0, it follows from Eq. (4.42) that the velocity node occurs when

$$\tan \frac{\omega}{c} (\ell - x) = \frac{\omega(x_1 + \ell)}{c} \tag{4.45}$$

or

$$\tan \frac{\omega}{c} (\ell - x) = \frac{\omega \ell}{c} \qquad \text{(when } d_1 = 0) \tag{4.45a}$$

The half-wave resonant length ℓ of a conical horn is determined by applying the boundary condition that

$$\left[\frac{\partial v}{\partial x}\right]_{x = 0} = 0.$$

By differentiating Eq. (4.42)

$$\tan \frac{\omega \ell}{c} = \frac{\omega \ell c}{\omega^2 x_1^2 + \omega^2 \ell x_1 + c^2} = \frac{\omega \ell c (d_2 - d_1)^2}{\omega^2 \ell^2 d_1 d_2 + c^2 (d_2 - d_1)^2} \tag{4.46}$$

or

$$\tan \frac{\omega \ell}{c} = \frac{\omega \ell}{c} \qquad \text{(when } d_1 = 0) \tag{4.46a}$$

For half-wave horns $\omega\ell/c$ lies in the third quadrant, i.e., $\pi < \omega\ell/c < 3\pi/2$.

The stress s at x is given by

$$s = j \frac{Y}{\omega} \frac{\partial v}{\partial x} = j \frac{YV_2}{\omega(x_1 + x)^2} \left[(x - \ell) \cos \frac{\omega}{c} (x - \ell) - \left(x_1^2 + x_1\ell + x_1 x + \ell x + \frac{c^2}{\omega^2}\right) \frac{\omega}{c} \sin \frac{\omega}{c} (x - \ell)\right] \tag{4.47}$$

or, when $d_1 = 0$

$$s = j \frac{YV_2}{\omega x^2} \left[(x - \ell) \cos \frac{\omega}{c} (x - \ell) - \left(\ell x + \frac{c^2}{\omega^2}\right) \frac{\omega}{c} \sin \frac{\omega}{c} (x - \ell)\right] \tag{4.47a}$$

Differentiating Eq. (4.47) and setting $\partial s/\partial x = 0$

$$\tan \frac{\omega}{c}(x - \ell) = \frac{\omega}{c}\left[\frac{2c^2(x - \ell) + \omega^2(x_1 + \ell)(x_1 + x)^2}{2c^2 + \omega^2 x_1^2 + 2\omega^2 x_1 \ell + 2\omega^2 \ell x - \omega^2 x^2}\right] \tag{4.48}$$

at the maximum stress position.

Therefore, since $Z = F/v = sS/v$. it follows from Eqs. (4.42) and (4.47) and the equation describing the area S at x, [Eq. (4.36)] that

$$Z = \frac{F}{v}$$

$$= j\frac{\pi Y d_1^2 (x_1 + x)}{4x_1^2[\omega(x_1 + \ell) + c \tan \frac{\omega}{c}(x - \ell)]} \left\{(x - \ell) - [x_1^2 + x_1 \ell + x_1 x + \ell x + \frac{c^2}{\omega^2}]\frac{\omega}{c} \tan \frac{\omega}{c}(x - \ell)\right\} \tag{4.49}$$

or when $d_1 = 0$

$$Z = j\frac{\pi d_2^2 Y x}{4\ell^2}\left[\frac{(x - \ell) - [\ell x + (\frac{c^2}{\omega^2})]\frac{\omega}{c}[\tan \frac{\omega}{c}(x - \ell)]}{\omega\ell + c[\tan \frac{\omega}{c}(x - \ell)]}\right] \tag{4.49a}$$

As Eqs. (4.45), (4.46), and (4.48) show, the design of a conical horn involves solving an algebraic function of x and a trigonometric function of x simultaneously to determine the location of the node, the half-wave length, and the position of maximum stress.

When $d_1 = 0$, the quantity $\omega\ell/c$ is constant regardless of frequency or velocity of sound and is approximately 4.5. Then $\ell = 4.5\ c/\omega$. The ratio of velocities and displacement amplitudes at the ends of this horn is

$$\frac{V_1}{V_2} = \sec 4.5 = -4.61$$

which is the maximum amplification possible for a conical horn. As Eq. (4.46) shows, for all values of $d_1 > 0$, $\pi \leq \tan \omega \ell / c < 4.5$. When $d_1 = d_2$, $\tan \omega\ell/c = \omega\ell/c = \pi$ and $\ell = \pi c/\omega$, which corresponds to Eq. (4.2) for the cylindrical horn.

III.F. CATENOIDAL HORNS

The equation of the catenary is

$$y = \frac{a}{2}(e^{x/a} + e^{-x/a}) = a \cosh \frac{x}{a} \tag{4.50}$$

(see Fig. 4.6).

If y corresponds to the radius of the horn at x, and a is the radius at x = 0, the area S_x is

$$S_x = \pi y^2 = \frac{\pi a^2}{4}(e^{2x/a} + 2 + e^{-2x/a}) \tag{4.51}$$

$$= \frac{S_1}{4}(e^{2x/a} + 2 + e^{-2x/a}) = \frac{\pi d^2}{4}$$

$$= S_1 \cosh^2 \frac{x}{a}$$

Differentiating Eq. (4.51) with respect to x gives

$$\frac{\partial S}{\partial x} = \frac{S_1}{2a}(e^{2x/a} - e^{-2x/a}) = \frac{2}{a} S_1 \cosh \frac{x}{a} \sinh \frac{x}{a} \tag{4.52}$$

Substituting Eq. (4.52) in the horn equation [Eq. (3.21)] and assuming simple harmonic motion for the displacement, one may write

$$\frac{\partial^2 v}{\partial x^2} + \frac{2}{a}\left[\frac{e^{x/a} - e^{-x/a}}{e^{x/a} + e^{-x/a}}\right]\frac{\partial v}{\partial x} + \frac{\omega^2}{c^2} v = 0$$

or

$$\frac{\partial^2 v}{\partial x^2} + \left[\frac{2}{a} \tanh \frac{x}{a}\right] \frac{\partial v}{\partial x} + \frac{\omega^2}{c^2} v = 0 \tag{4.53}$$

for the differential equation of the catenoidal horn (3). The solution to Eq. (4.53) is

$$v = \frac{1}{a \cosh x/a}[A \cos k'x + B \sin k'x] \tag{4.54}$$

where

$$k' = \sqrt{\frac{\omega^2}{c^2} - \frac{1}{a^2}}$$

At $x = 0$

$$\frac{\partial v}{\partial x} = 0 \text{ and } v = V_1$$

Therefore

$$V_1 = \frac{A}{a} \quad \text{or} \quad A = aV_1$$

$$\left.\frac{\partial v}{\partial x}\right|_0 = 0$$

and

$$Bk'/a = 0$$

Therefore, $B = 0$, and Eq. (4.54) becomes

$$v = \frac{V_1 \cos k'x}{\cosh \frac{x}{a}} = j\omega\zeta \tag{4.55}$$

By definition, $D_2/D_1 = \cosh \ell/a$, i.e., $D_1 = 2a$ and $D_2 = 2a \cosh \ell/a$ according to Eq. (4.50). Therefore

$$V_2 = \frac{V_1 \cos k'\ell}{(D_2/D_1)} \tag{4.56}$$

The length ℓ of a half-wave catenoidal horn is

$$\ell = \frac{\pi}{k'} = a \cosh^{-1} \frac{D_2}{D_1} = \frac{D_1}{2} \cosh^{-1} \frac{D_2}{D_1} \tag{4.57}$$

The velocity and displacement node occurs where $\cos k'x = 0$, i.e., where $k'x = \pi/2$. Therefore, the node occurs when

$$x \Big|_{v=0} = \frac{\pi}{2k'} = \frac{\pi}{2\sqrt{\frac{\omega^2}{c^2} - \frac{4}{D_1^2}}} \tag{4.58}$$

The stress along a catenoidal horn is given by Eq. (4.17)

and is

$$s = j\,\frac{Y}{\omega}\,\frac{\partial v}{\partial x}$$

$$= -j\,\frac{YV_1 \cos k'x}{\omega a \cosh \frac{x}{a}} \left[\tanh \frac{x}{a} + ak' \tan k'x\right] \tag{4.59}$$

At the point of maximum stress s_m, $\partial s/\partial x = 0$.

$$\frac{\partial s}{\partial x} = j\,\frac{YV_1 \cos k'x}{\omega a^2 \cosh \frac{x}{a}} \left\{\left[\tanh \frac{x}{a} + ak' \tan k'x\right]^2 - \left[\operatorname{sech}^2 \left(\frac{x}{a}\right) + a^2k'^2 \sec^2 k'x\right]\right\} = 0 \tag{4.60}$$

from which

$$\tanh \frac{x}{a} \left[\tanh \frac{x}{a} + ak' \tan k'x\right] = \frac{1 + a^2k'^2}{2} \tag{4.61}$$

where $a = D_1/2$ and $k'^2 = \omega^2/c^2 - 4/D_1^2$ where s is maximum.

The impedance at x of a catenoidal horn is $Z_x = s_x S_x / v_x$

$$Z_x = -j\,\frac{YS_1}{\omega a} \left[\tanh \frac{x}{a} + ak' \tan k'x\right] \cosh^2 \frac{x}{a} \tag{4.62}$$

IV. COMBINING SECTIONS OF DIFFERENT CONFIGURATIONS FOR PRACTICAL APPLICATIONS

The previous sections contain design formulas for the solid horns which are most commonly used in the production of high-intensity ultrasonic energy used for materials-forming and certain processing applications. In materials-forming applications, especially ultrasonic drilling, usually a tool is silver soldered to a horn. This horn is then attached by means of a stud to a transducer or to a second horn which is attached permanently to the transducer (see Fig. 4.7). Maximum displacement is obtained at the tip of the tool when the resonance frequencies of the transducer, the permanent horn, and the tool/horn combination are the same.

Toolwear during ultrasonic machining causes an upward shift in frequency. Temperature changes in any of the components cause inverse changes in resonance frequency of the corresponding component. For these reasons it is difficult to maintain a perfect match between elements of the system of Fig. 4.7.

The problem of frequency shift is minimized by using transducers with low mechanical Q, which produces a fairly broad tuning curve. The permanently attached horn is tuned to the center frequency of the transducer. The attached toolholder with a new tool attached resonates at a slightly lower frequency. As the tool wears, the frequency increases and may rise to a frequency slightly above the resonant frequency of the transducer.

A practical limit on the distance the resonant frequency of the toolholder may shift from the center frequency of the transducer is imposed by the geometry of the system. When the three sections (transducer, permanent horn and toolholder plus tool) of Fig. 4.7 are resonant at the same frequency, the impedance at each junction would be zero if there were no energy losses in the system. In the practical case, the stress level at the junctions is minimum, being sufficient to compensate for losses in the system and thus to maintain a constant level of vibration. Under this ideal condition, maximum energy is transferable from the transducer to the free end of the tool. Each section terminates at a velocity antinode.

When the various sections are not matched, if a velocity antinode appears at the free end of the tool, the opposite end of the toolholder is not at an antinode. Energy is reflected back to the transducer out of phase with the transmitted wave. The resulting cancellation effect reduces the energy available to the toolholders. In addition, the junction is no longer at a minimum stress level. Driving at high intensities can result in fatigue failure in the coupling stud.

When two members of different materials, or different geometry, or both, are combined to form a single half-wave resonant member (as the toolholder combination, Fig. 4.7), matching is accomplished by equating the magnitudes of the mechanical impedances of the two sections at the junction and at the desired frequency, i.e., $|Z_h| = |Z_\ell|$. The length of the horn ℓ_h should always be greater than $\lambda/4$ and that of the tool (a uniform bar) always less than $\lambda/4$ for maximum displacement amplification. The junction between the tool and the toolholder should never occur at the point of maximum stress if the tool is to be driven at high intensity. The junction would fail in fatigue within a very short time.

V. EFFECT OF DAMPING ON THE OPERATION OF HORNS

The horn equation may be corrected for damping by inserting the term R_m/ρ as follows:

$$\frac{1}{c^2}\frac{\partial^2\zeta}{\partial t^2} + \frac{R_m}{\rho}\frac{\partial\zeta}{\partial t} - \frac{1}{S}\frac{\partial S}{\partial x}\frac{\partial\zeta}{\partial x} - \frac{\partial^2\zeta}{\partial x^2} = 0 \tag{4.63}$$

or

$$\frac{\partial^2 v}{\partial x^2} + \frac{1}{S}\frac{\partial S}{\partial x}\frac{\partial v}{\partial x} + \left(\frac{\omega^2}{c^2} - j\frac{R_m}{\rho}\right) v = 0 \tag{4.64}$$

where ω is the undamped angular frequency and c is the longitudinal velocity of sound in the medium.

The solution to Eq. (4.64) depends upon S. For instance, for the exponentially tapered horn, where $S = S_o e^{-\gamma x}$, Eq. (4.64) reduces to

$$\frac{\partial^2 v}{\partial x^2} - \gamma\frac{\partial v}{\partial x} + \left(\frac{\omega^2}{c^2} - j\frac{R_m}{\rho}\right) v = 0 \tag{4.65}$$

The solution to Eq. (4.65) is

$$v = e^{\gamma x/2}\left[K_{d1}\cos\frac{\omega x}{c_{d'}} + K_{d2}\sin\frac{\omega}{c_{d'}}\right] \tag{4.66}$$

which is similar to Eq. (4.11). However, $c_{d'}$ contains the addional term $-jc^2R_m/\omega^2\rho$ due to damping. Therefore, the apparent phase velocity of an exponentially tapered horn with losses, c_d', is

$$c_d' = \frac{c}{\sqrt{1 - \dfrac{\gamma^2 c^2}{4\omega^2} - j\,\dfrac{c^2 R_m}{\omega^2 \rho}}} \tag{4.67}$$

Applying the correction due to damping for the design equations of the various other configurations is left to the reader.

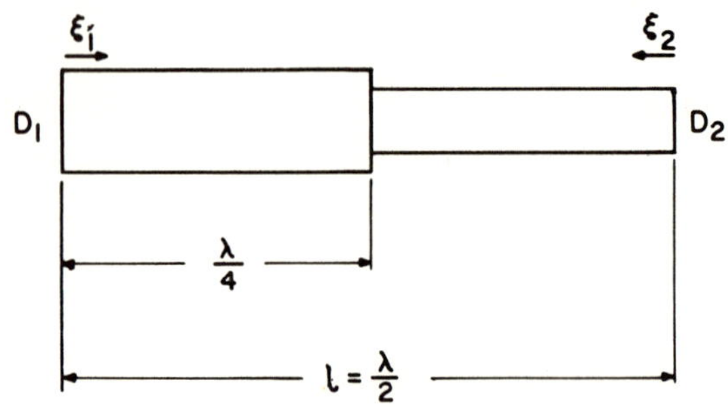

FIG. 4.1. Stepped horn.

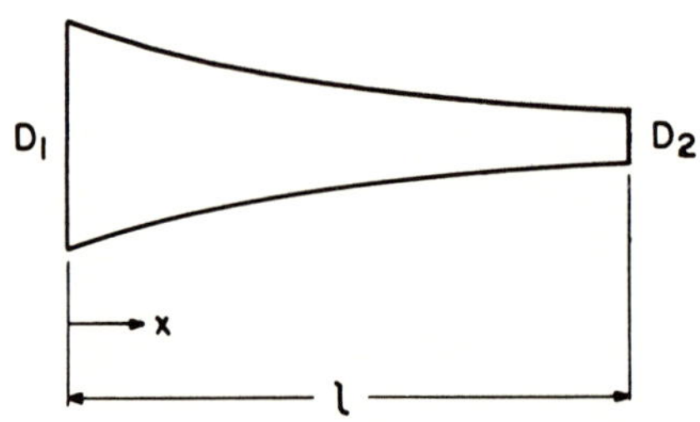

FIG. 4.2. Exponential horn.

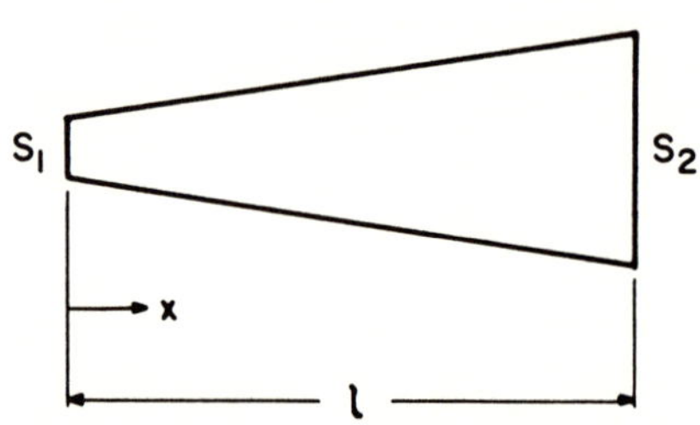

FIG. 4.3. Wedge-shaped horn.

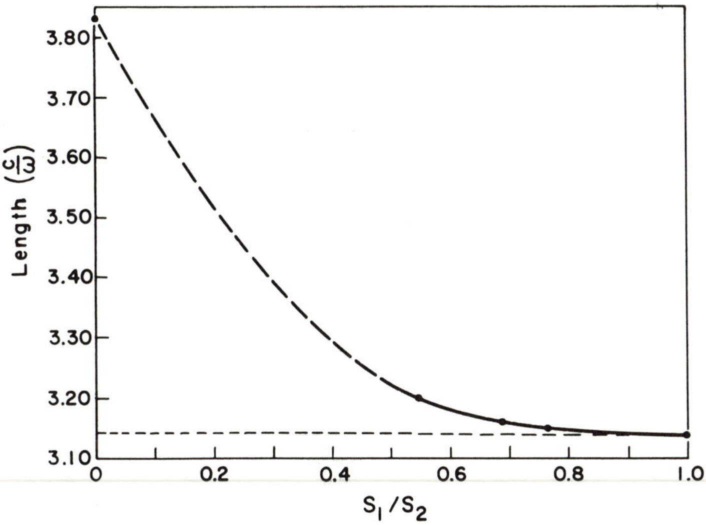

FIG. 4.4. Wedge length as a function of taper.

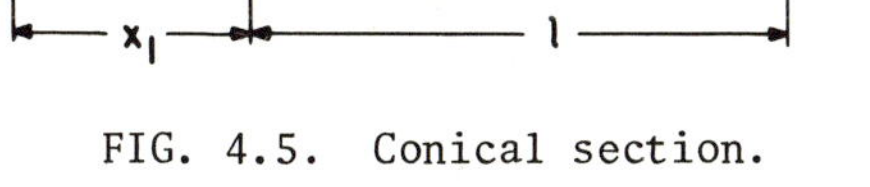

FIG. 4.5. Conical section.

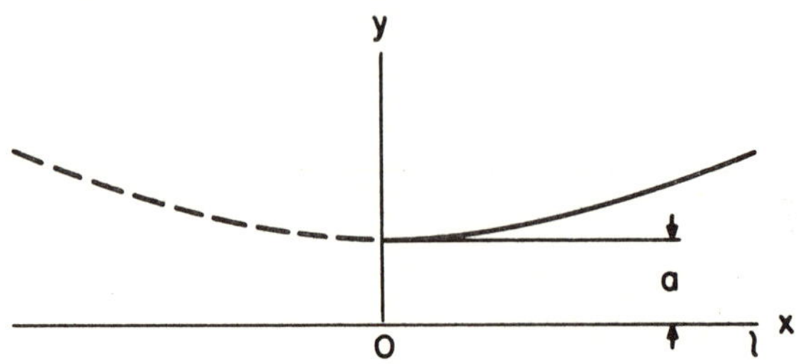

FIG. 4.6. The catenary.

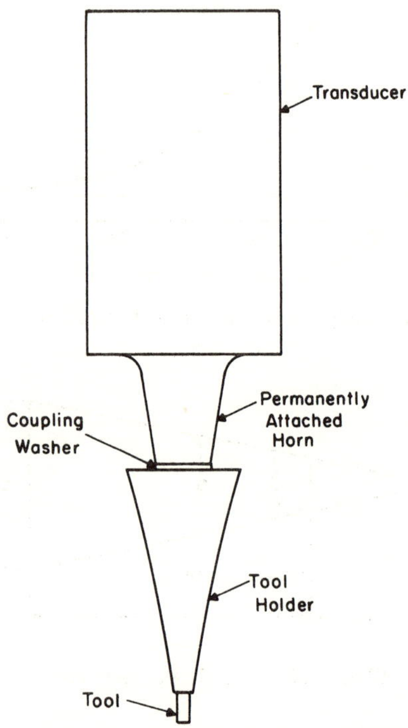

FIG. 4.7. Vibratory components of conventional ultrasonic machining system.

REFERENCES

1. W. P. Mason and R. F. Wick, Acoust. Soc. Am., 23, 209-214 (1951).

2. D. Ensminger, Acoust. Soc. Am., 32, 194-196 (1960).

3. L. G. Merkulov, Soviet Phys.-Acoust., 3, 246-255 (1957).

Chapter 5

BASIC DESIGN OF ULTRASONIC TRANSDUCERS

I. INTRODUCTION

Ultrasonic energy is generated and detected by devices called transducers. By definition a transducer is "a device that is actuated by power from one system to supply power to a second system"; i.e., a transducer converts energy from one form to another. In ultrasonics, the most typical conversions are electrical to ultrasonic energy (transmitters) or ultrasonic to electrical energy (receivers). Transducers most often used for generating ultrasonics are piezoelectric, magnetostrictive, electromagnetic, pneumatic (whistles), and mechanical devices. Transducers most often used for detecting or receiving ultrasonic energy are piezoelectric and magnetostrictive devices. A few basic principles of transducer design are presented in this chapter. Manufacturers of piezoelectric or magnetostrictive materials usually make available specific information on the characteristics of their products which is helpful in designing and fabricating transducers.

Piezoelectric, magnetostrictive, and electromagnetic transducers are electrical devices; i.e., they convert electrical energy to ultrasonic energy or ultrasonic energy to electrical energy. They also are mechanical devices in that mechanical oscillations are involved in the generation and detection of ultrasonic energy. The transducer designer must be concerned with the electrical characteristics of the transducer in relation to associated electrical equipment such as electronic power sources or receiver amplifiers. He must be concerned with the mechanical properties of the transducer in relation to its vibrational performance and its radiation characteristics. The electrical design of these "electromechanical" transducers is emphasized in this chapter. Mechanically, the transducers incorporate characteristics of ultrasonic transmission lines with various terminations, including reflections and transmissions of energy across boundaries, for which adequate equations are provided in Chapts. 2 through 4. A little investigation will reveal that a large number of possible configurations include bars vibrating in longitudinal modes, plates vibrating in thickness modes, bars vibrating in flexure, and so on.

In the transfer of energy from the electrical system to the mechanical system and vice versa, there must be an interaction between the two systems; i.e., the impedance measured at the electrical terminals is influenced by the motional characteristics of the transducer and by the type of mechanical loading that it may experience. In other words, the transducer is equivalent to an electrical transformer in which the electrical components, i.e., the true electrical resistances and reactances, appear in the primary circuit and the mechanical components appear in the secondary circuit. With respect to their influence on the electrical properties of the transducer, the mechanical components are real impedances which can be computed by using certain transformation constants to be discussed later. These properties are convenient for use in the initial electrical design of transducers; and it must be the initial design in the case of transducers for high-intensity applications because these devices are nonlinear and their loads may also exhibit nonlinear characteristics as functions of intensity. They also provide means for predicting and measuring efficiencies of transducers on the basis of the electrical characteristics of the transducers. To facilitate accomplishing these tasks, the transducers are represented by "equivalent" electrical circuits which contain components representing all of the factors known to influence the electrical characteristics of the transducer. These circuits are similar to equivalent circuits of electrical transformers. Representative equivalent circuits for piezoelectric and magnetostrictive transducers are described in this chapter.

A brief discussion of the design of each of the following types of transducers is presented in this chapter: piezoelectric, magnetostrictive, electromagnetic, pneumatic, and mechanical.

Piezoelectric transducers are used at all ultrasonic frequencies for generating and detecting ultrasonic energy at all levels of intensity. These transducers utilize piezoelectric components, such as plates or other suitable configurations which generate a charge on preferred surfaces under the influence of stresses, or which change dimensions when they are subjected to an electric field. Applications for piezoelectric transducers range from ultrasonic velocity measurements and nondestructive testing at low intensities to cleaning and welding at high intensities.

Magnetostrictive transducers are used at the lower ultrasonic frequencies, primarily for generating high-intensity ultrasonic energy. They are used at times for detecting ultrasonic energy. Magnetostrictive transducers utilize materials which exhibit the magnetostrictive effect, i.e., that change dimensions under the influence of a magnetic field. Typical uses of this type of transducer are driving ultrasonic cleaners and ultrasonic machine tools.

Electromagnetic transducers make use of the attractive forces of electromagnets to generate vibrations. They are used to obtain high-amplitude vibrations at frequencies usually below the ultrasonic range.

Pneumatic transducers, such as whistles, are useful at frequencies extending into the lower ultrasonic range for producing small-particle sprays of liquids for use in burners and in certain types of bulk cleaners. They have been used to dry materials and to break up foams resulting from various chemical processes.

Mechanical transducers are devices which are actuated mechanically. Their applications are more correctly classed as macrosonic. They are used to obtain high-amplitude, often high-intensity, vibrations at low sonic frequencies. A typical example is the sonic pile driver, in which counterbalanced eccentrics are used to generate vibrations in a direction parallel to the axis of the pile.

The designs for two types of transducers, sirens and combustion types of generators, are not discussed in this chapter.

Sirens have been made to operate at low ultrasonic frequencies. These transducers employ rotating perforated disks or cylinders which effect intermittent opening and closing of ports through which air (or other suitable gas) passes to generate the sound. Mechanical tolerances must be very close for efficient operation at ultrasonic frequencies.

Transducers using combustion and detonation principles have been used for specific applications. Combustion can be effective in generating continuous waves by synchronizing the fuel injection and ignition with the reverberating pressure wave in a resonating gas column. A combustion type of transducer is also a low-frequency device, but extremely high intensities can be generated in air at frequencies to 1000 Hz. Detonation-type transducers are useful in generating high-pressure pulses. The blast can be "shaped" into a pulse of a few cycles by using reflectors spaced at regular intervals from the detonation source. The pulse repetition rate is necessarily low, if detonating mixtures of gas are used, because time is required to refill the cavity with the proper mixture to accomplish detonation. A 50/50 mixture of acetylene and oxygen is a typical detonating mixture.

II. EQUIVALENT CIRCUITS

As mentioned previously, the electrical characteristics of transducers are influenced by their mechanical properties and by the types of mechanical loading to which they are subjected. Since these influences have the characteristics of electrical impedances, they may be represented in a circuit diagram which also includes all of the electrical parameters of the transducer. A circuit diagram of this type is called an "equivalent" circuit. It is useful for the electrical design of the transducer, for predicting its efficiency and power-handling capabilities, and for analyzing the performance of a transducer after it is constructed.

An equivalent circuit is based on the assumption that the transducer operates in a linear manner and, therefore, all circuit values are constants. This assumption is true only to a limited extent. Nonlinearities occur in the stress-strain curve of materials used in the construction of the transducer, in the electrical-mechanical interaction at various driving potentials, and in variations in the acoustical load impedance at various levels of intensity. Although nonlinearities cause alterations in the constants of the circuits, they do not diminish its value for transducer design or analysis. Reliable circuits can be drawn for specific conditions of operation. If these conditions are changed, the parameters of the circuit must be changed accordingly.

III. PIEZOELECTRIC TRANSDUCERS

III.A. EQUIVALENT CIRCUIT OF A SIMPLE PIEZOELECTRIC TRANSDUCER

A simple piezoelectric transducer, consisting of a plate vibrating in the thickness mode, may radiate from both faces (symmetrical load) or from only one face (air backed). An air-backed transducer has only one-half the radiating surface of a symmetrically loaded transducer. Thus, if a transformation factor α is used to convert mechanical impedances to electrical impedances in a symmetrically loaded transducer, a factor 2α must be used for the air-backed transducer.

To develop the equivalent circuit, it is necessary to determine the quantities that influence the overall input impedance to the transducer. In a piezoelectric material, the imposed stress is in phase with the impressed voltage. The piezoelectric element is an electrical condenser of capacitance C_0 by virtue of the dielectric nature of the transducer material and the electroded surfaces. The electrical resistance of the condenser is negligible. The capacitance C_0 appears in parallel with a series branch that includes the converted mechanical impedances. These mechanical impedances consist of (1) a resistance R_ℓ corresponding to the losses in the transducer (negligible in high-Q piezoelectric transducers), (2) a load resistance or radiation impedance $Z_R = \rho_o c_o S$, where $\rho_o c_o$ is the characteristic acoustic impedance of the load and S is the area of the radiating surface; (3) an inductance M due to the mass of the transducer; and (4) a capacitance 1/K due to the compliance of the transducer. The equivalent electrical components assuming R_ℓ negligible, are:

For a symmetrically loaded transducer

$$R_R = Z_R/\alpha^2, \quad L = M/\alpha^2, \quad C = \alpha^2/K \tag{5.1a}$$

For an air-backed transducer

$$R_R = Z_R/4\alpha^2, \quad L = M/4\alpha^2, \quad C = 4\alpha^2/K \tag{5.1b}$$

Since the transducer is a simple plate, a simple equivalent circuit can be drawn, as shown in Fig. 5.1.

The remaining discussion applies only to air-backed transducers. Figure 5.1 transforms into Fig. 5.2 by virtue of the transformer ratio $N = \frac{1}{2}\alpha$.

It should be noted that the circuits of Figs. 5.1 and 5.2 apply only to a specific type of transducer, a simple piezoelectric plate. Additional components, such as mechanical horns, necessitate additions to the circuit and the values of these as well as those of the driving element alone depend upon the geometry of the additional element or elements, the mass, the elasticity, and the mode of operation of the overall system.

III.B. EFFICIENCY OF A SIMPLE PIEZOELECTRIC TRANSDUCER

The efficiency of any system is the ratio of power out to the total power into the system. Referring to Fig. 5.2, the total power dissipation due to the transducer is that which is radiated into the load through R_R; i.e., the efficiency of such a transducer would be 100 per cent. In actual practice, transducer losses are present, and in the piezoelectric elements they include dielectric losses and internal losses due to strain.

At resonance, $\omega L = 1/\omega C$, and, inserting the resistances R_D corresponding to dielectric losses and R_f corresponding to the internal strain losses, Fig. 5.2 may be modified according to Fig. 5.3.

Referring to Fig. 5.3, the power output of the transducer is

$$P_{out} = I_3^2 R_R \tag{5.2}$$

The total power input of the transducer is

$$P_{in} = I_1^2 R_D + I_3^2 (R_f + R_R) \tag{5.3}$$

The efficiency of the transducer is

$$\eta = \frac{P_{out}}{P_{in}} = \frac{R_D R_R}{(R_f + R_R)(R_d + R_R + R_D)} \tag{5.4}$$

or, in terms of acoustical radiation impedance

$$\eta = \frac{R_D \left(\frac{\rho_o c_o S}{4\alpha^2}\right)}{(R_f + \frac{\rho_o c_o S}{4\alpha^2})(R_f + R_D + \frac{\rho_o c_o S}{4\alpha^2})} \tag{5.5}$$

where S is the cross-sectional area of the transducer and also the area of the radiating surface.

III.C. MAXIMUM POWER TRANSFER BETWEEN ELECTRONIC POWER SOURCE AND SIMPLE PIEZOELECTRIC TRANSDUCERS

According to elementary electric-circuit theory, maximum power transfer from a generator to a load occurs as follows:

(1) When the internal impedance of the generator is resistive, the load is resistive and equal in magnitude to the internal generator resistance;

(2) When the internal generator impedance is complex, the load impedance is the conjugate of the generator impedance; and

(3) When neither (1) or (2) are possible, the load impedance is equal in magnitude to the generator impedance.

In the equivalent circuit of Fig. 5.3, C_o causes a low power factor which usually may be eliminated by "tuning out" C_o with an inductance in parallel with the transducer. Therefore, for maximum power transfer to the transducer

$$R_i = \frac{R_D\,(R_f + R_R)}{R_D + R_f + R_R} = \frac{R_D\left(R_f + \dfrac{\rho_o c_o S}{4\alpha^2}\right)}{R_D + R_f + \dfrac{\rho_o c_o S}{4\alpha^2}} \tag{5.6}$$

where R_i is the internal resistance of the generator.

Resistances R_D and R_f are determined by the material of the transducer and its dimensions. For the condition of Eq. (5.6) to exist

$$R_R = \frac{R_i\,R_D}{R_D - R_i} - R_f \tag{5.7}$$

The maximum transfer of energy to the acoustic load occurs when $R_R = R_f$, at which time

$$R_i = \frac{2R_D\,R_R}{R_D + 2R_R}$$

$$R_R = \frac{R_i\,R_D}{2(R_D - R_i)} = \frac{\rho_o c_o S}{4\alpha^2}$$

(air-backed transducer) (5.7a)

for maximum power transfer from the generator to the load. Substituting Eqs. (5.7a) in Eq. (5.5) gives an efficiency

$$\eta = \frac{R_D}{2(R_D + 2R_R)} \tag{5.8}$$

when the transducer is matched to the generator and the radiation impedance is equal in magnitude to the internal loss resistance of the transducer. If, in this special case, R_D is negligible, i.e., $R_D = \infty$, the transducer efficiency is 50 per cent and the overall efficiency of the ultrasonic system is 25 per cent.

Other equivalent circuits representative of more complex systems can be developed by using the procedures followed in these sections. The previously derived equations for efficiency and maximum power transfer are easily adapted to the special case of a resonant piezoelectric transducer radiating through a half-wave mechanical transformer (solid double cylinders, exponential tapers, etc.) without losses. In each formula, the value of R_R representing the actual radiation impedance is multiplied by the appropriate transformation ratio. Typical ratios are:

For the exponential taper, $N_1/N_2 = d_1/d_2$

For the double cylinder, $N_1/N_2 = d_1^2/d_2^2$

For the cone, N_1/N_2 depends upon the dimensions; goes to a limit of 4.6

For the wedge, N_1/N_2 depends upon the dimensions; solution is a Bessel's function

For example, the efficiency of a half-wave resonant piezoelectric transducer radiating through a half-wave lossless exponentially tapered horn is

$$\eta_{\text{exp.}} = \frac{R_D\left(\frac{d_1}{d_2}\right)^2 R_R}{\left[R_f + \left(\frac{d_1}{d_2}\right)^2 R_R\right]\left[R_f + \left(\frac{d_1}{d_2}\right)^2 R_R + R_D\right]} \tag{5.5a}$$

and maximum power transfer from the generator to the transducer occurs when

$$\left(\frac{d_1}{d_2}\right)^2 R_R = \frac{R_i\ R_D}{R_D - R_i} - R_f \tag{5.7b}$$

where S_2 is the radiating surface $\pi d_2^2/4$ and $R_R = \rho_o c_o S_2/4\alpha^2$. If d_1 of the exponentially tapered horn is equal to the diameter of the piezoelectric transducer

$$\left(\frac{d_1}{d_2}\right)^2 R_R = \frac{\rho_o c_o S_1}{4\alpha^2}$$

which is identical to the radiation impedance of the transducer working into the same load without the horn. The exponentially tapered horn is the only one, other than the straight cylinder, that produces this condition.

III.D. DETERMINING TRANSFORMATION FACTOR (α) FOR THE PIEZOELECTRIC TRANSDUCER

The transformation factor α relates mechanical impedances to electrical impedances for use in the design of ultrasonic systems. The constant α is a function of the electromechanical coupling factor k_c by the relationship

$$\alpha^2 = \frac{\text{clamped capacitance of the piezoelectric element}}{\text{mechanical compliance of the piezoelectric element}} \times k_c^{\,2}$$

$$= \frac{C_o}{(1/K)} k_c^{\,2} \tag{5.9}$$

where $k_c^{\,2}$ is defined as the energy transformed divided by the total input energy, regardless of whether electrical energy is transformed to mechanical energy or vice versa. The approximate value of $k_c^{\,2}$ may be obtained by measuring the frequency, f_r, at resonance and the frequency, f_a, at antiresonance. Then

$$k_c^{\,2} \simeq \frac{\pi^2}{4} \frac{(f_a - f_r)}{f_r} \tag{5.10}$$

III.E. QUALITY FACTOR (Q) OF PIEZOELECTRIC TRANSDUCERS

In electrical systems, the quality factor (Q) of an inductance or a capacitance is the ratio of the energy stored in the component to the energy dissipated in an associated resistance. Similarly, the Q of a transducer is the ratio of energy stored in its reactive components to the energy dissipated in its resistive components, which include the radiation resistance. The Q of a transducer is a measure of the sharpness of resonance and, therefore, the operating frequency bandwidth of the transducer is a function of the Q.

Neglecting all electrical dissipative factors, from Fig. 5.2 we see that the Q of a simple piezoelectric transducer with air backing is

$$Q = \frac{\omega L}{R_R} = \frac{(\omega M/4\alpha^2)}{(\rho_o c_o S/4\alpha^2)} = \frac{\omega M}{\rho_o c_o S} \tag{5.11}$$

where M is the equivalent mass of the transducer. ($M = \rho_1 S\ell/2$, where ℓ is the thickness of the transducer.) Then

$$Q = \frac{\pi\rho_1 f\ell}{\rho_o c_o} = \frac{\pi\rho_1 \quad (\frac{\lambda f}{2})}{\rho_o c_o} = \frac{\pi\rho_1 c_1}{2\rho_o c_o} \tag{5.11a}$$

or, for the nth harmonic

$$Q_n = \frac{n\pi}{2}\frac{\rho_1 c_1}{\rho_2 c_2} \qquad \text{(air backed)} \tag{5.11b}$$

Equations (5.11), (5.11a), and (5.11b) apply to ideal lossless transducers in which R_D and R_f are negligible.

III.F. PIEZOELECTRIC TRANSDUCERS FOR HIGH-INTENSITY APPLICATIONS

Piezoelectric transducers for power applications sometimes consist of plates or blocks of piezoelectric materials resonant at the desired frequency epoxy bonded to the surface of a treatment chamber--for example, the bottom of a cleaning tank. A more desirable technique is to clamp the element between two blocks which become a part of the resonant system. A typical half-wave transducer is shown schematically in Fig. 5.4 and is reminiscent of the Langevin sandwich structure.

Here, two piezoelectric elements C_1 and C_2 are located between two identical metal blocks A and B. The elements C_1 and C_2 are separated by an electrode E connected to the high voltage lead. The electrode is, therefore, located at a node and the elements C_1 and C_2 must be polarized in opposite directions for optimum activity.

If the mating surfaces are perfectly flat and highly polished, coupling without bonding between the piezoelectric elements and the metal end pieces may be accomplished fairly well by applying high pressure (bias pressure) across the elements. Since these elements are ceramic, usually with tensile strengths within the range of 5,000 to 10,000 psi, a bias pressure provides a safety factor against fatigue failure. In Fig. 5.4, the elements are located on either side of the plane of maximum stress, which can easily exceed the tensile strength of the elements or of the adhesive which bonds them together, if one is used.

Bonding the elements together on clamping improves coupling and permits operation at lower and more effective bias pressures (7,000-11,000 psi compared to 13,000-14,000 psi without bonding in lead-ziconate titanate elements).

Clamping may be done in one of two ways: (1) a bolt may be run through the center of the elements, but insulated from the piezoelectric elements; or (2) a flange may be attached to the metal blocks and clamped externally by a series of bolts.

The proper torque to apply to the bolts may be determined by (1) calculating the total force required to produce the desired stress, (2) determining the force F that each bolt must hold, and (3) applying this value of F in the following equation:

$$T = \frac{Fp}{2\pi e} \tag{5.12}$$

where T is the torque (lb-in.);
F is the force per bolt (lb),
p is the pitch of the bolt thread (in.), and
e is the efficiency of the threaded combinations.

The assembly just described will resonate at a frequency that is below that of either piezoelectric element when it is unloaded by an amount controlled by the dimensions and acoustical characteristics of the metal blocks and of the piezoelectric elements.

The dimensions of the transducer elements can be calculated for a given operating frequency, using equations given in Chap. 4 and the known elastic properties of the materials to be employed.

Piezoelectric transducers are high-impedance units and the voltage used in driving power units may exceed 20,000 V. These voltages do not pose a difficult problem, but must be taken into consideration in the design of the assembly.

Structures used to support transducers for high-intensity applications provide paths for conducting energy from the transducers. Part of this energy is dissipated and the remainder reverberates within the structure. The efficiency of the system suffers both by the energy dissipated and by the interference of the reverberating waves in the structure with the oscillations of the transducer. It is good practice to acoustically isolate the transducer from all structures not intended as a part of the vibratory system in order to conserve as much energy as possible for processing. This is usually done (1) by mounting the transducer at nodes through O-ring assemblies, (2) by using the radiating surface as the mounting plate (for example, the bottom of ultrasonic cleaning tanks), or (3) by floating or pinning the transducer in its housing for those applications which do not require an applied external force of any significance.

III.G. PULSE-TYPE TRANSDUCERS FOR LOW-INTENSITY APPLICATIONS

Transducers generally used in nondestructive testing are of the piezoelectric type. Piezoelectric elements have high Q values and, as a result, a short electrical impulse will cause them to vibrate or "ring" for a long period of time. With few exceptions, in pulse-type nondestructive testing such ringing is definitely undesirable. The length of the pulse, which is the length of time required for the transducer to stop ringing, corresponds to a distance of pulse travel in the medium under investigation. In pulse-echo testing, resolution of defects is a function of the pulse length. If an echo is received before the initial pulse dies down, its electrical indication becomes lost in the electrical indication of the initial pulse. Therefore, the ability to detect defects near the surface of a material or to resolve two defects of only slightly different depths is a function of the pulse length.

Shorter pulse lengths also give better defect sensitivity. Shorter pulses contain a wider band of frequencies. The increased sensitivity to small defects is due to the shorter wavelengths of the higher frequencies associated with the wide bandwidth of the short pulse.

The bandwidth of a transducer is a function of its Q as follows:

$$Q = f_c/(f_2 - f_1) \tag{5.13}$$

where f_c is the center frequency of the pulse, and f_1 and f_2 are the half-power points on each side of f_c where the amplitude is 0.707 times the amplitude at f_c.

The high Q generally obtained in the piezoelectric materials used to produce high-frequency, short-pulse transducers is offset by bonding a material with high damping characteristics to one face of the piezoelectric plate. Ideally, these two materials should have identical characteristic acoustic impedances to eliminate reflections from the interface. The opposite, or radiating, face of the piezoelectric plate may be bonded to a low-loss window, wedge (for generating modes other than longitudinal), lens, or other useful structure.

Various materials have high absorption characteristics and are suitable for damping purposes. Suitable mixtures can be formulated with graphite, powdered metals, and metal oxides of random grain sizes either fused or adhesively bonded together.

Tranducers used for nondestructive testing are identified by frequency, the frequency usually being the fundamental resonance frequency of the unloaded crystal and supposedly designating the frequency at the center of the bandwidth, f_c. Often after the back-up material has been added, the center frequency drops significantly below that desired. The proper center frequency can be maintained by properly controlling the dimensions of the backup material. Or, the frequency rating may be determined empirically. Transducers that produce extremely short pulses are not designated by a frequency.

Piezoelectric materials have high impedances and are driven at high voltages. The two major (and opposite) faces are electroded and connected to the electrical leads. The radiating face usually is at ground potential and the high-potential lead is connected to the opposite face. Proper spacing and insulation between high-potential leads and grounded structures is essential to avoid electrical arcing and damage to the piezoelectric plate.

IV. MAGNETOSTRICTIVE TRANSDUCERS

An extensive treatise on the design of magnetostrictive transducers is included in a Summary Report of the National Defense Research Committee and is available through the Library of Congress (2).

Magnetostriction is a term applied to the change in dimensions of a magnetic material when the impressed magnetic field is varied in magnitude. This phenomenon was discovered by Joule and it was described by him in a paper published in 1847 (3).

The earliest high-power ultrasonic transducers operating in the low-kilohertz frequency range (below 100 kHz) were made of magnetostrictive materials, high-frequency generation being left to piezoelectric transducers. The trend today is toward piezoelectric ceramic materials for both low- and high-frequency applications, but certain applications for ultrasonic energy may continue to favor certain characteristics peculiar to magnetostrictive devices. One such characteristic is the fact that, if the temperature of a magnetostrictive transducer should accidentally exceed the Curie point (temperature above which it becomes inactive), the core is usable upon cooling. However, the thermal history of the core does influence the magnetostrictive properties of the material. When the temperature of a ceramic piezoelectric material exceeds the Curie temperature, it must be polarized anew before it can be used again for ultrasonic generation or reception.

The electrical components of a magnetostrictive transducer consist of a coil wrapped about a core of magnetic material. The coil has an inductance corresponding to the number of turns N, the permeability of the core material μ_o, the dynamic permeability of the core μ_i, the cross-sectional area of the core S, and the length of the magnetic path ℓ, as follows:

$$L_c = \mu_i \mu_o SN^2/\ell \tag{5.14}$$

when the magnetic path is a closed loop. If an air gap is present, the reluctance of the air gap must be taken into account.

The coil exhibits the electrical resistance common to any inductor with equivalent length and size of conductor, and, also, the magnetic circuit is subject to hysteresis and eddy-current losses.

Consider the mechanical components of the magnetostrictive transducer. These are the elastic properties of the material, the mass of the material, the load characteristics (i.e., radiation impedance or the mechanical characteristics of a transition member such as a horn and its radiation impedance), and the mechanical losses of the material.

The electrical and the mechanical quantities may be represented in an equivalent electrical circuit as shown in Fig. 5.5.

In Fig. 5.5, $\alpha = -(\Lambda/\omega N)e^{-j\phi}$

where
- Λ is the magnetostrictive constant,
- $e^{-j\phi}$ is a phase factor accounting for hysteresis,
- N is the number of turns on the core,
- ω is $2\pi f$, and
- f is the frequency.

The magnetostrictive constant Λ and the transformation coefficients are related to the electromechanical coupling coefficient k_c as follows:

$$R_c = \sqrt{\mu_i \mu_o \Lambda^2/Y_o}$$

$$= \frac{\Lambda}{N}\sqrt{\frac{L_c \ell}{S\,Y_o}} = -\omega\alpha\sqrt{\frac{L_c \ell}{S\,Y_o}}\; e^{j\phi} \tag{5.15}$$

or

$$\alpha^2 = \frac{-jSk_c^{\,2}\,Y_o\,e^{-j2\phi}}{\omega\ell Z_o} \tag{5.16}$$

where $Z_o = j\omega L$.

The circuit of Fig. 5.5 may be drawn in its equivalent form, adding the internal impedance of the generator Z_i, as in Fig. 5.6.

Certain facts must be kept in mind in using such an idealized circuit as that shown in Fig. 5.6. First, the use of the circuit implies linearity in the operation of the transducer and that each component in the circuit has a constant value. Actually this assumption approaches validity only when very small driving magnetomotive force (mmf) is applied. Second, the circuit applies only to closed loops as the elements are defined; however, the circuit can be modified to accommodate other magnetic configurations. Third, frequency f must be lower than a critical value given by

$$f_c = \frac{\rho_c}{2\pi^2 \mu_1 \mu_o d^2} \tag{5.17}$$

where ρ_c is the resistivity of the core (Ωm), and

d is the lamination thickness, (m).

Therefore, since the transducer element provides a nonlinear impedance/power function, equivalent circuits must be used with care.

Keeping these statements in mind, one may proceed to derive equations for maximum power transfer and efficiency for a magnetostrictive transducer operating at low amplitude. The equations derived for efficiency will be the maximum that can be expected from a magnetostrictive device. At high amplitudes, where the characteristics become nonlinear, the efficiency is always lower.

IV.A. MAXIMUM POWER TRANSFER TO THE MAGNETOSTRICTIVE TRANSDUCER

Referring to Fig. 5.6, at resonance the reactive components of the mechanical branch of the circuit are "tuned out". By adding a tuning capacitor at the output of the generator, $j\omega L_o$ also may be tuned out.

Letting the internal impedance of the generator now be R_i, the condition for maximum power transfer from the generator to the load is

$$R_i = R_o + \left[\frac{(R_m + Z_R)}{\alpha^2} Z_c\right] \Bigg/ \left[\frac{R_m + Z_R}{\alpha^2} + Z_c\right] \tag{5.18}$$

or

$$Z_R = \frac{R_o R_m + \alpha^2 Z_c R_o + R_m Z_c - R_i R_m - \alpha^2 R_i R_c}{R_i - R_o - Z_c} \tag{5.19}$$

for a transducer radiating from one end, where $Z_R = (\rho c)_\ell S$ and $(\rho c)_\ell$ is the acoustic impedance of the load medium.

IV.B. EFFICIENCY OF THE MAGNETOSTRICTIVE TRANSDUCER

The efficiency of the magnetostrictive transducer is

$$\eta = \frac{\text{power output}}{\text{power input}} = \frac{W_o}{W_{in}} \tag{5.20}$$

The power output, W_o, is the power delivered to the acoustic load, or

$$W_o = I_m^2 \frac{(\rho c)_\ell S}{\alpha^2} = I_m^2 R_R/\alpha^2 \tag{5.21}$$

where I_m is the current through the output branch of Fig. 5.6, and is given by

$$I_m = \frac{Z_c I_T}{Z_c + \dfrac{(R_m + Z_R)}{\alpha^2}}$$

The power input, W_{in}, is

$$W_{in} = I_T^{\,2}\left[R_o + \frac{Z_c\,(R_m + Z_R)}{\left(Z_c + \dfrac{R_m + Z_R}{\alpha^2}\right)\alpha^2}\right] \tag{5.22}$$

where I_T is the total current delivered to the transducer. Then

$$\eta = \frac{\alpha^2\,(\rho c)_\ell S Z_c^{\,2}}{[\alpha^2 Z_c + R_m + (\rho c)_\ell S][\alpha^2 R_o Z_c + R_o R_m + R_o(\rho c)_\ell S + R_m Z_c + (\rho c)_\ell S Z_c]} \tag{5.23}$$

IV.C. MAGNETOSTRICTIVE TRANSDUCERS FOR HIGH-INTENSITY APPLICATIONS

Magnetostrictive transducers are low-impedance, high-current devices. The impedance of a 20-kHz, 1000-W transducer of A-nickel is approximately 10 Ω if the coil resistance is negligible.

Several materials exhibit magnetostrictive properties. Nickel, nickel-iron, and iron-cobalt (Permendur) alloys are most commonly used for transducers.

Transducers of A-nickel (commercial, high purity) are rugged. They are usually made in laminated stacks that are one-half wavelength long, cut so that the rolling direction is in the direction of motion.

The bar velocity of sound in nickel is 188,500 in./sec. Therefore, a half-wavelength stack of nickel with uniform cross section vibrating at 20 kHz is approximately 4.7 in. long.

The thickness of the laminations used in a magnetostrictive stack is determined by the flux penetration, which is limited by eddy currents in the materials. The total thickness of each lamination is approximately twice the effective penetration. The penetration depth is determined by

$$\delta = 1.98\sqrt{\rho_e/\mu_\Delta f} \tag{5.24}$$

where ρ_e is electrical conductivity of the magnetic material, (Ω-cm),

μ_Δ is relative magnetic permeability of the magnetic material, and

f is frequency (Hz).

Annealing A-nickel in air at 1400°F for 1 hr and allowing it to cool very slowly in the furnace produces good magnetostrictive characteristics for ultrasonic transducers. The oxide layer thus produced is tough and has good electrical resistance, so that the laminations can be stacked in direct contact with one another.

Magnetostrictive stacks are commonly formed as rectangular stacks with windows, thus giving a closed magnetic flux path. The energizing coils are wound through the windows. The coils are protected from mechanical damage to the insulation by slabs of insulators. The oxide layer may be removed from one end of the stack and the stack silver brazed to a solid half-wave horn which is used to couple the energy to the work.

A-nickel saturates at a magnetomotive force slightly above 100 Oe and a flux density of approximately 6000 G. At these levels of magnetization in a dc electric field, the length of the stack shortens by approximately 33 μ in./in. However, at resonance, the maximum strain is a function of the mechanical Q of the transducer times the dc strain corresponding to the peak ac magnetization.

The efficiency of a magnetostrictive transducer is affected by the magnetic bias. At zero bias, the transducer vibrates at twice the frequency of the driving current. Operation at zero bias is inefficient. The highest efficiency is obtained with A-nickel when it is biased to approximately 4000 G.

Magnetostrictive transducers are also low-Q devices. This low-Q is due partly to electrical eddy currents and hysteresis losses and partly to internal friction losses. These losses also produce heat which must be dissipated. Nickel transducers usually are water cooled to maintain a temperature at which the transducer will operate at its best.

Magnetostrictive devices may be used at frequencies as high as 100 kHz; however, for practical power-handling capabilities, they are limited to frequencies below 50 kHz.

V. ELECTROMAGNETIC DEVICES

Electromagnetic devices for generating vibratory energy may be produced in many forms: for example, loudspeakers, foundry-flask vibrators, marking tools, and hair clippers. They are fairly simple devices to design and build.

A simple device may consist of an armature mounted on a spring in a magnetic circuit (Fig. 5.7).

The frequency at which the armature will vibrate is determined by the equation for the mass-loaded spring or

$$f = \frac{1}{2\pi} \sqrt{\frac{k}{M + m_s/3}} \tag{5.25}$$

where M is the mass of the armature (lb-sec^2/in.),
m is the mass of the spring (lb-sec^2/in.), and
k is the spring constant (lb/in.).

VI. PNEUMATIC DEVICES (WHISTLES)

Whistles have been used to generate high-intensity sound both in air and in liquids. Although whistles are capable of generating ultrasonic energy in air to 500 kHz, using helium or hydrogen, a practical upper frequency for power applications of whistles in air appears to be approximately 30 kHz.

Several types of whistles are possible. Most people are familiar with the common whistle based on the organ-pipe effect. A jet of air impinges on a sharp edge and generates a noise which is shaped into a fairly pure tone through a resonant column. The efficiencies of such whistles are low (below 5 per cent).

A more efficient whistle (η = 5 per cent) is based upon a design credited to Hartmann (Fig. 5.8).

A jet of gas of diameter D_1 is directed into a cavity of diameter D_2. The impact of the jet on the bottom of the cavity causes a rise in pressure which in turn causes a counterflow of gas toward the nozzle. The pressure is released at the open end of the cavity which causes a rarefaction in the cavity, when the jet again enters the cavity to complete the pressure cycle. The wavelength of the resulting sound is approximately (see Fig. 5.8)

$$\lambda = 4(b + 0.3D_2) \tag{5.26}$$

where b is the depth of the cavity and the quantitiy $0.3D_2$ is an end correction factor.

In the Hartmann whistle, the diameter of the cup D_2 is equal to the diameter of the jet D_1. The spacing a between the end of the nozzle and the open end of the cavity is critical as is the axial alignment between the two sections. According to Hartmann (4) the relationship between wavelength λ, nozzle and resonator diameter $D_1 = D_2 = b$, excess gas pressure p, and nozzle-to-resonator spacing a is

$$\frac{\lambda}{D_1} = \frac{\lambda}{D_2} = 5.8 + 2.5 \left\{ \frac{a}{D_1} - [1 + 0.041\ (p - 0.9)^2] \right\} \tag{5.27}$$

where p is in kilograms per square centimeter, and
a, b, D_1, and D_2 are in millimeters.

Hartmann shows that a may vary within limits a_o and a_m given by

$$a_o = [1 + 0.041\ (p - 0.9)^2]\ D_2 \tag{5.28}$$

and

$$\frac{a_m - a_o}{D_2} = 0.13\ (p - 0.9)^2 \tag{5.29}$$

Within the range a_o to a_m, the wavelength increases nearly linearly at a rate

$$d\lambda/da = 2.5$$

The permissible range over which the efficiency is maximum is somewhat less than that given by Eq. (5.29) and is more nearly

$$\frac{a_2 - a_o}{D_2} = 0.44 \sqrt{p - 1.8} \tag{5.30}$$

According to Hartmann (4) the radiated power P_o, in watts is

$$P_o = 3D_2^{\,2} \sqrt{p - 0.9} \qquad (W) \tag{5.31}$$

where p is now given in atmospheres.

The efficiency η, is

$$\eta = P_o/P_j = \frac{3D_2^{\,2} \sqrt{p - 0.9}}{P_j} \tag{5.32}$$

where P_j is the theoretical power required to maintain the jet. The 5 per cent efficiency claimed by Hartmann is based upon Eq. (5.32) and, therefore, it does not include the efficiency of the compressor.

Boucher (5) has developed a modification of the Hartmann whistle for which he claims efficiencies of 14 per cent. In Boucher's whistle, the resonator is mounted on a stem extending axially through the nozzle (Fig. 5.9).

Boucher claims to verify Hartmann's equation with his "stem-jet" whistle; however, he finds an improvement in efficiency by making the ratio $D_2/D_1 \geq 1.3$. He also finds that the wavelength-to-distance ratio $d\lambda/da$ over the interval a_o to a_m is more nearly

$$d\lambda/da = 1.88$$

for the stem-jet whistle.

The depth, b, of the resonator is not critical at low frequencies. Cleanliness is important. A thin film of grease, dirt, or liquid in the cup can lower the efficiency by a considerable amount.

Figure 5.10 is a photograph of two modified Hartmann whistles. These whistles operate at 10 kHz and deliver 100 acoustical watts at an airline pressure of 90 psig. The smaller reflector is equipped with a series of holes through which liquids are passed to be nebulized.

Whistles for generating sound in liquid have been developed for industrial use. The Hartmann whistle required jet velocities equal to the velocity of sound. The velocity of sound is considerably higher in liquids than in gases. Flow velocities equal to the velocity of sound are impractical in liquids. Therefore, the liquid whistles must operate on a different principle.

The whistles of Janovsky and Pohlmann (6) operate on a jet-edge principle. A high pressure jet of the liquid or liquids is forced into the edge of a thin plate which is mounted at the nodes. The plate vibrates in flexure at resonance. The resulting cavitation in a multiliquid mixture causes the liquids to emulsify more effectively than is accomplished by a similar jet impinging on a stationary plate. Again, liquid whistles are low-frequency devices, 5,000 Hz being typical.

VII. MECHANICAL DEVICES

High-amplitude vibrations at low frequencies are sometimes generated by rotating, counterbalanced weights. Two sets of weights are mounted to balance out forces in all directions but one. The two sets are rotated in opposite directions (Fig. 5.11.

In Fig. 5.11, the centrifugal force due to the rotation of each weight is given by

$$F = mr_g\omega^2 \tag{5.33}$$

where F is the force (lb),

m (= W/g) is the mass of the weight ($\text{lb-sec}^2/\text{in.}$),

W is the weight of the mass (lb),

g is gravity, 386 in./sec^2,

r_g is the radius of gyration of the weight (in.),
ω is $2\pi n$, and
n is the number of revolutions the weight makes in 1 sec.

Devices of this type are limited to low frequencies by mechanical restrictions--restrictions common to the use of bearings and gears.

From Fig. 5.11, it is evident that the sum of forces in the horizontal direction is zero, but that the vertical components are additive and the maximum vertical force equals the sum of the centrifugal forces. Forces about a vertical axis also must cancel out to avoid rotation. This is done by using more than one weight for each direction and balancing these for zero rotation.

VIII. GENERAL REMARKS

Design of ultrasonic transducers has been treated in an elementary manner in this chapter. A more extensive treatment is not within the scope of this book. Since an ultrasonic transducer is essentially a device for producing ultrasonic vibrations or for detecting ultrasonic energy, any device that is capable of performing these functions is a potentially useful transducer, so that the list of types is not limited to those mentioned in this chapter.

The potential designs of transducers incorporating the principles discussed in this chapter are probably infinite. This might be recognized by considering the transducer elements presented (piezoelectric plates, magnetostrictive stacks, etc.) as motors that energize a piece of machinery and comparing the possible combinations with the number of ways in which electric motors are used.

A distinction should be made between the terms effectiveness, efficiency, maximum power transfer, and quality factor (Q) of a system which can affect the designers' approach to the design of transducers.

The effectiveness of an ultrasonic transducer or system is determined by the adequacy of the transducer to perform the required task. An effective transducer is not necessarily an efficient transducer. In fact, efficiency may have to be sacrificed to obtain the desired effect, especially in certain high-intensity applications of ultrasonics.

The efficiency of a transducer is the ratio of radiated power to power delivered into the transducer. The efficiency of the electronic generator is the ratio of the power delivered to the transducer to the power into the generator. These combined efficiencies (product of the two) gives the efficiency of the total system.

Maximum power transfer occurs when the load impedance matches the internal impedance of a power source. When this match exists, the generator is capable of delivering the maximum amount of power to the load. This condition is often mistakenly considered to be the condition of maximum efficiency.

The quality factor Q of a transducer is the ratio of the total inductive reactance (at resonance) to the total resistance, or the ratio of the energy stored in the system to the energy dissipated. The amplitude of vibration of transducers increases with Q. If, under load, the Q of a transducer continues to be high, it is an indication that little energy is dissipated in the load and, therefore, the efficiency of the system is correspondingly low. There are applications where high-Q operation is required, such as where transducers are used for accurately controlling the frequency of an oscillator. Many applications require low-Q operation under load, for example, ultrasonic cleaning wherein high-efficiency operation is very important.

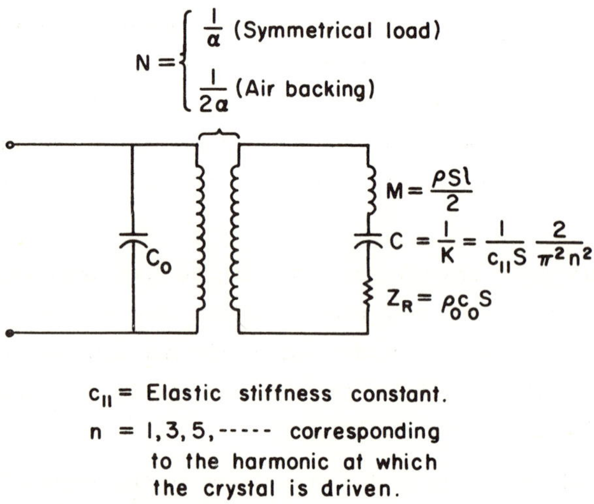

FIG. 5.1. Equivalent circuit of a simple piezoelectric transducer.

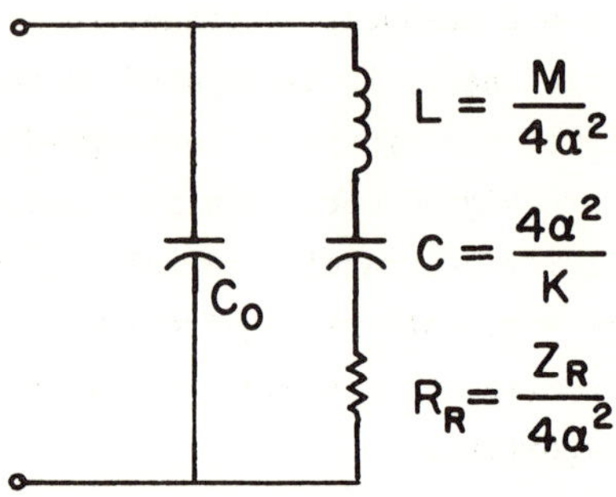

FIG. 5.2. Equivalent circuit of a simple air-backed transducer.

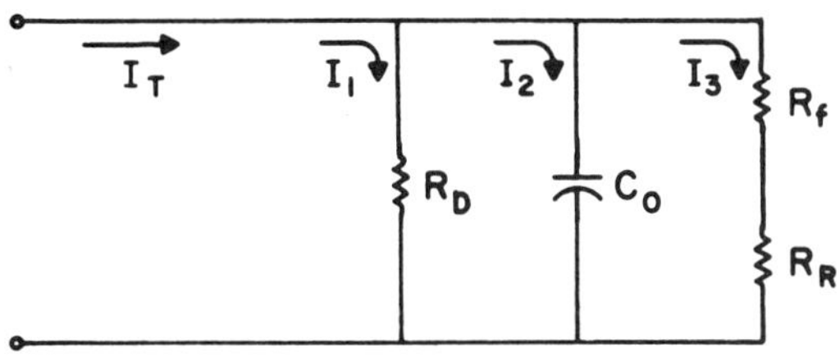

FIG. 5.3. Equivalent circuit of a simple resonant piezoelectric transducer with losses.

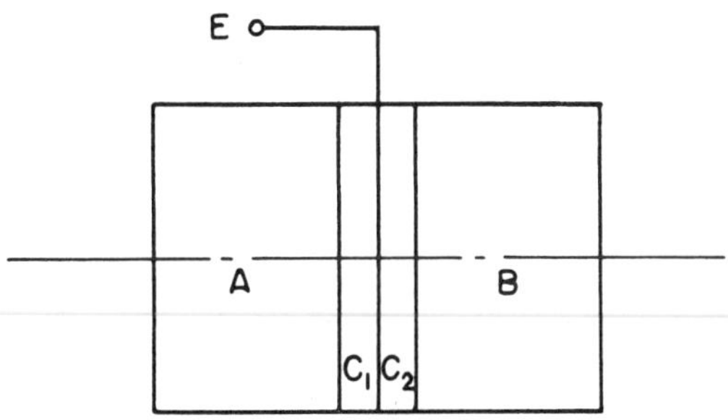

FIG. 5.4. Typical half-wave piezoelectric transducer.

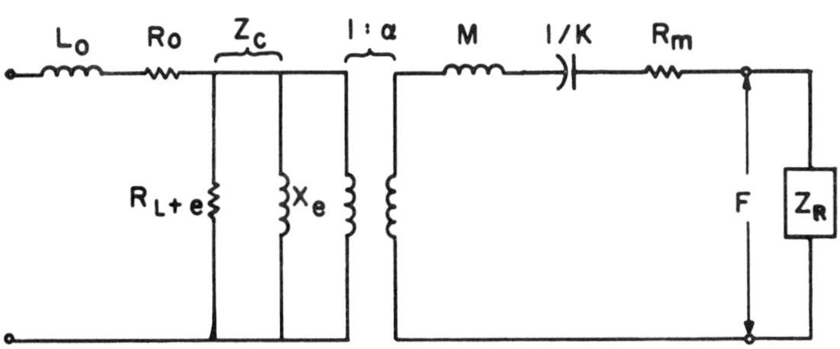

FIG. 5.5. Equivalent circuit of a magnetostrictive transducer.

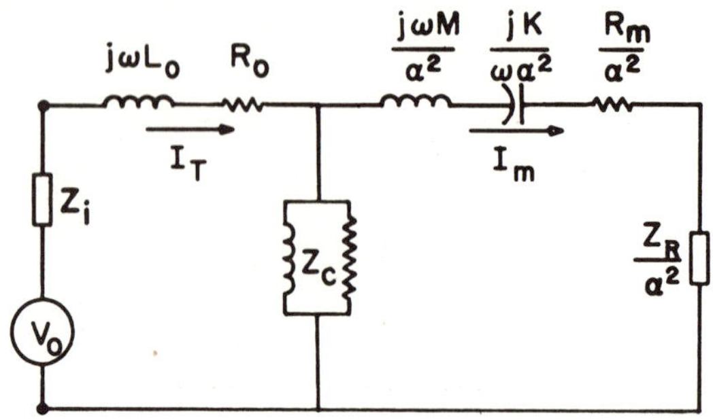

FIG. 5.6. Equivalent electrical circuit of a magnetostrictive transducer.

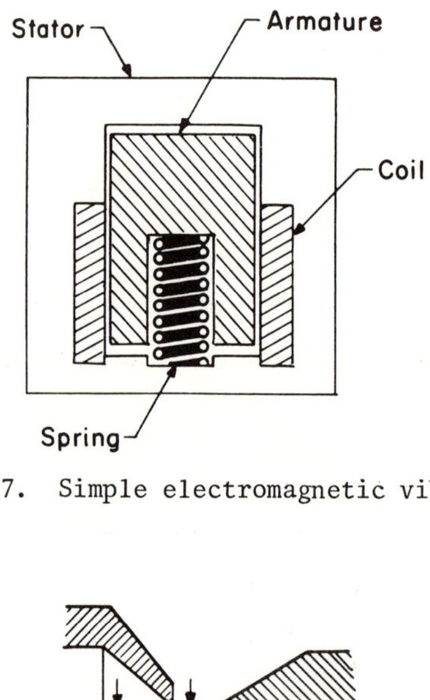

FIG. 5.7. Simple electromagnetic vibrator.

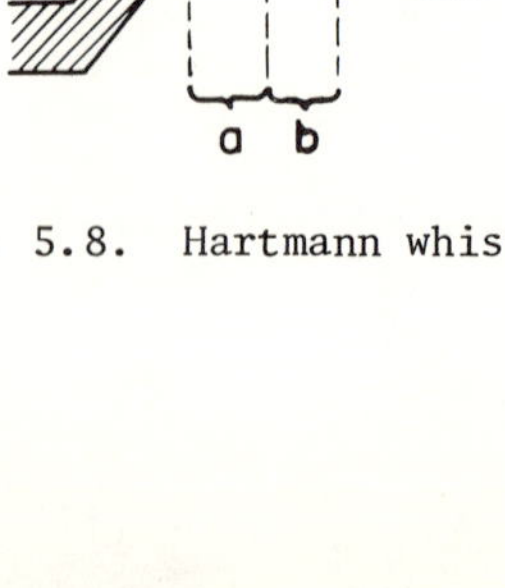

FIG. 5.8. Hartmann whistle.

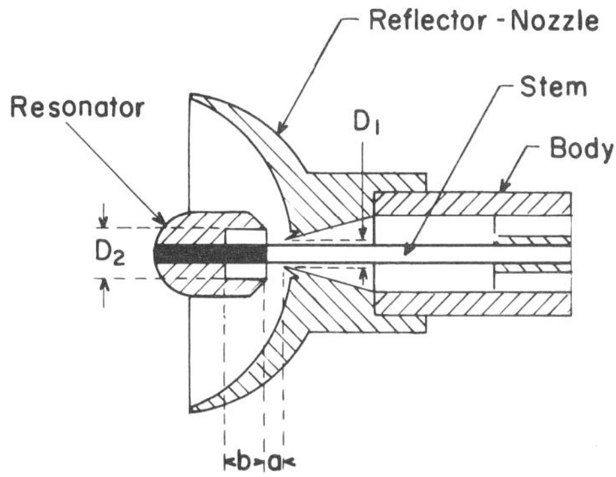

FIG. 5.9. Modified Hartmann whistle.

FIG. 5.10. Modified Hartmann whistles for high-intensity applications.

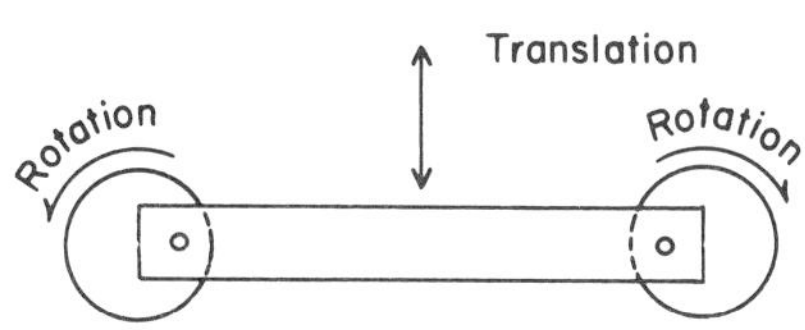

FIG. 5.11. Counterbalanced mechanical vibration generator.

REFERENCES

1. T. F. Hueter and R. H. Bolt, Sonics, Wiley, New York; Chapman and Hall, London, 1955.

2. Summary Technical Report of the National Defense Research Committee, The Design and Construction of Magnetostriction Transducers, 1946, available through Library of Congress, Washington, D. C.

3. J. P. Joule, Phil. Mag., III, 30, 76 (1847).

4. J. Hartmann, "Construction, Performance, and Design of the Acoustic Air-Jet Generator," J. Sci. Instr., 16, 140-149 (1939).

5. E. Brun and R. M. G. Boucher, J. Acoust. Soc. Am., 29, 573-583 (1957).

6. W. Janovsky and R. Pohlmann, Z. Angew. Phys., 1, 222 (1948).

Chapter 6

DETERMINING PROPERTIES OF MATERIALS

I. INTRODUCTION

In the previous chapters, it was shown that many factors influence the propagation characteristics of ultrasonic waves. In this and succeeding chapters (through Chap. 10), methods of measuring and evaluating propagation characteristics are presented, demonstrating many varied and practical low-intensity applications of ultrasonic energy. Low-intensity applications are those wherein the primary purpose is to transmit the energy through a medium. The objective is never to change the state of the medium; i.e., the energy level is so low that the medium remains unchanged from its original state after the wave has passed through it. Presented in this chapter are methods of measuring propagation characteristics for obtaining engineering data for materials.

Propagation characteristics involve both velocity of sound and attenuation. The elastic properties and density of a medium control the velocity of sound. Mechanisms producing losses of energy control the attenuation. Both velocity and attenuation are measurable quantities.

There are many methods of measuring velocity of sound. The choice of method depends upon the nature of the material and its environment, and upon the accuracy required. Discussed in this chapter are the interferometer, the resonance, the "sing-around" and pulse-superposition methods. Methods used at high temperatures and those used at high pressures are presented as special cases of interest.

Measurements of flexural, longitudinal, and torsional resonant frequencies of bars provide means of determining the elastic moduli and Poisson's ratio of materials, which are quantities of engineering importance. These measurements have been valuable in determining the quality, including strength, of ceramic materials and of powder-metal compacts.

Ultrasonic methods of determining viscosity and of determining the properties of plastics and high polymers are also presented.

The acoustical properties of materials are presented in tabular form at the end of this chapter.

II. METHODS OF MEASURING VELOCITY OF SOUND

Measurements conducted over a period of many years by numerous investigators have resulted in an extensive accumulation of data on velocity of sound. However, there is a continued interest in such measurements in new materials and in materials to be used in hostile or abnormal environments--high temperatures and high pressures, for example.

There are basically three methods of measuring the velocity of sound: (1) by determining the length of a continuous wave at a known frequency, (2) by determining the time required for a progressive wave to travel a known distance, and (3) by determining the angle of refraction of a beam of sound incident on an interface between two media in one of which the velocity of sound is known. These methods have been implemented in numerous ways. A few of these are described in the following sections.

II.A. INTERFEROMETER METHOD

The interferometer is a continuous-wave device which has been used for measuring velocity and attenuation of sound in liquids and gases. It consists of a piezoelectric transducer and a reflector. The reflector is moved toward or away from the transducer by a micrometer adjustment mechanism. As the reflector moves, the reflected wave becomes periodically in phase and out of phase with the transmitted wave, thus causing a corresponding constructive and destructive interference. The reaction of the interference on the crystal influences the load impedance reflected into the electronic system. The plate current of the electronic amplifier is caused to fluctuate accordingly. The wavelength of the sound is determined by the distance the micrometer moves the reflector during one cycle of plate-current fluctuation. The accuracy of the measurement depends upon the accuracy of the micrometer readings, the parallelism between reflector and transducer surfaces, and the accuracy of the frequency determination. The velocity of sound is given by multiplying the wavelength times the frequency.

II.B. RESONANCE METHOD

The resonance method is similar to the interferometer method and is applicable to the measurement of velocity of sound in gases, liquids, or solids.

The resonance method involves the use of a fixed transducer and a fixed reflector or two transducers spaced a known distance apart. The transducer is driven through a range of frequencies to determine successive resonances. The difference between two successive resonant frequencies is equal to the fundamental resonant frequency of the medium or

$$c = 2\ell\Delta f \qquad (6.1)$$

where ℓ is the distance between the transducer and the reflecting surface, and Δf is the difference between successive resonant frequencies.

The accuracy of the resonance method depends upon the accuracy of the frequency determination and of the measurement of the distance between reflecting surfaces. Phase shifts at reflecting interfaces must be taken into account.

II.C. "SING-AROUND" METHOD

The "sing-around" method of measuring the velocity of sound involves the use of two piezoelectric transducers, one at each end of the specimen. One transducer receives an impulse from the electronic source and converts it into an ultrasonic pulse in the specimen. The pulse of ultrasonic energy, after passing through the specimen, is detected by the receiving transducer. The received pulse triggers the electronic generator to initiate a succeeding pulse. The velocity is related to the pulse repetition rate as follows:

$$c = \frac{\ell \times \text{P.R.R.}}{(1 - e \times \text{P.R.R.})} \approx \ell \times \text{P.R.R} \qquad (6.2)$$

where P.R.R. is pulse repetition rate per second,

ℓ is the length of the specimen, and

e is a correction factor for delays in coupling between the transducers and the specimen and in the electrical components.

II.D. PULSE-SUPERPOSITION METHOD

The pulse-superposition method (1,2) is capable of measuring the velocity of sound to accuracies of 1 part in 5,000 or better. A piezoelectric transducer initiates pulses of ultrasonic energy into a specimen. These pulses echo back and forth within the specimen. A continuous-wave oscillator is used to control the pulse repetition rate. When the repetition rate is adjusted so that the initiation of a pulse coincides with the return of the first echo from the preceding pulses, the change in the signal amplitude indicates superposition. The pulse repetition rate is a measure of the travel time within the specimen (twice the thickness of the specimen). The velocity is twice the specimen thickness times the pulse repetition rate after correcting for phase shift at the bond areas.

II.E. MEASUREMENTS AT HIGH TEMPERATURES

The velocity of sound in materials can be measured at elevated temperatures by means of transmission lines (or buffer rods). Wall effects and phase shifts due to coupling media must be taken into consideration.

When the attenuation in the rod is low and the material is plentiful, the specimen can be used to provide its own transmission line from the transducer located outside the furnace to the constant temperature zone of the furnace. A short length on the end of the rod may be machined to provide a shoulder from which energy is reflected back into the transmission line to the transducer. Echoes from this shoulder and from the end of the rod return to the transducer at different times corresponding to the length of the specimen section and the velocity of sound in the reduced section. Machined notches in the rod at the high temperature section may provide additional reverberations for more accurate measurements of both velocity and attenuation (2).

The delay time may be measured by comparator methods in which a pulse of energy is passed through a suitable calibrated delay line and displayed on an oscilloscope to coincide consecutively with each echo from the specimen area. Time-interval counters are available which make possible accurate measurements of the delay times.

Similar measurements are possible in high-loss materials by attaching the specimen to low-loss buffer rods. McSkimin (2) discusses these techniques in greater detail.

The thermal expansion of the specimen and phase shifts at bond areas must be taken into account in these measurements. In addition, wall effects, causing mode conversion and consequently interferences between modes of different velocities, can cause problems and should be prevented. One method often used to reduce wall effects in the buffer rods is to roughen or thread the surfaces of the transmission-line segments of the rods.

Similar techniques can be used to measure the velocity of sound in liquids at high temperatures. Figure 6.1 is a method used by Higgs and Litovitz (3) to measure ultrasonic absorption and velocity in molten salts. The buffer rods are fused quartz to which are bonded piezoelectric transducers. The upper end of the lower rod contains a machined shoulder to provide mechanical support for the sample cup. The sample cup is sealed to the quartz rod at the shoulder by means of Saureisen cement.

The sample cup consists of a 24/40 standard taper joint (fused quartz) mated with a cup made of 35 mm-ID quartz tubing. Parallelism between the adjacent ends of the rods is important for accuracy.

Measurements of absorption and velocity were made by comparator methods (4). The amplitude of the echoes for absorption measurements was measured by feeding the calibrated pulsed output of a standard signal generator into the receiver and delaying it in time so that it is adjacent to the echo under consideration.

Velocity measurements were made by comparing the velocity of sound through the specimen with the velocity of sound in water in which the temperature was controlled to ± 0.1°C. By making the travel time in the water identical with that in the specimen and determining the path length in each, it was possible to determine the velocity of sound as follows:

$$c_x = (\ell_x/\ell_w)c_w \tag{6.3}$$

where c_x is the velocity of sound in the specimen,

ℓ_x is the distance the pulse travels in the specimen in time T,

ℓ_w is the distance the pulse travels in the water in time T, and

c_w is the velocity of sound in the water, or comparator fluid.

The accuracy of the velocity measurement depends upon the accuracy of the micrometer readings on the comparator fluid path and the specimen path and also upon the accuracy to which the velocity of sound in the comparator fluid is known.

Another factor to be considered in measuring velocity and absorption of sound in liquids at high temperatures is the compatibility of the materials used with both the environment and the specimen materials. For example, fused quartz cannot be used at temperatures above which it devitrifies. Also, certain salts rapidly attack fused quartz at elevated temperatures.

Figure 6.2 is a modification of Higgs and Litovitz's (3) method of measuring velocities and attenuations of sound in molten salts that react with fused quartz. Dense alumina rods are used instead of quartz. The cup is also alumina attached to the lower rod by means of an alumina cement. An alumina collar is fitted near the upper end of the rod and cemented to it to provide additional strength to support the weight of the specimen. A hole is machined through the center of the bottom of the cup which rests on and is cemented to the collar. The cement is slowly and carefully cured to avoid damage from vapor-pressure buildup when the temperature is raised.

A pulse generator provides an impulse to excite the transmitter transducer and simultaneously trigger one sweep of a dual-beam oscilloscope. (Adjustable Delay No. 1 is optional and may be eliminated if Adjustable Delay No. 2 has sufficient delay capability.) The transmitted pulse travels through the buffer rods and the specimen and is detected by the receiver. After amplification, the received pulse is displayed on the oscilloscope screen at a time corresponding to the total delay time.

The same electrical impulse that excites the transmitter also triggers a time-interval counter and afterward passes through a variable delay. The output from the variable delay is used to stop the counter and to trigger the sweep of the second trace on the dual-beam oscilloscope. Adjusting the delay time on the second trace to coincide with the delay through the rods and the specimen gives a reference time reading T_1. Changing the path length through the specimen material and repeating the procedure gives a new time T_2. Both T_1 and T_2 are presented on the digital counter. The difference in path length, $\Delta \ell$, is the difference in the two micrometer readings, and the velocity of sound is determined by the ratio

$$c = \Delta \ell / \Delta T \qquad (6.4)$$

where $\Delta T = T_1 - T_2 \; (T_1 > T_2)$, and
$\Delta \ell = \ell_1 - \ell_2 \; (\ell_1 > \ell_2)$.

The attenuation is obtained by comparing the calibrated output signal from a continuous-wave oscillator with the received pulse height corresponding to each of the positions at which T_1 and T_2 are obtained. The attenuation is given by

$$\alpha = \frac{\ln A_1 - \ln A_2}{\Delta \ell} \qquad (A_1 > A_2) \qquad (6.5)$$

where A_1 and A_2 are the amplitudes of the received signal corresponding to positions 1 and 2, respectively. Correction for diffraction is necessary for absolute attenuation measurements.

The thermal coefficient of expansion of alumina must be considered in making measurements with the setup of Fig. 6.2. Sufficient spacing can be maintained by running the two rods together at each temperature and adjusting the carriage to the minimum spacing desired.

Rugged micrometers accurate to the nearest 10 μ in. are available. A less expensive micrometer accurate to the nearest 25 μ in. and a total range of 1 in. will provide accuracies that are more than adequate compared with the capabilities of the rest of the system.

II.F. MEASUREMENTS AT HIGH PRESSURES

Velocity of sound and absorption can be measured at high pressures by techniques similar to those used at high temperatures. Dimensional changes due to the compressibilities of the materials must be taken into consideration. Also, the design of the equipment must be such that there is no interference between sound bypassing the specimen and waves passing through the specimen.

Peselnick, Meister, and Wilson (5) use the phase-comparison method of McSkimin (6) to measure the velocity of sound in fused quartz to pressures of 10 kbars. Figure 6.3 is the pressure-vessel apparatus used.

With this setup of Fig. 6.3, separate transducers are used to generate the shear and the longitudinal pulses. A stable continuous-wave oscillator, periodically gated, drives the transducer. The resulting ultrasonic pulses are transmitted through the tool-steel cap and into the specimen, where they reverberate and return to the transducer where they are detected. When the reverberations are in phase, corresponding to a resonant condition within the specimen, the velocity of sound is

$$c = \frac{2\ell f}{N + (\gamma/360)} = \lambda f \qquad (6.6)$$

where

c is the velocity of sound, longitudinal or shear wave,

ℓ is the thickness of the specimen,

f is the frequency at which the in-phase condition occurs,

N is the number of half-wavelengths in ℓ, an integer,

γ is the phase shift due to the couplant between the cap and the specimen, and

λ is the equivalent wavelength at frequency f.

The fundamental frequency is the difference, Δf, between two successive in-phase frequencies. Therefore, $N = f/\Delta f$ at any given frequency.

Measurements in geological specimens are subject to hysteresis effects. As a result, making two series of measurements in order to obtain both the longitudinal-mode and the shear-mode data may be unsatisfactory and result in considerable error. These measurements can be made while subjecting the specimen to only one pressure cycle by a technique described by Gregory and Podio (7). Figure 6.4 shows the apparatus used by Gregory and Podio. The longitudinal wave is generated by a piezoelectric transducer vibrating in its thickness mode and attached to the vertical side of the aluminum block. The ultrasonic wave impinges on the 45° face of the aluminum block and is reflected downward and toward the specimen.

The shear wave is generated by mode conversion at the interface between the aluminum and an oil bath containing the shear-wave transducer. The angle of incidence is greater than the critical angle for longitudinal waves in the aluminum block. The shear wave is refracted at 45° to the sloping side of the block. This refraction angle also causes the shear wave to be directed toward the specimen.

Figure 6.5 is a schematic diagram of the electronic-acoustic system of Gregory and Podio.

III. LOW-FREQUENCY MEASUREMENTS OF ELASTIC MODULI AND POISSON'S RATIO

The elastic moduli--Young's modulus, shear modulus, and

Poisson's ratio--can be determined by fairly simple devices and equipment. In this section are described one method of obtaining Young's modulus, one for obtaining shear modulus, and the use of these data to obtain Poisson's ratio.

III.A. MEASURING FLEXURAL AND LONGITUDINAL RESONANT FREQUENCIES OF BARS

Figure 6.6 is a schematic diagram of a simple method of measuring the flexural resonance frequency of a uniform bar. In Chap. 3, section VI.B., it was shown that the fundamental resonant frequency of a free-free bar is

$$f_1 = \frac{\pi c \kappa}{8\ell^2}(3.0112)^2 \tag{3.62}$$

where κ is the radius of gyration of the cross section,

c is the bar velocity of sound (thin bar compared to the length, and

ℓ is the length of the bar.

At this frequency, the nodal points occur 0.224 ℓ from either end. Therefore, if the knife edges are located at 0.224 ℓ from either end of the bar (Fig. 6.6), a simple electromagnetic, piezoelectric, or magnetostrictive transducer driving a stylus located at the center of the bar can set the bar in flexural resonance. The true resonant frequency can be determined by scanning the length of the bar with the phonograph pickup. At the nodes, the signal amplitude will be minimum, and at the antinodes, the signal amplitude will be maximum. When the minima coincide with the positions of the knife edges, the maxima occur at the center and at each end, and no other maxima or minima can be identified by the pickup, then the frequency as shown by the counter is the fundamental resonant frequency of the bar. The Young's modulus can be calculated from the following equation:

$$c = \sqrt{Y/\rho} = (8\ell^2 f_1)/\pi\kappa(3.0112)^2 \tag{6.7}$$

The radius of gyration, κ, depends upon the geometry of the cross section of the bar. For a square or rectangular section

$\kappa = 0.289$ a, where a is the thickness of the bar in the direction of vibrational motion. For a circular cross section

$$\kappa = 0.5 \, r$$

where r is the radius of the cross section.
For a hollow cylinder

$$\kappa = 0.5\sqrt{r^2 + r_o^2}$$

where r is the radius of the outer circle and r_o is the radius of the inner circle.

Longitudinal resonant frequencies of bars may be measured by a similar method, except that the exciter is located at one end of the bar and the bar is laid on soft rubber, or a similar material. A microphone with wide response is located above the bar. The driving potential and the amplified signal from the microphone are applied, respectively, to the horizontal and to the vertical sweep terminals of an oscilloscope. The resulting Lissajous patterns can be used to locate the fundamental longitudinal frequency of the bar. The exciter generates a wide band of frequencies in striking the bar, so that, for example, if the driver oscillates at one-half the resonance frequency, a double loop is formed which indicates that the microphone detects a frequency which is double that of the exciter. Increasing the frequency of the exciter by a factor of two should verify the resonance frequency.

Bars used in this manner to determine the elastic modulus should be slender, uniform in dimension, and homogeneous. On the other hand, the method is often used to detect differences in homogeneity and quality of materials because variations in density and strength and the presence of defects affect the vibrational characteristics and the resonance frequencies.

III.B. MEASURING TORSIONAL RESONANT FREQUENCIES OF ISOTROPIC BARS

The torsional method has been used for measuring shear modulus of isotropic bars such as glass (8). The elastic relations for torsional vibrations in crystalline specimens are complicated by the anisotropy of the crystals. In a circular bar of length ℓ, the fundamental frequency at torsional resonance is

$$f_r = c_s/2\ell = (\mu/\rho)^{\frac{1}{2}}/2\ell \tag{6.8}$$

where μ is shear modulus of elasticity,

ρ is the density of the material, and

c_s is the velocity of a shear wave in the material of the bar.

The device used by Spinner (8) to excite both flexural and torsional vibrations in glass rods consisted simply of an electromagnetic driver which excited the bar from a point located near one end and offset from the axis of the bar. The bar was mounted at the center. The modes were identified by scanning the rod with a phonograph pickup.

Several precautions should be observed in making torsional-resonance measurements and most of these apply to flexural measurements as well:

1. Identification of the mode of vibration should be done with care to be certain that the vibration is torsional
2. Interferences due to structural resonances should be prevented to avoid erroneous interpretations.
3. It is desirable to mount the specimen so that it approaches a free state or so that any necessary clamping occurs at the preferred locations of the nodes and these nodal clamps approach line-contact conditions. The objective of the measurement is to identify a quarter-wavelength of the proper mode in the specimen bar at resonance.

It may be desirable to bond the specimen to another rod which may be more readily vibrated in a torsional mode. If, at resonance, a quarter-wavelength in the specimen can be identified, one may easily calculate c_s and, therefore, μ.

III.C. DETERMINING POISSON'S RATIO, YOUNG'S MODULUS, AND SHEAR MODULUS FROM FLEXURAL AND TORSIONAL RESONANCE DATA

Having determined the bar velocity c, by Eq. (6.7), and shear-wave velocity c_s, by Eq. (6.8), Poisson's ratio may be calculated by

$$\sigma = [(c/c_s)^2 - 2]/2 \tag{6.9}$$

Also from Eqs. (6.7) and (6.8), Young's modulus is

$$Y = \rho c^2 \tag{6.7a}$$

and the shear modulus is

$$\mu = \rho c_s^{\ 2} \tag{6.8a}$$

IV. VISCOSITY MEASUREMENTS BY ULTRASONICS

Methods of measuring viscosity by means of ultrasonics have been implemented. One commercial device makes use of the damping effect of a viscous liquid on a pulsed, resonant, magnetostrictive strip (9). The system consists of the magnetostrictive probe and an electronic computer. The probe is about the size of a fountain pen. Its operating frequency is approximately 28 kHz.

The computer is an analog type and its function is to produce an electrical output which is a function of the viscosity times the density.

The viscometer measures viscosity by comparing the damping effects of the viscous medium with that of an unloaded probe. The transducer is energized with a pulse. Between pulses, the vibration amplitude decays at a rate proportional to $e^{-\alpha t}$ where α is a function of viscosity and is given by

$$\alpha = (n\pi c_m \rho\eta/8L\rho_m^{\ 2}d^2)^{\frac{1}{2}}, \quad n = 1,3,\ldots \tag{6.10}$$

where c_m is the bar velocity of the probe strip,
ρ is the density of the viscoelastic material,
η is shear-viscosity coefficient;

L is the length of the probe strip,

ρ_m is the density of the material of the probe strip, and

d is the thickness of the strip.

The probe is calibrated for the true viscosity of Newtonian liquids. However, for non-Newtonian liquids, which exhibit a shear modulus, the indicated viscosity is in error by a function of the shear modulus of the material under test.

A method developed by Mason and McSkimin (10) is based upon measuring the acoustic shear-load impedance of a film of the test liquid on the surface of a transmission line (Fig. 6.7). The crystals are Y-cut quartz oriented to produce a shear wave in which the particle motion is parallel to the surface of the bar on which the fluid is located. The shear impedance of the test liquid is measured by comparing signals transmitted through two identical rods, the one containing the liquid load and the other remaining empty. The load impedance offered by the liquid is

$$Z_L = Z_q \left[\frac{1 - r^2 + 2jr \sin \phi}{1 + r^2 + 2r \cos \phi}\right] \cos \theta \tag{6.11}$$

where Z_q is the shear-wave impedance of the rod material,

r is the ratio of the amplitude of the attenuated signal to the amplitude of the unattenuated signal,

ϕ is the phase shift between the two branches, and

θ is the angle of incidence of the shear wave on the surface containing the liquid.

This method is applicable to measurements from 3 MHz to 100 MHz. At lower frequencies, Mason and McSkimin used shear loading on torsional crystals to measure dynamic viscosities ranging from 10 to 10^3 p to an accuracy of approximately 10%.

The complex impedance of the liquid, whether measured by the shear-wave-reflection method of Fig. 6.4 or the torsional-wave method, may be written in the usual form for complex impedances:

$$Z_L = R + jX = j\omega\rho\eta = j\omega\rho(\eta_1 - j\eta_2) \tag{6.12}$$

where $\bar{\eta}$ is the complex viscosity of the liquid, which is

$$\bar{\eta} = \eta_1 - j\eta_2 \tag{6.12a}$$

For a simple Newtonian liquid, $\eta_2 = 0$, and the real and imaginary parts of Eq. (6.12) are equal and

$$Z = (1 + j)(\omega\rho\eta/2)^{\frac{1}{2}} \tag{6.12b}$$

For a viscoelastic liquid

$$\eta_1 = 2RX/\omega\rho \tag{6.12c}$$

$$\eta_2 = (R^2 - X^2)/\omega\rho \tag{6.12d}$$

V. DETERMINING PROPERTIES OF PLASTICS AND HIGH POLYMERS

Sonic techniques are used to determine certain mechanical properties of high polymers and to locate flaws in a variety of manufactured plastic products. By measuring the velocity of sound, or resonance characteristics of beams under controlled conditions, it is possible to determine the physical constants characteristic of the viscoelastic behavior of the material. Lohr, et al. (11) claim that long-time mechanical and thermal stability of polymeric materials can be predicted from experiments conducted over relatively short periods of time. The materials that they tested included Plexiglass, polyvinyl chloride, polystyrene, Delrin, and Lexan.

The velocity of sound in polymers may be used to determine the degree of polymerization or cure of plastics, which is a measure of the molecular weight (12).

Pulse-echo and through-transmission techniques have both been used in nondestructive testing of plastics and plastic laminates. Pulse-echo techniques are those methods involving the detection of pulses of ultrasonic energy reflected from impedance discontinuities. Through-transmission techniques are those methods involving the use of a transmitter at one end or side of a specimen and a receiver at the other end whereby it is possible to evaluate the material on the basis of the nature of the effect produced on the propagating wave. Properties evaluated include delamination, bond condition, material thickness, and bond strength (13).

Zurbrick (14) has used ultrasonics to determine the performance reliability of glass-reinforced laminates designed for structural, primary-load-bearing components. Bobrikov and Medredev (15) discuss the possibility of using ultrasonic energy for determining the binder content and the porosity of monodirectional glass-fiber-reinforced plastic materials. They give semiempirical relations between the propagation rate of longitudinal ultrasonic oscillations and the binder content or porosity of such material. Schultz (16) shows correlations between ultrasonic longitudinal bulk-wave velocity and the bulk density, and between the velocity and the elastic modulus for laminates approximately 1 in. thick, used as heat shields. The material studied was carbon fabric impregnated with phenolic resin. The correlations exhibited variabilities (attributed to fabrication processing) of a few percent over the density range 0.28 to 1.44 g/cm^3 and elastic modulus range 1.20 to 1.75 times 10^6 psi. From a velocity-density correlation, the variability of a velocity-modulus correlation can be determined, and, if an approximate value of Poisson's ratio can be estimated, the velocity-modulus correlation itself can be determined. If the velocity-density and velocity-modulus correlations are known, Poisson's ratio can be determined from equations describing the propagation of ultrasonic waves.

VI. ACOUSTICAL PROPERTIES OF MATERIALS

Because it is most convenient for the reader, the acoustical properties are presented in tabular form. Tables 6.1, 6.2, 6.3, and 6.4, respectively, present these properties for solids, common liquids and molten materials, organic liquids, and gases.

TABLE 6.1

Acoustical Properties of Solids

Materials	Velocities, 10^5 cm/sec Longitudinal		Shear	Density, g/cm^3	Characteristic impedance 10^6 g/cm^2-sec	References
	Bar (c_o)	Bulk (c_B)	(c_s)	(ρ)	(ρc_B)	
			Metals			
Aluminum 2SO	5.10	6.35	3.10	2.71	1.73	17-21
Aluminum 17ST	5.08	6.25	3.10	2.80	1.75	17,18
Antimony	3.40	--	--	6.70	--	21
Beryllium	12.75	12.80	8.71	1.82	2.33	17,18
Beryllium	--	12.89	8.88	1.80	2.32	20
Bismuth	1.79	2.18	1.10	9.80	2.14	21
Brass 70-30	3.40	4.37	2.10	8.50	3.70	20
Brass (naval)	3.49	4.43	2.12	8.10	3.61	17,18,21
Bronze (phosphor 5%)	3.43	3.53	2.23	8.86	3.12	17,18
Cadmium	2.40	2.78	1.50	8.60	2.40	21
Constantan	4.30	5.24	2.64	8.80	4.60	21
Copper	3.71	4.66	2.26	8.90	4.18	17,18
Copper	3.71	4.70	2.26	8.90	4.18	21
Copper	3.65	4.80	2.33	8.90	4.25	20
Copper	3.60	4.60	--	3.20	4.10	19

Gold	2.03	3.24	1.20	19.30	6.26	18,20,21
German silver	3.58	4.76	2.16	8.40	4.00	21
Hastelloy X	--	5.79	2.74	8.23	4.77	22
Hastelloy C	--	5.84	2.90	8.94	5.22	22
Iridium	4.79	--	--	22.40	--	21
Iron	5.18	5.96	2.05	7.90	4.68	20
Iron	5.17	5.85	3.23	7.80	4.56	21
Iron, cast	3.0-4.7	3.50-5.60	2.2-3.2	7.20	2.50-4.00	20
Iron, cast	--	4.50	2.40	7.70	2.96	18
Iron, cast	3.70	4.35	--	7.70	3.35	19
Iron, wrought	5.00	5.90	--	7.80	4.60	19
Lead, pure	1.20	2.16	0.70	11.40	2.46	17,18,21
Lead, pure	1.25	2.40	0.79	11.30	2.72	20
Lead, pure	1.20	2.20	--	11.30	2.50	19
Lead (antimony 6%)	1.37	2.16	0.81	10.90	2.36	17
Magnesium	4.90	5.74	3.08	1.70	0.99	20
Magnesium (Am 35)	5.00	5.79	3.10	1.74	1.01	17,18
Manganin	3.83	4.66	2.35	8.40	3.90	21
Molybdenum	5.45	6.29	3.35	10.09	6.35	17,18
Molybdenum	--	6.25	3.35	10.20	6.37	20
Nickel	4.90	5.60	--	8.80	4.90	19
Nickel	4.79	5.63	2.96	8.80	4.95	17,18,21
Inconel X	--	5.94	3.12	8.30	4.93	22

Table 6.1

Acoustical Properties of Solids (con't)

Materials	Velocities, 10^5 cm/sec Longitudinal Bar (c_o)	Bulk (c_B)	Shear (c_s)	Density, g/cm^3 (ρ)	Characteristic impedance 10^6 g/cm^2-sec (ρc_B)	References
Nickel (con't)						
Inconel (wrought)	5.08	7.82	3.02	8.25	6.45	17
Monel (wrought)	4.52	6.02	2.72	8.83	5.31	17
Silver-nickel (18%)	3.83	4.62	2.32	8.75	4.03	17
Platinum	2.80	3.96	1.67	21.40	8.46	20,21
Silver	2.67	3.70	1.70	10.50	3.90	19,20
Silver	2.64	3.60	1.59	10.50	3.80	18,21
Steel	5.05	6.10	--	7.70	4.70	19
Steel	5.17	5.85	3.23	7.80	4.56	17,18
Steel, mild	5.05	6.10	3.24	7.90	4.82	20
Steel, stainless 302	4.90	5.66	3.12	8.03	4.55	17,18
Steel, stainless 304L	--	5.64	3.07	7.90	4.45	22
Steel, stainless 410	5.03	7.39	2.99	7.67	5.67	17,18
Steel, stainless 416	--	6.02	3.23	7.70	4.64	22
Tantalum	3.35	--	--	16.60	--	21
Tin	2.73	3.32	1.67	7.30	2.42	18,21

Tin	2.74	3.38	1.61	7.30	2.47	20
Titanium	--	5.99	3.12	4.50	2.70	20
Titanium (Ti 150A)	5.08	6.10	3.12	4.54	2.77	17,18
Tungsten	4.60	5.18	2.87	19.25	9.98	17,18
Tungsten	4.31	5.46	2.62	19.10	10.42	21
Tungsten	--	5.17	2.88	19.30	10.00	20
Uranium	--	3.37	2.02	18.70	6.30	20
Zinc	3.81	4.17	2.41	7.10	2.96	18,20,21
Zirconium	--	4.65	2.30	6.40	2.98	20
			Nonmetals			
Concrete	3.10	--	--	2.60	--	19
Cork	0.50	--	--	0.24	--	19,21
Glass, Crown	5.30	5.66	3.42	2.50	1.41	20,21
Flint	4.00	4.26	2.56	3.60	1.54	21
Plate	--	5.77	3.43	2.51	1.45	17
Pyrex	--	5.57	3.44	2.23	1.24	17
Quartz	5.37	5.57	3.51	2.60	1.45	21
Soft	5.00	5.40	--	2.40	1.30	19
Granite	3.95	--	--	2.75	--	20
Ice	--	3.98	1.99	0.90	0.36	20
Ice	3.20	--	--	0.92	--	19

[For an extensive lising of velocities of sound in rocks at various pressures, see Birch (23).]

TABLE 6.1

Acoustical Properties of Solids (con't)

Materials	Velocities, 10^5 cm/sec Longitudinal		Shear	Density, g/cm^3	Characteristic impedance 10^6 g/cm^2-sec	References
	Bar (c_o)	Bulk (c_B0	(c_s)	(ρ)	(ρc_B)	
Ice	3.28	3.98	1.99	1.00	0.40	17,21
Nylon	--	1.8-2.2	--	1.1-1.2	0.2-0.27	20
Paraffin, hard	--	2.20	--	0.83	0.18	20
Plastic, acrylic resin	--	2.67	1.12	1.18	0.32	17
Plexiglas	1.80	2.68	1.32	1.20	0.32	20
Plexiglas	--	2.67	1.12	1.18	0.32	21
Polystyrene	--	2.67	--	1.06	0.28	20
Polystyrene	--	2.35	1.12	1.06	0.23	21
Porcelain	4.88	5.34	3.12	2.41	1.29	21
Quartz, natural	--	5.73	--	2.65	1.52	17
Quartz, fused	5.37	5.57	3.52	2.60	1.45	20
Quartz, fused	--	5.93	3.75	2.20	1.30	17
Quartz, X-cut	5.45	5.75	--	2.65	1.52	19
Rubber, hard	1.45	--	--	1.10	--	19
Rubber, India	--	1.48	--	0.90	0.14	21

Rubber, soft	0.07	--	----	0.95	--	19
Teflon	--	1.35	--	2.20	0.30	20
Tungsten carbide	--	6.66	3.98	10.0-15.0	6.65-9.85	20
Wood, oak	4.10	--	--	0.72	--	19
Wood, oak	4.10	--	--	0.80	--	20
Wood, pine	3.65	--	--	0.45	--	19

[For an extensive table of acoustic properties of biological tissues, see Goldman and Hueter (24).]

TABLE 6.2

Acoustical Properties of Common Liquids and Molten Materials

Materials	Temperature, C	Density (ρ), g/cm^3	Velocity (c), 10^5 cm/sec	Characteristic impedance (ρc) 10^5 g/cm^2-sec	References
		Common liquids			
Alcohol (ethyl)	20	0.79	1.17	0.92	19
Glycerine	20	1.26	1.92	2.40	19, 20
Mercury	20	13.546	1.451	19.72	19, 21
Mercury	50	--	1.440	--	21
Oil, transformer	20	0.92	1.38	1.27	17
Oil, castor	20	0.95	1.54	1.45	19

TABLE 6.2

Acoustical Properties of Common Liquids and Molten Materials (con't)

Materials	Temperature, C	Density (ρ), g/cm^3	Velocity (c), 10^5 cm/sec	Characteristic impedance (c) 10^5 g/cm^2-sec	References
Oil, SAE 20	--	0.87	1.74	1.50	20
Turpentine	20	0.87	1.33	1.15	19
Water, fresh	4	--	1.422	--	25
Water, fresh	10	--	1.448	--	25
Water, fresh	15	--	1.466	--	25
Water, fresh	20	0.998	1.483	1.48	19,25
Water, fresh	50	--	1.543	--	25
Water, fresh	75	--	1.555	--	25
Water, fresh	90	--	1.551	--	25
Molten materials (low melting point)					
Bismuth	271.0	--	1.635	--	21
Cadmium	321.0	--	2.222	--	21
Cesium	28.5	--	0.967	--	21

[For an extensive list of the velocities of sound in fresh water over a wide temperature range in both metric and English units, see Greenspan and Tschiegg (25).]

Gallium	29.5	--	2.740	--	21
Indium	156.0	--	2.215	--	21
Lead	327.0	--	1.790	--	21
Potassium	64.0	--	1.820	--	21
Rubidium	39.0	--	1.260	--	21
Sodium	98.0	--	2.395	--	21
Sulfur	115.0	--	1.350	--	21
Thallium	302.0	--	1.625	--	21
Tin	232.0	--	2.270	--	21
Zinc	419.0	--	2.790	--	21
85% Sodium carbonate +15% sodium sulfide	896	--	2.274	--	Battelle
85% Sodium carbonate +15% sodium sulfide	1015	--	2.180	--	Battelle
70% Sodium carbonate + 30% sodium sulfide	860	--	2.250	--	Battelle
70% Sodium carbonate + 30% sodium sulfide	1000	--	2.148	--	Battelle
Sodium carbonate	903	--	2.300	--	Battelle
Sodium carbonate	1001	--	2.220	--	Battelle
Sodium chloride	864	--	1.726	--	Battelle
Sodium chloride	998	--	1.593	--	Battelle
Sodium sulfide	960	--	2.037	--	Battelle
Sodium sulfide	1020	--	2.000	--	Battelle

TABLE 6.2

Acoustical Properties of Common Liquids and Molten Materials (con't)

Materials	Temperature, C	Density (ρ) g/cm^3	Velocity (c), 10^5 cm/sec	Characteristic impedance (ρc) 10^5 g/cm^2-sec	References
Sodium sulfide	1060	--	1.963	--	Battelle
50% Sodium carbonate + 25% Sodium chloride + 25% Sodium sulfide	860	--	2.070	--	Battelle
50% Sodium carbonate + 25% Sodium chloride + 25% Sodium sulfide	947	--	1.988	--	Battelle
50% Sodium carbonate + 25% Sodium chloride + 25% Sodium sulfide	1002	--	1.941	--	Battelle
70% Sodium carbonate + 25% Sodium sulfide + 5% Sodium aluminate	840	--	2.280	--	Battelle
70% Sodium carbonate + 25% Sodium sulfide + 5% Sodium aluminate	1000	--	2.160	--	Battelle

70% Sodium carbonate + 25% Sodium sulfide + 5% Sodium hydroxide	840	--	2.278	--	Battelle
70% Sodium carbonate + 25% Sodium sulfide + 5% Sodium aluminate	840	--	2.280	--	Battelle
70% Sodium carbonate + 25% Sodium sulfide + 5% Sodium aluminate	1000	--	2.160	--	Battelle
70% Sodium carbonate + 25% Sodium sulfide + 5% Sodium hydroxide	840	--	2.278	--	Battelle
70% Sodium carbonate + 25% Sodium sulfide + 5% Sodium hydroxide	998	--	2.161	--	Battelle
		Liquid gases			
Argon	-186.0	1.404	0.837	1.175	21
Helium	-269.1	0.125	0.1798	0.0225	21
Helium	-271.5	0.146	0.2314	0.0338	21
Hydrogen	-252.7	0.355	1.127	0.400	21
Nitrogen	-197.0	0.815	0.869	0.708	21
Oxygen	-183.6	1.143	0.911	1.041	21

TABLE 6.3

Acoustical Properties of Organic Liquids[a]

Liquid	Chemical formula	Temperature, C	Density (ρ), g/cm^3	Velocity of sound (c), 10^5 cm/sec	Characteristic impedance (ρc) 10^5 g/cm^2-sec
Acetic anhydride	$(CH_3CO)_2O$	24	1.087	1.384	1.51
Acetone	$CH_3 \cdot CO \cdot CH_3$	20	0.792	1.192	0.94
Acetonylacetone	$CH_3CO(CH_2)_2COCH_3$	20	0.970	1.416	1.38
Acetophenone	$C_6H_5 \cdot CO \cdot CH_3$	20	1.026	1.496	1.54
Acetylacetone	$CH_3COCH_2COCH_3$	20	0.976	1.383	1.35
Acetylchloride	C_2H_3OCl	20	1.105	0.060	1.17
Acetylene dichloride	$CHCl{=}CHCl$	25	1.265	1.025	1.30
Acetylene tetrabromide	$CHBr_2 \cdot CHBr_2$	20	2.963	1.041	3.09
Acetylene tetrachloride	$CHCl_2 \cdot CHCl_2$	28	1.578	1.155	1.82
Acrolein	$CH_2{:}CHCHO$	20	0.841	1.207	1.02
Amyl acetate	$CH_3COO(CH_2)_4CH$	26	0.879	1.168	1.03
tert-Amyl alcohol (dimethylethylcarbinol)	$CH_3CH_2C(CH_3)OHCH_3$	28	0.809	1.204	0.97
n-Amyl bromide	$CH_3(CH_2)_3CH_2Br$	20	1.246	0.981	1.22
Amyl ether	$[CH_3(CH_2)_3CH_2]_2O$	26	0.774	1.153	0.89
Amyl formate	$HCOO(CH_2)_4CH_3$	26	0.893	1.201	1.07
Aniline	$C_6H_5NH_2$	20	1.022	0.656	1.69

Biacetyl	$CH_3COCOCH_3$	25	0.990	1.236	1.22
Benzaldehyde	C_6H_5CHO	20	1.050	1.479	1.55
Benzene	C_6H_6	20	0.879	1.326	1.17
Benzylacetone	$C_6H_5CH_2CH_2COCH_3$	20	0.989	1.514	1.50
Benzyl alcohol	$C_6H_5CH_2OH$	20	1.050	1.540	1.62
Benzyl chloride	C_7H_7Cl	20	1.1026	1.420	1.57
Bromal	$CHBr_3CHO$	20	2.30	0.966	2.22
Bromoform	$CHBr_3$	20	2.890	0.928	2.68
1-Bromonaphthalene	$C_{10}H_7Br$	20	1.487	1.372	2.04
2-Butanone (ethyl methyl ketone)	$CH_3COC_2H_5$	20	0.805	1.207	0.97
n-Butyl alcohol	$CH_3(CH_2)_2CH_2OH$	20	0.810	1.268	1.03
sec-Butyl alcohol	$CH_3CH_2CHOHCH_3$	20	0.808	1.222	0.99
tert-Butyl alcohol	$(CH_3)_3COH$	20	0.789	1.155	0.91
n-Butyl bromide	$CH_3(CH_2)_2CH_2Br$	20	1.299	0.990	1.29
n-Butyl chloride	C_4H_9Cl	20	0.884	1.133	1.00
Butyl formate (butyl methanoate)	$HCOOC_4H_9$	24	0.911	1.199	1.09
n-Butyl iodide	$CH_3(CH_2)_2CH_2I$	20	1.617	0.977	1.58

TABLE 6.3

Acoustical Properties of Organic Liquids (con't)

Liquid	Chemical formula	Temperature, C	Density (ρ), g/cm^3	Velocity of sound (c), 10^5 cm/sec	Characteristic impedance (ρc) 10^5 g/cm^2-sec
Butyric acid	C_3H_7COOH	20	0.959	1.203	1.15
Caprylic acid	$CH_3(CH_2)_6COOH$	20	0.910	1.331	1.21
Carbon disulfide	CS_2	20	1.263	1.158	1.46
Carbon tetrachloride	CCl_4	20	1.595	0.938	1.50
Carvacrol	$CH_3(C_3H_7)C_6H_3OH$	20	0.976	1.475	1.44
Chlorobenzene	C_6H_5Cl	20	1.107	1.291	1.43
Chloroform	$CHCl_3$	20	1.498	1.005	1.51
1-Chlorohexane	$CH_3(CH_2)_4CH_2Cl$	20	0.872	1.221	1.06
1-Chloronaphthalene	$C_{10}H_7Cl$	20	1.194	1.481	1.77
o-Chlorotoluene	$ClC_6H_4CH_3$	20	1.0817	1.344	1.45
m-Chlorotoluene	$ClC_6H_4CH_3$	20	1.072	1.326	1.42
p-Chlorotoluene	$ClC_6H_4CH_3$	20	1.0697	1.316	1.41
Cinnamaldehyde	C_9H_8O	25	1.112	1.554	1.73
Citral b	$C_{10}H_{16}O$	20	0.888	1.442	1.28
o-Cresol	$CH_3C_6H_4OH$	25	1.046	1.506	1.58
Crotonaldehyde	C_4H_6O	20	0.858	1.344	1.15

Cumene	$C_6H_5CH(CH_3)_2$	20	0.862	1.342	1.16
Cyclohexane	C_6H_{12}	20	0.779	1.284	1.00
Cyclohexanol	$C_6H_{12}O$	20	0.945	1.493	1.41
Cyclohexanone	$CO(CH_2)_4CH_2$	20	0.948	1.449	1.37
Cyclohexene	$CH_2CH_2CH_2CH_2CH:CH$	20	0.811	1.305	1.06
Cyclohexylamine	$C_6H_{11}NH_2$	20	0.819	1.435	1.18
Cyclohexyl chloride	$C_6H_{11}Cl$	20	1.016	1.319	1.34
1,3-Cyclopentadiene	$CH:CHCH:CHCH_2$	20	0.805	1.421	1.14
Cyclopentanone	$COCH_2CH_2CH_2CH_2$	24	0.948	1.474	1.40
p-Cymene (p-isopropyltoluene)	$C_6H_4CH_3CH(CH_3)_2$	28	0.857	1.308	1.12
1-Decene	$C_{10}H_{20}$	20	0.741	1.250	0.93
n-Decyl alcohol	$C_{10}H_{21}OH$	20	0.829	1.402	1.16
2-Dibromoethylene	$CHBr \cdot CHBr$	20	2.271	0.957	2.17
m-Dichlorobenzene	$C_6H_4Cl_2$	28	1.285	1.232	1.58
o-Dichlorobenzene	$C_6H_4Cl_2$	20	1.305	1.295	1.69
1,1-Dichloroethane	$CH_3CH_2Cl_2$	20	1.174	1.034	1.21
Diethyl aldehyde	$CH_3CH(OC_2H_5)_2$	24	0.825	1.378	1.14
N,N-diethyl aniline	$C_6H_5N(C_2H_5)_2$	20	0.935	1.482	1.39
Diethylene glycol	$O(CH_2CH_2OH)_2$	25	1.132	1.586	1.80
Diethyl ester malonic acid	$CH_2(COOC_2H_5)_2$	22	1.055	1.386	1.46
2,3-Dimethylbutanoic acid	$C_5H_{11}COOH$	20	0.928	1.280	1.19

TABLE 6.3

Acoustical Properties of Organic Liquids (con't)

Liquid	Chemical formula	Temperature, C	Density (ρ), g/cm^3	Velocity of sound (c), 10^5 cm/sec	Characteristic impedance (ρc) 10^5 g/cm^2-sec
Di-n-propyl ether	$(CH_3CH_2CH_2)_2O$	20	0.736	1.112	0.82
Diphenyether	$C_6H_5OC_6H_5$	24	1.0728	1.469	1.57
Diphenylmethane	$C_6H_5-CH_2C_6H_5$	28	1.001	1.501	1.50
n-Dodecyl alcohol	$CH_3(CH_2)_{11}OH$	30	0.831	1.388	1.15
Ethyl acetate	$CH_3COOC_2H_5$	20	0.901	1.176	1.06
Ethyl-aceto-acetate	$CH_3 \cdot CO \cdot CH_2 \cdot COOC_2H_5$	25.5	1.025	1.417	1.45
Ethyl acetylmalonate	$CO \cdot (CH_2COOC_2H_5)_2$	22.5	1.085	1.348	1.46
Ethyl adipate	$CH_2 \cdot CH_2 \cdot COOC_2H_5$ | $CH_2 \cdot CH_2 \cdot COOC_2H_5$	22	1.009	1.376	1.39
Ethyl alcohol	C_2H_5OH	20	0.789	1.180	0.93
Ethyl bromide	C_2H_5Br	28	1.428	0.892	1.27
Ethyl butyrate	$C_3H_7COOC_2H_5$	23.5	0.879	1.1171	0.98
Ethyl caprylate	$CH_3(CH_2)_6COOC_2H_5$	28	0.867	1.263	1.10
Ethyl carbonate	$(C_2H_5)_2CO_3$	28	0.975	1.173	1.14
Ethyl chloroethanoate	$CH_2ClCOOC_2H_5$	25.5	1.159	1.234	1.43
Ethylene bromide	$C_2H_4Br_2$	20	2.1701	1.009	2.19

Ethylene chloride	$CH_2Cl \cdot CH_2Cl$	23	1.257	1.240	1.56
Ethylene glycol	CH_2OHCH_2OH	20	1.115	1.616	1.80
Ethyl ester propionic acid	$CH_3CH_2COOC_2H_5$	23.5	0.8846	1.185	1.05
Ethyl ether	$C_4H_{10}O$	20	0.714	1.008	0.72
Ethyl formate	$H \cdot COOC_2H_5$	24	0.924	1.721	1.59
Ethyl iodide	CH_3CH_2I	20	1.940	0.869	1.69
Ethyl-o-tolyl ether	$CH_3C_6H_4OC_2H_5$	25	0.959	1.315	1.26
Ethyl oxalate	$(COOC_2H_5)_2$	22	1.084	1.392	1.51
Ethyl phenyl ketone (propiophenone)	$C_2H_5COC_6H_5$	20	1.012	1.498	1.51
Ethyl p-phthalate	$C_6H_4(COOC_2H_5)_2$	23	1.123	1.471	1.65
Ethyl succinate	$(CH_2-COOC_2H_5)_2$	22	1.040	1.378	1.43
Formamide	$HCONH_2$	20	1.134	1.550	1.76
Formic acid	$HCOOH$	20	1.226	1.287	1.58
Furfuryl alcohol	$C_4H_3O \cdot CH_2OH$	25	1.1296	1.450	1.64
Geranyl acetate	$C_{12}H_{20}O_2$	28	0.917	1.328	1.22
Glycerine	$CH_2OHCHOHCH_2OH$	20	1.261	1.923	2.42
n-Heptane	C_7H_{16}	20	0.684	1.162	0.78
Heptanoic acid	$CH_3(CH_2)_5COOH$	20	0.913	1.312	1.20
1-Heptene	$CH_2 \cdot CH(CH_2)_4CH_3$	20	0.697	1.128	0.79
n-Heptyl alcohol	$CH_3(CH_2)_5CH_2OH$	20	0.823	1.341	1.10
Hexane	C_6H_{14}	20	0.660	1.083	0.71
n-Hexyl alcohol	$CH_3(CH_2)_4CH_2OH$	20	0.820	1.322	1.08

TABLE 6.3

Acoustical Properties of Organic Liquids (con't)

Liquid	Chemical formula	Temperature, C	Density (ρ), g/cm^3	Velocity of sound (c), 10^5 cm/sec	Characteristic impedance (ρc) 10^5 g/cm^2-sec
Hexylmethyl ketone	$CH_3COC_6H_{13}$	24	0.818	1.324	1.08
Indan (hydrindene)	$C_6H_4CH_2CH_2CH_2$	20	0.965	1.403	1.35
Indene	$C_6H_4CH_2CH{:}CH$	20	1.006	1.475	1.48
Iodobenzene	C_6H_5I	20	1.832	1.113	2.04
1-Iodohexane	$CH_3(CH_2)_4CH_2I$	20	1.441	1.081	1.56
α-Ionon	$C_{13}H_{20}O$	20	0.930	1.432	1.33
d-Linalool	$C_{10}H_{17}OH$	20	0.863	1.341	1.16
Mesityl oxide	$(CH_3)_2C{:}CHCOCH_3$	20	0.854	1.310	1.12
3-Methoxytoluene	$CH_3C_6H_4OCH_3$	26	0.976	1.385	1.35
Methyl alcohol	CH_3OH	20	0.796	1.123	0.89
Methyl acetate	CH_3COOCH_3	25	0.928	1.154	1.07
N-Methyl aniline	C_6H_5NHCH	20	0.986	1.586	1.56
Methyl chloroethanoate	$CH_2ClCOOCH_3$	26	1.227	1.331	1.64
Methyl cyanide	CH_3CN	20	0.783	1.304	1.02
Methylcyclohexane	$CH_3C_6H_{11}$	20	0.7864	1.247	0.98
2-Methylcyclohexanol	$CH_3C_6H_{10}OH$	25.5	0.937	1.421	1.33
3-Methyl-1-cyclohexanol	$CH_3C_6H_{10}OH$	25.5	0.914	1.406	1.28

4-Methylcyclohexanol	$CH_3C_6H_{10}OH$	25.5	0.913	1.387	1.27
2-Methylcyclohexanone	$C_7H_{12}O$	25.5	0.924	1.353	1.25
4-Methylcyclohexanone	$C_7H_{12}O$	25.5	0.913	1.348	1.23
Methylene bromide	CH_2Br_2	24	2.4953	0.971	2.42
Methylene chloride	CH_2Cl_2	20	1.336	1.092	1.45
Methylene iodide	CH_2I_2	24	3.325	0.977	3.25
Methyl ester salicylic acid	$HOC_6H_4COOCH_3$	28	1.184	1.408	1.67
Methyl iodide	CH_3I	20	2.279	0.834	1.90
Methyl phenol ether (anisole)	$C_6H_5OCH_3$	26	0.9988	1.353	1.35
Methyl propionate	$CH_3CH_2COOCH_3$	24.5	0.915	1.215	1.11
Methyl propyl ketone (diethyl ketone)	$C_2H_5COOC_2H_5$	24	0.816	1.314	1.06
Monoethyl ether	$C_2H_5OCH_2CH_2OCH_2CH_2OH$	25	0.990	1.458	1.44
Morpholine	$OCH_2CH_2NHCH_2CH_2$	25	1.000	1.442	1.44
Nicotine	$C_{10}H_{14}N_2$	20	1.009	1.491	1.50
Nitrobenzene	$C_6H_5NO_2$	20	1.199	1.473	1.77
2-Nitroethanol	$NO_2C_2H_4OH$	20	1.270	1.578	2.00
Nitromethane	CH_3NO_2	20	1.130	1.346	1.52
o-Nitrotoluene	$NO_2C_6H_4CH_3$	20	1.163	1.432	1.67
m-Nitrotoluene	$NO_2C_6H_4CH_3$	20	1.164	1.489	1.73
Nonane	$CH_3(CH_2)_7CH_3$	20	0.718	1.248	0.89
1-Nonene	$CH_3(CH_2)_6CH:CH_2$	20	0.729	1.218	0.89

TABLE 6.3

Acoustical Properties of Organic Liquids (con't)

Liquid	Chemical formula	Temperature, C	Density (ρ), g/cm^3	Velocity of sound (c), 10^5 cm/sec	Characteristic impedance (ρc) 10^5 g/cm^2-sec
n-Nonyl alcohol	$CH_3(CH_2)_7CH_2OH$	20	0.828	1.391	1.15
n-Octane	$CH_3(CH_2)_6CH_3$	20	0.703	1.197	0.84
1-Octanol	$CH_3(CH_2)_6CH_2OH$	20	0.825	1.358	1.12
n-Octyl bromide	$CH_3(CH_2)_6CH_2Br$	20	1.116	1.182	1.32
n-Octyl chloride	$CH_3(CH_2)_6CH_2Cl$	20	0.875	1.280	1.12
Oleic acid	$C_8H_{17}CH{:}CH(CH_2)_7COOH$	45	0.873	1.333	1.16
Paraldehyde	$C_6H_{12}O_3$	20	0.994	1.204	1.20
Pentachloroethane	$CHCl_2CCl_3$	20	1.709	1.113	1.90
Pentane	C_5H_{12}	20	0.626	1.008	0.63
1-Pentanol	$C_5H_{11}OH$	20	0.815	1.294	1.06
Perchloroethylene	$CCl_2{:}CCl_2$	20	1.623	1.066	1.73
Phenetole	$C_2H_5OC_6H_5$	26	0.967	1.153	1.12
Phenylethane	$C_6H_5 \cdot C_2H_5$	20	0.867	1.338	1.16
Phenylhydrazine	$C_6H_5NHNH_2$	20	1.098	1.738	1.91
Phenyl mustard oil	C_6H_5NCS	27	1.131	1.412	1.60
α-Picoline	$CH_3C_5H_4N$	28	0.951	1.453	1.38
β-Picolene	$CH_3C_5H_4N$	28	0.961	1.419	1.36
dl-Pinene	$C_{10}H_{16}$	24	0.858	1.247	1.07

Piperidine	$(CH_2)_5NH$	20	0.862	1.400	1.20
Propionic acid	CH_3CH_2COOH	20	0.992	1.176	1.17
n-Propionitrile	CH_3CH_2CN	20	0.783	1.271	1.00
n-Propyl acetate	$CH_3COOC_3H_7$	26	0.887	1.182	1.05
n-Propyl alcohol	$CH_3CH_2CH_2OH$	20	0.804	1.223	0.98
n-Propyl chloride	$CH_3CH_2CH_2Cl$	20	0.890	1.091	0.97
n-Propyl iodide	C_3H_7I	20	1.747	0.929	1.62
Pseudocumene	$(CH_3)_3C_6H_3$	20	0.876	1.368	1.20
Pyridine	C_6H_5N	20	0.982	1.445	1.42
Quinaldine	$CH_3C_9H_6N$	20	1.1013	1.575	1.73
Quinaline	$C_6H_4N:CHCH:CH$	20	1.095	1.600	1.75
Resorcinol dimethyl ether	$C_6H_4(OCH_3)_2$	26	1.080	1.460	1.58
Salicylaldehyde	HOC_6H_4CHO	27	1.166	1.474	1.72
1,1,2,2-Tetrabromoethane	$CHBr_2CHBr_2$	20	2.963	1.041	3.08
1,1,2,2-Tetrachloroethane	$C_2H_2Cl_4$	20	1.6	1.171	1.87
Tetrachloroethylene	$CCl_2:CCl_2$	28	1.623	1.027	1.67
Tetraethylethylene glycol	$(C_2H_5)_2COHCOHC(C_2H_5)_2$	25	1.123	1.586	1.78
1,2,3,4-tetrahydronaphthalene (Tetralin)	$C_{10}H_{12}$	20	0.971	1.492	1.45
Tetranitromethane	$C(NO_2)_4$	20	1.650	1.039	1.71
Thiolacetic acid	CH_3COSH	20	1.074	1.168	1.25
Thiophene	$SCH:CHCH:CH$	20	1.065	1.300	1.38
Toluene	$C_6H_5CH_3$	20	0.867	1.328	1.15

TABLE 6.3

Acoustical Properties of Organic Liquids (con't)

Liquid	Chemical formula	Temperature, C	Density (ρ), g/cm^3	Velocity of sound (c) 10^5 cm/sec	Characteristic impedance (ρc) 10^5 g/cm^2-sec
o-Toluidine	$CH_3C_6H_4NH_2$	20	1.004	1.634	1.64
m-Toluidine	$CH_3C_6H_4NH_2$	20	0.989	1.620	1.60
1,2,4-Trichlorobenzene	$C_6H_3Cl_3$	20	1.456	1.301	1.89
Trichloroethylene	$CHCl:CCl_2$	20	1.477	1.049	1.55
Tridecylene	$C_{13}H_{26}$	20	0.798	1.313	1.05
Triethylene glycol	$(CH_2OCH_2CH_2OH)_2$	25	1.125	1.608	1.81
Trimethylene bromide	$BrCH_2CH_2CH_2Br$	23.5	1.979	1.144	2.26
Triolein	$(C_{18}H_{33}O_2)_3C_3H_5$	20	0.915	1.482	1.36
o-Xylene	$C_6H_4(CH_3)_2$	20	0.880	1.360	1.20
m-Xylene	$C_6H_4(CH_3)_2$	20	0.864	1.340	1.46
p-Xylene	$C_6H_4(CH_3)_2$	20	0.861	1.330	1.15

[a]In general the velocities used in this table are from Bergmann (21). Densities are from Handbook of Chemistry and Physics (26) and ρc is corrected to these values of ρ.

TABLE 6.4

Acoustical Properties of Gases

Gas	Temperature, C	Density (ρ), 10^{-3} g/cm^3	Velocity (c), 10^4 cm/sec	Characteristic impedance (ρc) g/cm^2-sec	References
Air	0	1.293	3.316	42.8	19
Air	20	1.21	3.43	41.5	19
Argon	20	1.72	3.21	55.2	21
Carbon dioxide (low frequency)	0	1.98	2.58	51.0	19
Carbon dioxide (high frequency)	0	1.98	2.68	53.0	19
Carbon dioxide	16.6	1.935	2.77	53.6	21
Carbon monoxide	18.7	1.215	3.49	42.4	21
Helium	17.5	0.174	9.97	17.35	21
Hydrogen	0	0.09	12.65	11.5	19
Hydrogen	19.9	0.0868	13.29	11.54	21
Neon	19.0	0.872	4.50	39.25	21
Nitric oxide	16.3	1.311	3.34	43.8	21
Nitrogen	19.9	1.207	3.49	42.1	21
Oxygen	0	1.43	3.14	45.0	19
Oxygen	19.6	1.38	3.27	45.1	21
Steam	100	0.6	4.75	28.5	19

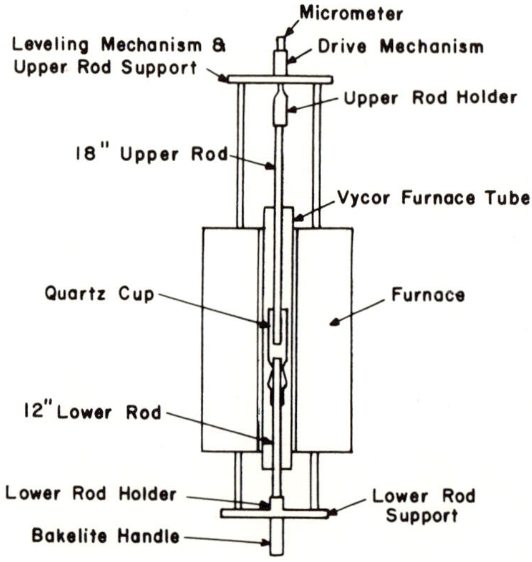

FIG. 6.1. Equipment assembly for measuring the ultrasonic absorption and velocity in molten salts (3).

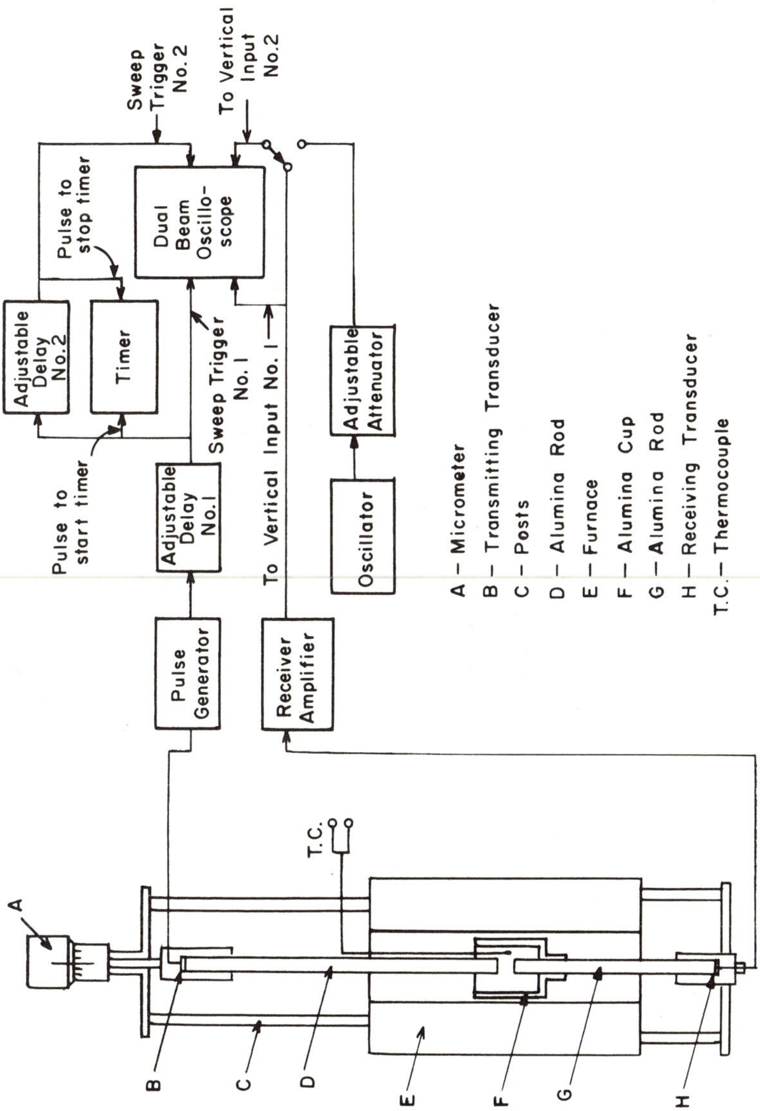

FIG. 6.2. Schematic diagram of apparatus for ultrasonic measurements in molten salts.

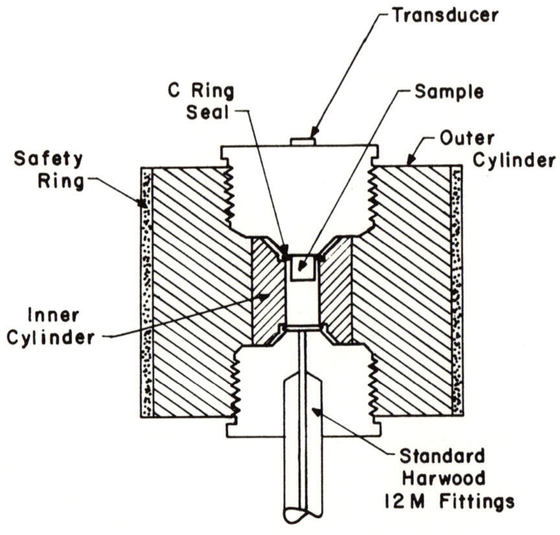

FIG. 6.3. Pressure vessel apparatus for measuring velocity of sound in fused quartz to pressures of 10 kbars (5).

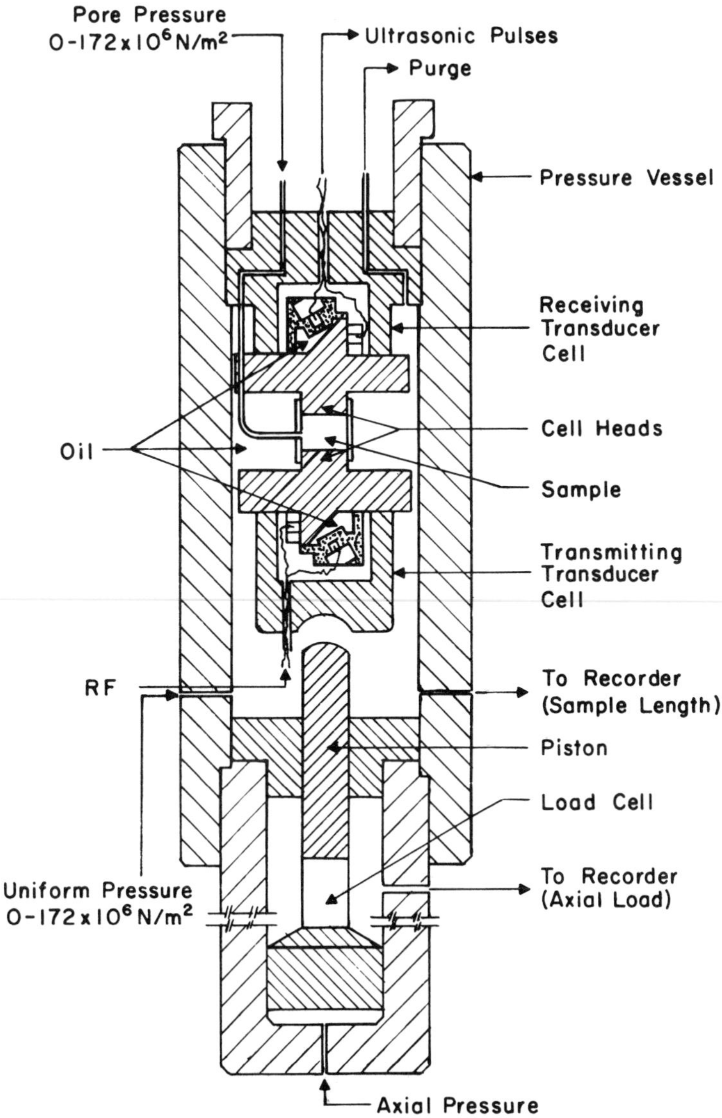

FIG. 6.4. Dual-mode ultrasonic apparatus for measuring velocity of longitudinal and shear waves in specimens under pressures (7).

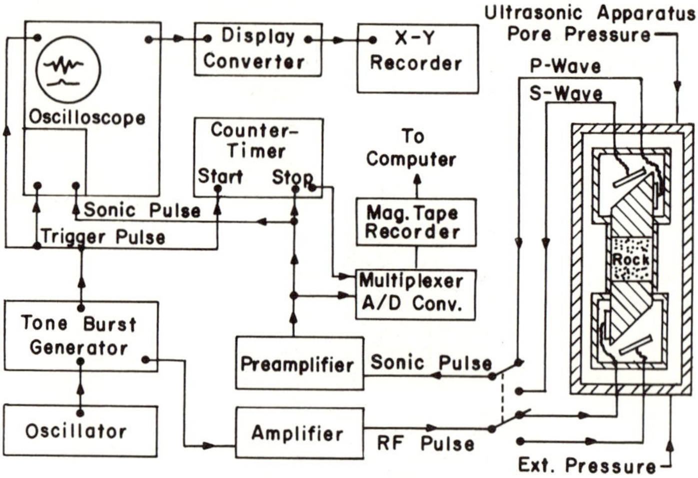

FIG. 6.5. Schematic diagram of the electronic-acoustic system of Gregory and Podio (7).

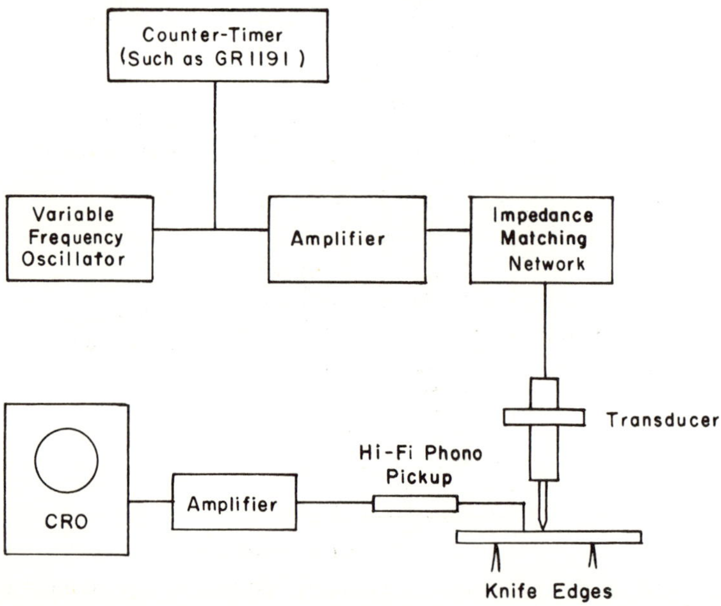

FIG. 6.6. Block diagram of circuit for measuring flexural resonance frequency in bars (fundamental frequency only).

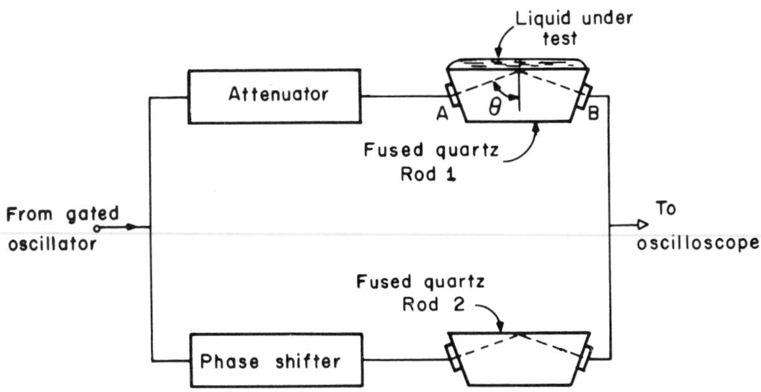

FIG. 6.7. Shear wave reflection technique of McSkimin (10).

REFERENCES

1. H. J. McSkimin, J. Acoust. Soc. Am., 33, 12 (1961).

2. H. J. McSkimin, J. Acoust. Soc. Am., 33, 606-615 (1961).

3. R. W. Higgs and T. A. Litovitz, J. Acoust. Soc. Am., 32, 1108-1115 (1960).

4. T. A. Litovitz, T. Lyon, and L. Peselnick, J. Acoust. Soc. Am., 26, 566 (1954).

5. L. Peselnick, R. Meister, W. H. Wilson, J. Phys. Chem. Solids, 28, 635-639 (1967).

6. H. J. McSkimin, IRE Transactions, PGUE-5, 25 (1957).

7. A. R. Gregory and A. L. Podio, Paper D-10, 1969 IEEE Ultrasonics Symposium, St. Louis, Missouri, September 24-26, 1969.

8. S. Spinner, J. Am. Ceram. Soc., 39, 113-118 (1956).

9. W. Roth and S. R. Rich, J. Appl. Phys., 24(7), 940-950 (1953).

10. W. P. Mason and H. J. McSkimin, Bell System Tech. J., 31, 122-171 (1952).

11. J. J. Lohr, D. E. Wilson, F. M. Hamaker, and W. J. Stewart, J. Spacecraft Rockets, 5(1), 68-74 (1968).

12. Anon., Tech. Rev., 53, 467 (1951).

13. W. C. Hitt and J. B. Ramsay, Rubber Plastics Age, 44, 411-413 (1963).

14. J. R. Zurbrick, Mater. Res. Stnd., 8(7), 25-36 (1968).

15. L. P. Bobrikov and M. Z. Medredev, Mekhan. Polimerov, 4, 547-554 (1968).

16. A. W. Schultz, Mater. Res. Std., 7(8), 341-345 (1967).

17. R. C. McMaster, ed., Nondestructive Testing Handbook, Vol. II, Ronald Press, New York, 1959.

18. General Dynamics Convair Division, Nondestructive Testing - Ultrasonic Testing, Classroom Training Handbook, San Diego, California, 1967.

19. L. E. Kinsler and A. R. Frey, Fundamentals of Acoustics, Wiley, New York, 1950.

20. J. R. Frederick, Ultrasonic Engineering, Wiley, New York, London, Sydney, 1965.

21. L. Bergmann, Der Ultraschall, Hirzel Verlag, Stuttgart, 1954.

22. W. D. Jolly, Materials Characterization Study--Ultrasonic Longitudinal and Shear Velocities of Candidate Isotope Encapsulation Materials, USAEC Report BNWL-547, Pacific Northwest Laboratories, Richland, Washington; available from Clearinghouse for Federal Scientific and Technical Information, 1967.

23. F. Birch, J. Geophys. Res., 65(4), 1083-1102 (1960).

24. D. E. Goldman and T. F. Hueter, J. Acoust. Soc. Am., 28(1), 35-37 (1956).

25. M. Greenspan and C. E. Tschiegg, J. Acoust. Soc. Am., 31(1), 75-76 (1959).

26. C. D. Hodgman, Ed. in chief, Handbook of Chemistry and Physics, 37th ed., Chemical Rubber Publishing Co., Cleveland, Ohio (1955).

Chapter 7

NONDESTRUCTIVE TESTING--BASIC METHODS AND GENERAL CONSIDERATIONS

I. INTRODUCTION

Ultrasonic methods are perhaps the most versatile of the techniques for nondestructive testing available to industry. As a result, nondestructive testing provides one of the largest, if not the largest, areas of applications for low-intensity ultrasonic energy. Books that cover this subject are included in the reference list at the end of this chapter (1-3). There are numerous good references in the technical literature on the use of ultrasonics for nondestructive testing.

The common methods of ultrasonic testing may be categorized under two headings: (1) resonance methods and (2) pulse methods. Other methods that do not fit into either category show promise in some areas of inspection. Discussed in this chapter are the ultrasonic methods of nondestructive testing, factors that affect ability to detect defects, and instrumentation.

II. BASIC METHODS

II.A. Resonance Methods

Resonance methods are useful for measuring thicknesses of walls or of delaminated layers when only one surface is available to the transducer. The method involves determining the fundamental frequency calibrated in terms of thickness. The thickness is one-half wavelength at the fundamental frequency, and at the overtones, the thickness is a multiple of half-wavelengths. A piezoelectric transducer used to excite ultrasonic waves in the specimen material is driven by a frequency-modulated electrical signal through a range of frequencies usually covering one octave. Changes in impedance due to the in-phase or out-of-phase conditions within the specimen are reflected through the transducer into the electronic system. The effects are indicated by a suitable display unit (usually an oscilloscope). When an oscilloscope is used, the sweep triggering and the sweep rate are synchronized with the variations in frequency of the oscillator. Therefore, resonance conditions are indicated on the oscilloscope screen as peaks at positions that correspond to multiples of half-wavelengths in the material. The thickness of the material is given, within the accuracy of the test conditions, by

$$\ell_T = c/2f_1 \tag{7.1}$$

where f_1 is the fundamental resonant frequency of the test material. Usually, the oscilloscope screen is equipped with scales calibrated in units of length corresponding to half-wavelengths within the material being tested. The important factor in this type of measurement is to identify or determine the fundamental resonance frequency. When resonance occurs at a harmonic frequency, more than one peak occurs on the oscilloscope screen. The difference between two successive resonant frequencies is the fundamental frequency. Since the calibration is in linear dimensions, it is possible to determine the actual dimension from the indicated dimensions on the oscilloscope screen, from the relationship

$$f_1 = f_{n+1} - f_n \tag{7.2}$$

where n is the order of the harmonic. If two successive peaks occur on the oscilloscope screen at ℓ_n and ℓ_{n+1} corresponding to f_n and f_{n+1}, the thickness ℓ is

$$\ell = \frac{\ell_n \ell_{n+1}}{\ell_n - \ell_{n+1}} \tag{7.3}$$

within the accuracy of the measurement technique. Some manufacturers supply calibration screens containing scales corresponding to the various harmonics, and by aligning the various peaks with corresponding values on each scale, it is possible to read the thickness directly without resorting to Eq. (7.3).

Since the velocity of sound is different for different materials, the instrumentation must be calibrated using calibration blocks made of the same material as that to be tested.

II.B. PULSE METHODS

Ultrasonic pulse methods of nondestructive testing are used more extensively than resonance methods in industry. Waves of various modes are used, including longitudinal, shear, Rayleigh, and Lamb waves. Ultrasonic pulses can penetrate to considerable depths--several feet in low-carbon steel, aluminum, etc.

Pulse methods may be classed under three categories: (1) pulse-echo, (2) pitch-catch, and (3) through transmission.

Pulse-echo methods involve the use of transducers which act first as emitters of ultrasonic pulses and then as receivers to detect echoes from defects or other interfaces.

Pitch-catch methods involve the use of two identical piezoelectric elements, often, but not necessarily, mounted in the same holder, with one element serving to emit ultrasonic pulses and the other to receive the reflected pulses.

Through transmission methods involve the use of two transducers located relative to each other and to the specimen in such a manner that one transducer receives the energy transmitted from the other after the energy has passed through a region of interest. Discontinuities cause losses of energy characterized by drops in output voltage from the receiver. When the longitudinal mode is used, the transmitter and receiver are located on opposite sides of the specimen. When shear waves are used, the transmitter and receiver may be located on the same face of the specimen, or on the opposite sides, but at opposite ends of the test area. When surface waves or Lamb waves are used, the transducers are located on the same surface, but at opposite ends of the test area.

III. FACTORS AFFECTING RESOLUTION AND SENSITIVITY

The sensitivity of an ultrasonic transducer refers to the relationship between the amplitude of the electrical voltage generated by the transducer and the magnitude of an ultrasonic signal impinging on its surface. It, therefore, is a determining factor in how small a defect can be detected. The sensitivity of transducers used for nondestructive testing is measured by the amplitude of their response to energy reflected from a standard discontinuity (for instance, a hole in a standard reference block or a round rod in a liquid bath). Several factors influence the sensitivity of a transducer and thus not all of a given type and make have similar sensitivities.

Resolution refers to the ability to separate echoes from two discontinuities located at only slightly different depths. A long pulse has poor resolving power. When a second echo arrives at the receiver before the first one dies out sufficiently for the second one to be observed, the second signal is masked by the first. Very short pulses are desirable for high resolution. This requires broad-bandwidth, low-Q transducers, where

$$Q = f_o/(f_2 - f_1) \tag{7.4}$$

where f_o is the center frequency of the transducer,

$(f_2 - f_1)$ is the bandwidth, and

f_1 and f_2 are the frequencies below and above f_o, respectively, where the amplitude of vibration is only 0.707 times the maximum amplitude.

Resolving power of broad-bandwidth transducers decreases with distance in materials, and the rate of decrease depends upon the attenuation characteristics of the material. For example, reflections at grain boundaries cause attenuation owing to scattering. The amount of scattering is a function of the anisotropy of the grains, of the metallurgical conditions at the grain boundaries, and of the relationship between the wavelength and the grain dimensions. Higher frequencies are attenuated more rapidly than are the lower frequencies, and this preferred attenuation tends to reduce the bandwidth and lower the center frequency of the pulse, thus causing the loss in resolving power.

III.A. NEAR-FIELD EFFECTS

Diffraction in the near field of a transducer produces

variations in the distribution of sound pressure along and parallel to the axis and also in a direction normal to the axis. In nondestructive testing, the amplitude of waves reflected from defects located within the near field is affected by the diffraction pattern and depends upon the location of the defect within the near field. In the far field, the amplitude of the reflected wave decreases approximately exponentially as the distance from the source increases. However, in the near field, the amplitude of the reflected wave may experience at least one maximum and one minimum as the distance from the source increases to a point corresponding to the start of the far field, where maximum amplitude usually is obtained. For this reason, it is difficult to determine the sizes of defects located in the near field by the amplitude of reflected waves alone.

III.B. PROPERTIES OF THE MATERIALS

The properties of the materials that influence defect resolution and sensitivity include geometry of the part, metallurgical conditions such as grain size and grain-boundary conditions, anisotropy, surface roughness, and the presence of internal scatterers and their scattering cross section.

Scattering from grain boundaries may completely obscure conditions within a material if the frequency chosen exceeds that for which the wavelength is greater than three times the grain dimensions. Therefore, grain size and the conditions that exist at the grain boundaries are determining factors in the choice of frequency and the size of defect that can be detected ultrasonically. Defects of sizes corresponding to the grain dimensions may be impossible to detect unless these defects appear in clusters such that the combination of scattering cross sections appears as a unit which is larger than the average grain sizes. For example, shrink porosity in wrought-iron castings is easily detected by pulse-echo and through-transmission methods because the clusters of small pores produce echoes and scattering that are similar to those produced by larger defects with continuous boundaries.

The geometry of the part determines the accessibility of the ultrasonic beam to the defects. It also is a determining factor in the choice of modes for the inspection. The surface contour in particular affects the resolution and sensitivity of the inspection by its effect on the incidence angle at any given position within the cross section of the beam. For instance, when a plane wave approaches by means of a liquid couplant, a curved surface which has a radius of the same order of magnitude as the cross section of the beam, the incident angle of approach may range from 0, where the radius of the surface is parallel to the direction of propagation, to a maximum of 90 where the radius is normal to the direction of propagation. Angles of reflection and refraction vary accordingly. Consequently, both shear and longitudinal waves are generated within the material. The effect of this mode conversion is to partition and scatter the energy within the part, and, since the different modes travel at different velocities, to decrease the sensitivity to defects because of the confusion that results.

Surface roughness affects the resolution and sensitivity of an ultrasonic nondestructive test in much the same manner that grain conditions do. Scattering from rough surfaces can be very bad if the wavelength is similar to the dimensions of the asperities. A review of the literature will show that analyses of scattering from rough surfaces and internal scatterers can be quite complicated (2-11). Not only does a rough surface cause scattering and mode conversion on entry, but it also may affect the coupling of energy into the part.

It is possible to define a critical surface roughness which corresponds to a condition of maximum destructive interference in the wave within the part. This condition is brought about by the difference in the velocity of sound in the liquid couplant and in the solid. The velocity of sound in a solid is greater than in a liquid. Waves entering the part at a roughness peak travel the distance to the bottom of the valley at a higher velocity than the waves which travel through the liquid and enter the part at the bottom of the valleys. When the phase difference between the two components after both enter the part is 180°, or an odd multiple thereof, maximum cancellation occurs. To obtain complete cancellation would require a unique situation that is not obtained with broad-band, short pulses. However, refraction, diffraction, and the difference in phase change the shape and affect the pattern of the pulse within the specimen.

The theoretical critical roughness, R_c is given by the equation

$$R_c = \frac{\lambda_1 c_2}{2(c_2 - c_1)} = \frac{\lambda_2 c_1}{2(c_2 - c_1)} \tag{7.5}$$

where λ_1 is the wavelength in the liquid,
c_1 is the velocity of sound in the liquid,
λ_2 is the wavelength in the solid, and
c_2 is the velocity of sound in the solid.

Anisotropy in a multicrystalline material contributes to scattering losses. The grain direction is distributed randomly with respect to the direction of wave propagation. Since the elastic constants vary with orientation, scattering occurs as a result of reflections at intersections between grain boundaries.

IV. UNCONVENTIONAL TECHNIQUES USED FOR NONDESTRUCTIVE TESTING

IV.A. EDDY-SONIC INSPECTION METHOD

In 1912, Thomas Fessenden developed the Fessenden oscillator for producing underwater sound signals. The oscillator was based upon the principle that the magnetic field developed by eddy currents in a conducting surface reacts with the source magnetic field which produces the eddy currents in the first place, and as a result, a force is applied to the conductor. Both the eddy currents and the mechanical force are time variant, and thus a sound wave is generated.

Botsco and his co-workers (12) have made use of the same principal to generate sound in materials to be tested, and the method has been applied successfully in testing adhesive-bonded honeycomb composites for the Saturn space vehicle (13). The method has been named eddy-sonic testing. The principal advantage of the method is that it requires no liquid couplant. Some constituent of the material to be inspected must be electrically conductive. High conductivity is preferred; however, low conductivity in materials which are magnetic, such as ferritic stainless steel, can be inspected successfully. Since the structure being examined is essentially the dynamic element of a transducer, any change in its condition affects the character of the sonic energy generated.

Detection of the radiated sound offers no difficulty because inspection is in the audible range--2 to 4 kHz being typical frequencies.

Commercial eddy-sonic units are available with a meter readout. However, C-scan recording also is possible.

Present equipment is practical for large panels such as honeycombs but impractical for small panels. Resonant patterns across small composite samples cause confusion when their signals are similar to those produced by defects.

An evaluation of the eddy-sonic method for inspecting adhesive-bonded honeycomb structures has been made by General American Research Division of General American Transportation Corporation (14). Their results showed that the technique is capable of resolving near-side unbonded areas approximately 2½ in. in diameter in test panels with a 0.020-in. thick aluminum facing sheet over an aluminum core, but far-side unbonds of the same size could not be detected in panels with a core thickness exceeding ¼ in. The actual sensitivity depends upon the test conditions--especially the dimensions of the panel, the conductivity of the conductive part of the panel, and the nature of the defect. The method has low sensitivity to tight delaminations.

Types of defects that can be detected are nonbonds, crushed core, fracture core, and voids in the adhesive.

IV.B. SONIC ANALYSIS

Sonic analysis in the present sense is the analysis of acoustic signals emitted by a device or system to detect the presence of undesirable conditions such as the development of defects that are the first stages of serious failure. Every person who can hear has applied sonic analysis mentally without being conscious of the mechanism. Identification of a person's voice, recognition of a jet aircraft flying overhead, and recognition of a discord in one's piano playing are typical of the many ways people use a mental analysis of sounds.

Similar principles may be applied in industry, but here the received signal must be processed in such a manner that identification of an anomaly may be made on the basis of a visual rather than an aural data presentation, if necessary. Any moving system generates a sound. If the system is operating in a normal manner, the signal emitted may be called the "acoustic signature" of that system. The acoustic signature, therefore, identifies the normal condition. Any abnormality will affect the character of the emitted sound. For instance, a mechanic can readily distinguish between the pounding of a broken connecting rod in an internal combustion engine and the sound of a properly functioning engine.

Sonic signature analysis is performed by comparing the known normal signature with the sound emitted by the operating system. The potential methods of instrumentation are numerous--but each has a sound- or vibration-sensing device (microphone or accelerometer). For instance, in the case of leak detection, the only equipment necessary may be a listening device capable of detecting frequencies within a range where the background noise is minimum. The Heco ultrasonic detector (15) is one such device. Gases escaping through a small opening produce a broad band of white noise, i.e., noise having a wide range of frequencies. The Heco device detects through a microphone those components of ultrasound falling within the range 35,000 to 45,000Hz. The detected signal is beat down to the audible range and detected through earphones or through an external amplifier and speaker. Orifice-type gas leaks may be detected at distances greater than 100 ft.

Another type of leak test makes use of the noise generated by a leaking liquid which escapes under conditions of temperature and pressure that cause the liquid to vaporize within the length of the crack. Still another method is based upon the detection of bursting microscopic bubbles produced by slowly escaping gas into a fluid coating applied over the area under test. The bursting bubbles release ultrasonic energy that can be detected by an ultrasonic probe. This method can be sensitive to leaks as small as 0.002 cc/sec.

Ultrasonic signature analysis is used in monitoring the condition of bearings. Defects in the inner or outer races or in a ball in a ball bearing produce distinguishable sounds as the balls move in the races. The equations for the rotational frequencies, as given by Martin (16), are

$$f_o = \frac{n}{2}\,\phi\left(1 - \frac{2r}{D}\cos\beta\right) \qquad (7.6a)$$

(flaw on outer race)

where n is the number of balls,

ϕ is the relative speed between inner and outer race (rps),

r is the radius of the ball,

D is the pitch diameter, and

β is the contact angle.

$$f_i = \frac{n}{2}\,\phi\left(1 + \frac{2r}{D}\cos\beta\right) \qquad (7.6b)$$

(flaw on inner race)

$$f = \frac{D}{2r}\,\phi\left[1 - (2r/D)^2\cos^2\beta\right] \qquad (7.6c)$$

(flaw on ball bearing)

Each time a moving ball contacts a crack or pit, for instance, it generates a shock wave that can drive the races and also the balls into resonance. The vibrations that are most easily detected are those that are generated in the races. The races are rings, and resonant frequencies of rings are given by the equation

$$f_r = \frac{w(w - 1)}{2(w^2 + 1)^{\frac{1}{2}}} \frac{1}{a^2} (YI/m)^{\frac{1}{2}} \tag{7.7}$$

where w is the number of waves around the circumference of a ring,
a is the radius of the race (to the neutral axis),
I is the moment of inertia of the cross section of the race about the neutral axis,
Y is the modulus of elasticity of the material of the race, and
m is the mass of the ring per linear foot.

When a defect such as a pit occurs in a ball bearing, the intensity of the radiated sonic, or ultrasonic, energy increases at frequencies corresponding to those determined by Eqs. (7.6) and (7.7). The amount of the increase in intensity depends upon the relationship between the rotational speed and the resonant frequencies of the races. It can be orders of magnitude higher than the intensity emitted at the same frequencies in the absence of defects.

Signature analysis and sonic detectors have been used to detect electrical noise in transformers, capacitors, or insulators. Other applications include detecting abnormal conditions in hydraulic systems, piston ring blow-by and cylinder scoring in internal combustion engines and compressors, incipient boiling in pressurized water reactors, and detecting broken blades and worn bearings in jet engines as well as monitoring machinery and transmission noises for indications of incipient failures.

IV.C. ACOUSTIC-EMISSION TECHNIQUE

Acoustic emission is the spontaneous emission of sound pulses from materials subjected to external stress as a result of sudden relaxation of stresses within the material. Assume that a material is subjected to tension and this material contains a point of stress concentration. As the average stress increases, the stress in the material about the point of stress concentration approaches the elastic limit before that of the remaining material. The material subjected to the concentrated stress is no longer capable of carrying its share of the load and the adjacent fibers suddenly experience an increase in stress. According to the development of the wave equation, in order for equilibrium to obtain, the stress distributes throughout the remaining material at the velocity of sound.

Stress relaxation producing acoustic emission can occur as a result of the nucleation and propagation of cracks or as a result of such elastic and plastic deformation processes as the slip of existing dislocations in metal, the activation of dislocation sources, twinning, and slip of grain boundaries (17).

The emitted energy consists of two components: (1) a low-level, high-frequency component which appears as an increasing noise as the stress continues to be applied and (2) a discontinuous, burst-type component or pulse. The high-frequency component is believed to be induced by dislocation pinning during slip deformation. The burst type of pulse is believed to be related to the formation of stacking faults and mechanical twins and to the formation of cracks.

The sound level of acoustic emission is very low and depends upon the stress level at which it is produced, neglecting strain rates and volume effects (18). The loading mechanism used to stress the material must be carefully designed so that it does not generate sufficient noise to mask the acoustic emission.

The frequency spectrum of acoustic emissions depends upon how the sound is produced. The frequencies range from the audible to the ultrasonic frequencies in the megahertz range. Sometimes frequency is used to denote rate of emission or the number of bursts or counts per second (17).

The microstructure and therefore the processing treatment that a metal has been subjected to affects the acoustic emission. Aging, either at zero stress or under an applied stress, precipitation hardening, and cold work affect the emission (16). Deformation may cause a random signal which is strain-rate dependent with an average frequency being discernible. Along with the random signal may occur sporadic bursts of higher frequency. Brittle fracture produces burst signals of much higher frequency (19).

When a metal is stressed to a value near or above the yield stress, unloaded, and then restressed to this same level again, if there had been emission during the first load cycle, it would be absent or very much reduced during the second application of a load up to the previous stress level. However, when the previous stress level is exceeded, acoustic emission is again evident (17). This phenomenon has been called the Kaiser effect.

Basic equipment required for applying acoustic emission to nondestructive testing comprises a highly sensitive microphone and an amplifier. If possible, detection is done at frequencies at which the background noise is minimum. This may be done by filtering or by choosing a transducer with response limited to a small range of frequencies within the range of interest.

Although most applications of acoustic emission for nondestructive testing evaluated have been in metals--pressure vessels, nuclear components, aerospace materials--Romrell and Bunnell (20) have indicated the feasibility of monitoring the growth of cracks in ceramic materials. They used a lead-zirconate-titanate transducer with a primary frequency response to acoustic emission at 600 kHz. They increased the signal-to-noise ratio by using a preamplifier with a 600kHz tuned circuit. For additional rejection to low-frequency noise, the signal conditioner included a high-pass filter adjusted to 450 kHz. Materials studied were thorium-yttrium oxide (tubes) and aluminum oxide rods containing magnesium oxide to allow sintering to nearly theoretical density while maintaining fine grain structure. The acoustic-emission signals were counted by both emission rate and summation circuitry which operated in parallel.

Other applications of acoustic emission include in-process inspection of welds (if cracks form during cooling they can be detected as they form), monitoring crack formation during hydrotesting of solid rocket chambers, and monitoring reactor pressure systems. The addition of brittle coatings to enhance the detection of high stress regions has been considered.

IV.D. ACOUSTIC-IMPACT TECHNIQUE

Testing a material by acoustic impact is not uncommon. When a banker drops a coin on the marble counter top to determine its genuineness, he is performing an acoustic-impact test. Thumping a melon to determine whether it is ripe is another acoustic-impact test. One may go on with many other illustrations, including the crewman with the hammer testing the wheels and other critical parts of a locomotive. In each of these examples, the brain processes and analyzes the data and from these arrives at an evaluation.

When a material is shocked by impact, a spectrum of frequencies is contained in the resulting pulse, and the wave constituents of the pulse depend upon the nature of the impact. As the slope of the wave-front increases, so does the width of the spectrum. Since the propagation of the waves depends upon the properties of the material, changes in these properties affect the propagation characteristics (velocity and attenuation).

An acoustic-impact technique has been developed by Schroeer, Rowand and Kamm (21), based upon the fact that the appearance of a defect changes the mechanical impedance of a structure or part and affects the vibrational response of the structure in the area of the defect. The instrumentation includes an impactor, a sensor, a preamplifier, a tunable band-pass filter, a signal processing circuitry, a tape recorder, an x-y or strip chart recorder, and an indicator (dial or oscilloscope). The impactor and sensor are included in one assembly called the probe. The impactor is the carbide-tipped core of a solenoid coil. It is capable of providing a single blow or a series of repetitive shocks as desired.

The sensor is a lightweight accelerometer capable of withstanding high acceleration forces. As would be expected, its position on the test specimen affects the accuracy of the test.

Schroeer, Rowand, and Kamm explain the effectiveness of the acoustic-impact technique on two mechanisms of damping of vibrations--hysteresis and shear damping. Hysteresis occurs in materials undergoing cyclic stressing, i.e., the stress-strain curve corresponding to decreasing stress does not coincide with the stress-strain curve of increasing stress and the dissipated energy is proportional to the area between the two curves. In a beam vibrating in flexure, the outer fibers experience the greatest stress and, therefore, the greatest hysteresis loss occurs in these fibers. When a crack forms at the surface in a direction normal to the stress, the stress in the outer fibers is relieved and the bulk of the load is carried by the outermost of the intact fibers. If the amplitude of vibration remains constant, the hysteresis losses drop. In addition, the effective thickness of the beam decreases, which causes a corresponding decrease in stiffness and, therefore, flexural resonant frequency. The mass of material between the surface and the depth of the cracks approaches a dead weight as the number of cracks increase. This fact, too, tends to cause a decrease in frequency. Therefore, the appearance of cracks in a flexing beam will lower its natural resonant frequencies and may decrease its damping by decreasing its hysteresis losses. On the other hand, if the mating surfaces of the cracks are subjected to shear stresses, energy is dissipated because of friction. The shear stress is equal to the product of the coefficient of sliding friction times the compressive stress normal to the mating surfaces. Frictional losses are greatest with the overtones. Therefore, cracks in a beam vibrating in flexure may cause a reduction in damping and the fundamental frequency, but may cause an increase in attenuation of higher frequency components.

Schroeer, Rowand, and Kamm developed their technique primarily for detecting cracks concealed under fastener heads but have shown its practicability in the inspection of composite materials and fatigue specimens. Many other applications are possible. In fact, similar techniques have been used to separate parts, for example, in the automotive industry, by ringing the test specimen and accepting or rejecting the part on the basis of its ringing frequency.

In order to follow the history of a failure, it is imperative that the probe be located exactly the same during each measurement. Variations in loading can affect the frequency content of the pulse detected by the sensor much as changing the position of the mass on a pendulum changes the frequency of the pendulum.

The acoustic-impact technique detects the presence of a crack by its influence on the vibrational characteristics of the specimen but it does not delineate the crack or cracks.

IV.E. FREQUENCY ANALYSIS

In most conventional applications of ultrasonics in nonstructive testing, evaluation is based primarily upon amplitude and time information. Transducers are selected according to a nominal test frequency that meets the requirements of the inspection task, but, in general, the relationship between the initial and received pulse spectra is not recorded. Additional information to be gained by analyzing pulses used in nondestructive testing becomes obvious when the frequency dependency of reflection, diffraction, and attenuation (especially in crystalline materials) is considered. For example, the components of a pulse in which the wavelengths are large compared to the dimensions of a discontinuity will be reflected spherically; those in which the wavelengths are small compared to the dimensions of the discontinuity are reflected according to the contour of the surface of the discontinuity upon which the pulse is incident. By diffraction, the longer waves pass the discontinuity producing no "shadow" effect while the shorter waves cast a shadow the nature of which depends upon the relationship between wavelength, cross section of the beam, and cross section of the discontinuity. These phenomena provide a basis for flaw analysis including size, geometry, and orientation. The frequency dependency of attenuation in crystalline materials can be used to determine certain properties of the grains, particularly average grain size of certain materials.

Procedures for ultrasonic frequency analysis have been developed and are being used in several areas. Early in 1964, Gericke (22,23) began developing equipment for ultrasonic spectroscopy. Gericke's application was to pulse-echo studies using the direct-contact methods. The essential components for such studies are a broadband transducer, rectangular pulse generators, time gate, broad band amplifiers, suitable detector, and a spectrum analyzer capable of analyzing typical pulses used in nondestructive testing.

The most important component is the transducer. Gericke uses lead-zirconate-niobate transducers that produce a pulse that contains various frequency components covering a band of approximately 3 to 10 MHz. The spectrum of the initial pulse is determined by the frequency response of the trnasducer and by the length of the rectangular voltage pulse used to excite the transducer. Gericke states that "if the excitation pulse length is properly matched to the transducer response curve, a sufficiently uniform ultrasonic spectrum can be generated to permit successful ultrasonic spectroscopy".

Commercially available transducers may exhibit characteristics that vary from those indicated by the nameplate (see Fig. 7.1). The nameplate designates a nominal center frequency. A spectrum analysis often shows peaks in sensitivity at frequencies other than the nominal frequencies. In general, commonly available transducers are unsuitable to use for ultrasonic spectroscopy. Desirable characteristics are very low Q, no significant resonances in the operating band, and good sensitivity.

Whaley and Cook (24) have developed means for ultrasonic frequency analysis which is composed of commercially available electronic equipment and can be used with conventional ultrasonic instruments operating in either a contact or immersion mode. The application goes beyond analyzing test material to include the evaluation of the characteristics of transducers, instrument types, transducer positioning, tuning devices, and collimation. All of these parameters are important, especially in comparing inspection data between instruments and also at various intervals of time. Differences in response and setup can result in serious problems of interpretation of inspection data. For their work, Whaley and Cook prefer an immersion setup.

Serabian (25) uses spectrum analysis to show that as an ultrasonic pulse propagates through highly attenuating, nondispersive materials, preferential attenuation of high frequency components causes the pulse to continuously change its shape as it propagates through the medium. The center frequency shifts toward the low-frequency end of the pulse, which also may be lower in intensity than the center frequency. As a result, sensitivity to defects decreases with distance in such materials not only because of absorption of energy but also because of the lower sensitivity to defects at lower frequencies. Thus, spectral changes in pulses propagating through such materials (for instance, graphite) impose a limitation on the use of ultrasonics for their inspection.

IV.F. CRITICAL-ANGLE ANALYSIS

Solids are capable of sustaining several wave modes. When a solid is immersed in a liquid and an ultrasonic wave impinges on the solid from the liquid, several wave modes may be generated in the solid by mode conversion. The critical angle for a given mode is the angle of incidence which produces an angle of refraction of 90° for that mode. According to Snell's law, the refraction angle is a function of the angle of incidence and of the relationship between the velocity of sound in the liquid and in the solid. The principle offers a sensitive means of analysis which has been used to a limited extent.

Critical-angle analysis makes use of phenomena which are easily observed. A reflected wave at a liquid/solid interface undergoes a shift along the interface which is dependent upon the angle of incidence. The amplitude distribution across the reflected beam differs from the distribution across a focused incident

incident beam when the axis of the incident beam coincides with the Rayleigh critical angle. The amount of amplitude redistribution depends upon the density and the elastic properties of the solid. The shift of the position of the reflected wave is the major factor in the redistribution of the amplitude in the reflected beam. The principle is illustrated in Fig. 7.2.

When a focused beam impinges on a liquid/solid interface so that the ray along the axis of the beam is at the Rayleigh critical angle, the beam also contains rays at incident angles which are either greater than or less than the critical angle. Those rays at angles exceeding the critical angle are totally reflected. Those rays at angles less than the critical angle are partially reflected and partially refracted into the solid. As a result, the reflected rays in the latter case are less intense than they would be if they were reflected at an angle greater than the critical angle. In addition, the varying amount of beam shift due to the multiple angles of incidence in a focused beam also causes variation in the interference pattern of the reflected beam, according to Huygens' principle.

In the application of the critical-angle technique, the focused beam is adjusted so that the ray along the axis is at the Rayleigh critical angle. The beam shift and intensity distribution of the reflected wave are then characteristic of the material under inspection. If the beam is caused to scan the surface of the metal while maintaining a constant angular relationship, the conditions within the reflected wave remain constant unless a change in the elastic properties or density of the solid occur (assuming a perfectly smooth surface). Since changes in elastic properties or density cause corresponding changes in the propagation velocity of ultrasonic waves, these former changes also cause a corresponding change in critical angle. Rays other than those on the axis of the beam are located at the new critical angle. Consequently, a change in amplitude distribution occurs which can be related to the new properties of the material.

The lateral shift of the reflected beam is given by

$$\Delta = -\frac{\lambda}{2\pi}\left(\frac{\partial \phi}{\partial \sin \alpha_1}\right) \tag{7.8}$$

where λ is the wavelength of sound in the liquid,

ϕ is the phase of the reflection coefficient, and

α_1 is the angle of incidence and the angle of reflection when α_1 is greater than the angle of total internal reflection.

Equation (7.8) is valid when the width of the incident beam is much larger than λ.

The phase shift is approximately 2π when the angle of incidence is nearly equal to the Rayleigh critical angle. The lateral shift for this condition is

$$\Delta = \lambda\left(\frac{2}{\pi}\right)\frac{\rho_m}{\rho_f}\sqrt{\frac{r}{s}\frac{(r-s)}{(s-1)}}\left\{\frac{1+6s^2(1-q)-2s(3-2q)}{s-q}\right\} \tag{7.9}$$

where ρ_f is the density of the fluid,

ρ_m is the density of the metal,

$r = (c_s/c_f)^2$; $s = (c_s/c_R)^2$; $q = (c_s/c_L)^2$

c_s is the shear-wave velocity in the metal,

c_f is the velocity of sound in the liquid,

c_R is the velocity of Rayleigh waves in the metal, and

c_L is the velocity of longitudinal waves in the metal.

In practice, a focused transducer is oriented with respect to the solid surface in an immersion system at the Rayleigh critical angle (see Fig. 7.3). A miniature receiving transducer is mounted on a support which is adjusted by micrometer. The scan of the receiver passes through a null region caused by destructive interference between reflected rays. The sensitivity of the miniature receiving transducer, which is approximately 1/8 in. in diameter, is enhanced by a collimator.

The technique thus described has been used by Willard (26) to detect hydrides in Zircaloy. The sensitivity was 600 ppm, being limited by the surface roughness of the specimens. The presence of hydrides in Zircaloy affects both the density and the elastic properties. Frequencies used by Willard were 1.0, 2.25, and 5.0 MHz in flat plate specimens 0.140 in. thick.

Hunter (27) has used the critical angle method to determine irradiation damage. Both shear and longitudinal waves are slowed by such damage, but, in some cases, specimens irradiated at high temperature exhibited increasing velocities of sound prior to the decreasing trend. Radiation damage in low-alloy steels appears as impact embrittlement.

The critical-angle method of analysis is potentially more useful than its limited application would indicate. The observed effects are attributable to boundary wave generation and these are potentially useful for determining near-surface conditions, such as measuring residual stresses, and detecting variations in elastic moduli, degree of cold-work, grain orientation, composition, and surface treatment. Replacing the miniature receiving transducer with an electronic-acoustic image converter might enhance the sensitivity of the method and permit more accurate evaluations. In addition, the electronic-acoustic image converter permits faster scanning than is possible by mechanical scanning.

V. INSTRUMENTATION

V.A. COUPLING ENERGY TO THE TEST OBJECT

A means must be provided for transferring ultrasonic energy between the piezoelectric transducers generally used in nondestructive testing and the test object. The great difference between the acoustic impedances of air and solids causes nearly 100 percent of the ultrasonic energy impinging on air/solid boundaries to be reflected. Maximum energy is transferred across an interface when the acoustic impedance of the two media are equal. The best condition exists when the densities are also equal. Although liquids seldom meet the conditions for maximum energy transfer, they are most generally used as the coupling medium for nondestructive testing because they do provide adequate energy-transfer capability and usually they are convenient to use.

Basically, three methods of coupling energy to a specimen material are used: (1) direct contact between the transducer and the test piece with a liquid or grease couplant between, (2) immersion of the transducer and the testpiece in a liquid bath and coupling through the intermediate liquid path, and (3) flooding a space separating the transducer and the testpiece with a liquid (gap method). Most other methods of coupling--liquid-filled tire, liquid-filled boot, etc.--are modifications or combinations of these techniques. If liquids are not permissible, sometimes thin films of soft pliable polymers can be used as coupling materials.

The eddy-sonic method (12) was developed in recent years with the objective of generating ultrasonic energy in a test object across an air gap. Here the testpiece actually becomes a part of the transducer, i.e., the electrical driving energy is actually transformed to mechanical energy within the testpiece.

Nearly all modes of ultrasonic waves used in nondestructive testing are produced by mode conversion at the boundaries between the couplant and the testpiece. The incident wave is generated by a transducer operating in a longitudinal mode. If testing is to be done using a longitudinal wave, the transducer is oriented so that the wave impinges at normal incidence on the surface of the testpiece. If a shear-wave technique is to be used, the incidence angle is made larger than the critical angle for longitudinal waves. When direct contact coupling is used with shear-wave testing, the transducer is mounted on a solid wedge (plastic) in which the velocity of sound is lower than that in the testpiece. The angle of incidence is designed to produce the desired refraction angle for the shear wave in the testpiece. When immersion coupling is used, the inspector can exercise greater freedom in choosing the incidence angle and therefore, the refraction angle within the testpiece.

In immersion testing, the distance between the transducer and the surface of the testpiece is chosen to avoid confusing the surface echo with reflections from within the testpiece. At normal incidence, the pulse will bounce back and forth between the transducer and the surface of the testpiece. Following the first return, time is allowed to receive all necessary information from the testpiece before the second reflection returns to the transducer. The spacing may be determined by the relative velocities in the two media. For example, ultrasonic pulses travel nearly four times as fast in steel as they do in water. If the ratio were 4:1, the recommended separation between the transducer and test material in a water bath would be at least 1 in. for every 4 in. the pulse travels in one direction in the test material.

V.B. RESONANCE METHODS

Generators designed for resonance testing consist of a CW oscillator, an amplifier, and a means, either manual or automatic, of modulating the frequency. Frequency modulation is obtained by varying either the inductance or the capacitance in the oscillator tank circuit.

Inductance modulation is obtained by an electrically variable saturable-reactor type of inductor in the tank circuit of the oscillator. The capacitance in the circuit is small. The upper limit on the frequency range usually is in the order of 30 MHz, which corresponds to the fundamental resonance of a steel or aluminum plate 0.004 in. thick.

Capacitance modulation is obtained either by means of a motor-driven variable condenser or by a manually controlled condenser.

When the oscillator frequency sweeps through a point of resonance in the material being inspected (i.e., the fundamental frequency or its harmonics), the amplitude of vibration of the material, under excitation by the transducer, increases characteristically. The energy dissipation increases and this is accompanied by an increase in plate current of the oscillator.

The information display is synchronized with the sweep of the frequency range. If an oscilloscope is used, the horizontal sweep is triggered to coincide with the start of the frequency sweep (starting at the high-frequency end of the range). The sweep rate is correlated with the frequency range. The vertical input is proportional to the plate current of the oscillator. Thus, resonance peaks occur on the oscilloscope screen at locations corresponding to resonant frequencies of the test material. Calibrated scales placed on the screen indicate the thickness directly in case of the fundamental frequency, or the thickness may be determined by means of Eq. (7.3) in case harmonics are used.

Correlation between frequency and indicator is assured in a stroboscopic type of thickness tester in which a disk with a neon lamp mounted on its outer edge rotates with the capacitor in the tank circuit of the oscillator. The increased plate current at resonance causes the neon lamp to flash, giving the appearance of a stationary light at positions corresponding to the fundamental frequency (or actual thickness) or the harmonics. In the case of harmonic indications, the thickness is determined by Eq. (7.3).

In some small, portable instruments which are battery operated, frequency is controlled manually. A plate-current meter indicates resonance. Thickness readings are obtained from the position of a pointer over a calibrated dial at resonance positions. Headphones or loudspeakers are often used, and resonance is indicated by a rise in the sound level in the devices. This type of unit is giving way to more modern equipment.

V.B.1. Transducers

Resonance measurements are dependent upon the change in impedance reflected into the oscillator circuit when the test material resonates. For this reason, the transducer must have a flat response over the frequency range of the test in order to avoid false indications. This is accomplished by using a transducer which resonates fundamentally at a frequency above the tuning range (usually 10 to 20 per cent higher). Piezoelectric elements which resonate at a fundamental frequency of 30 MHz are from 0.002 to 0.003 in. thick. The fragility of such thin elements is a limiting factor on the upper frequency that can be used for resonance testing.

The transducer consists of an undamped piezoelectric element. It may be flat for use on flat plates or contoured to match specific geometries.

Coupling between the transducer and the testpiece should be neither too tight nor too loose.

V.B.2. Data Recording

Automatic recording of thickness on a go-no-go basis is accomplished by gating circuits. The location and width of the gate are set to coincide with the thickness and allowable variations in thickness of the material being inspected. When indications occur outside the gate limits, the recorder receives no signal and, therefore, shows a zero reading. A number of gates can be used, each one supplying a signal to a separate recorder channel.

Actual thickness variations are recorded by means of an electronic circuit which produces a voltage which is proportional to the distance that the indication appears from the start of the gate. A similar technique can be used to produce a meter reading of thickness.

V.C. PULSE METHODS

The versatility of pulse methods of nondestructive testing is demonstrated by a variety of instruments available for generating ultrasonic pulses and recording test information. A basic system consists of a pulse generator, a pulse amplifier, a transducer for generating ultrasonic waves, a transducer for receiving the ultrasonic energy (which may or may not be the same one used for the transmitter), an amplifier for amplifying the received signal, and a means of displaying the test information.

Broad-band operation is desirable for most pulse-type applications. The pulse generator and amplifier must provide a "sharp" electrical pulse to excite the transmitter. The transmitting transducer and amplifier should have similar broad-band characteristics to avoid distorting the received signal.

The initial electrical impulse is produced by firing a pulse-generator device, such as a Thyraton tube, at the desired pulse repetition rate (500 to 1000 Hz is typical).

V.C.1. Transducers

In contrast to transducers used for resonance measurements which use undamped, high-Q, piezoelectric elements, pulse-type transducers are low-Q transducers. Resolution improves as the Q decreases.

Low-Q, broad-band characteristics are obtained by loading the nonradiating surface with a lossy material. The damping material preferably has the same characteristic impedance as the piezo-electric element. Sometimes the backing material is shaped to enhance its ability to absorb energy from the piezoelectric element by reflecting the energy in directions that increase its path length.

Commonly used piezoelectric materials are barium titanate, lead-zironate-titanate, lithium sulfate, and quartz.

Transducers for pulse-type measurements in nondestructive testing may be classified according to their methods of coupling to the testpiece--immersion, direct contact, gap, or coupling member (tire or sliding boot).

Contact types of transducers are available for producing the various types of waves used in nondestructive testing, i.e., longitudinal waves, shear waves, surface waves (Rayleigh and Love), and Lamb waves. Longitudinal waves are produced by a transducer element vibrating in its thickness mode and in a direction normal to the surface of the testpiece. Other vibrational modes are produced by mode conversion. To generate modes other than longi-tudinal, a piezoelectric element vibrating in the thickness mode is bonded to a surface of a wedge of material in which the velocity of sound is low compared with that in the test material. The surface to which the element is bonded is inclined at such an angle to the surface of the testpiece that the angle of incidence exceeds the critical angle of all modes which travel in the test material at velocities higher than the velocity of the desired mode.

Immersion types of transducers for nondestructive testing are sealed liquidtight. The beam may be plane or focused. Focusing is done through a lens attached to the face of the piezoelectric element. The transducers are mounted on immersion tubes, the mounts sometimes containing means for adjusting the incidence angle of the ultrasonic beam.

The immersion tubes are attached to manipulators located on a traversing mechanism which in turn is mounted on a bridge. The bridge travels the full length of an immersion tank and the traversing mechanism is designed to scan the width of the tank. The manipulator also contains angle adjustments to facilitate obtaining the proper incidence angle and inspection coverage.

Sometimes collimators are attached to immersion-type transducers (1) to improve the sensitivity to and location of defects, (2) to decrease the effects of surface roughness that tend to increase the length of the pulse reflected from the interface and thus reduce the resolving power, and (3) to produce a beam of limited size or controlled cross-sectional dimensions to prevent generating several modes in materials with irregular surfaces. The cross section of the beam may be round or rectangular. The walls of collimators are designed to absorb and reflect energy that does not pass through the opening so that it does not return to interfere with the received signal.

Two or more transducer elements may be located in one head. One example is a dual-element transducer used for thickness measurements. Two identical transducers are mounted on separate acoustical delay lines separated by an absorbent layer to avoid cross-talk. The transducers operate in a "pitch-catch" fashion, one being a transmitter and the other a receiver. The delay lines prevent echoes from returning to the receiver before the initial pulse has damped out, thus enhancing resolution. Thicknesses of less than 0.050 in. can be measured by this method; however, careful consideration must be given to the path that the beam follows in thin specimens to avoid errors due to multiple reflections within the testpiece.

V.C.2. Data Recording

Data from nondestructive tests using pulse methods may be presented in many ways. A few of the basic techniques will be discussed here.

The most familiar data presentation is the A-scan, shown in Fig. 7.4, which displays the various significant signals on an oscilloscope screen. The sweep is triggered by the impulse that energizes the transmitting transducer. Thus, the two events are synchronized. The sweep may be delayed and the sweep rate controlled to enable thorough examination of a desired region. This latter feature is especially useful in immersion testing. The settings may be adjusted so that front-surface reflection is indicated at the left of the oscilloscope screen, a back-surface reflection occurs at the right side of the oscilloscope screen, and reflections from discontinuities within the material occur between the front- and back-surface reflections.

B-scan presentations are pictorial displays of the cross section of a testpiece. They are more commonly used in medical diagnosis than in commercial nondestructive testing applications. The intensity of the echo signal is indicated by the brightness of the spot on the oscilloscope. Usually, the sweep connections on the oscilloscope are made to the vertical axis of the cathode-ray tube, and position signals, received from a power supply and whose amplitude is determined by a potentiometer, control the horizontal axis. The image of the test cross section may be retained on a long-persistence screen or directly on a photographic film exposed to the screen of an ordinary cathode-ray tube with provisions for Z-axis control. The camera lens is left open through the scanning period. A permanent photographic record of the cross section may be obtained by either method.

Pulse-type nondestructive-testing instruments may be equipped with gating circuits and recorder amplifiers which permit the retention of a test record on strip charts or on a plan-view type of recording. This type of display is especially useful in high-speed automatic scanning. Digital systems, such as counters that sum the amount of time the received signal is either above or below a reject level, and warning devices also may be employed.

A plan-view record of an ultrasonic nondestructive test is called a C-scan. The presence, but not the depth, of discontinuities is indicated. If indications of discontinuities within certain depth limits are desired, the gate settings may be adjusted to these limits and the signals from the discontinuities obtained by pulse-echo means. Multiple gates would permit obtaining plan views at various depths.

C-scan information may be recorded on electrosensitive paper or heat-sensitive paper. Recorders for C-scan operation are available commercially, but special apparatus can be made in any shop to suit specific needs. The location of the recording stylus to correspond to the position of the transducer may be controlled by mechanical linkages or by servomechanisms.

Printing may correspond either to the presence of defects or to the absence of defects, whichever is preferred. The recorder amplifier receives its signal from the gating circuit.

The spectrum of an ultrasonic pulse often may provide valuable information regarding a test specimen. Techniques have been developed for analyzing pulses before and after passing through a specimen, and by special data-processing methods it is possible to determine specific qualities that are not readily recognized by signal amplitude alone.

V.D. VIDEO-TAPE RECORDING OF ULTRASONIC TEST INFORMATION

An interesting method of recording ultrasonic data has been

developed by Youshaw, Dyer, and Criscuolo (28) at the United States Naval Ordnance Laboratory. A video-tape recorder is used to record the A-scan from a commercial ultrasonic tester along with the operator's voice giving the location, transducer position, and interpretation of test data. A differential amplifier is used between the ultrasonic flaw detector and the video recorder to match the systems and avoid excessive voltage at the input of the video recorder. Electronics are also provided for applying a marker wave to the video recorder.

V.E. COMMONLY USED SPECIFICATIONS AND STANDARDS (29)

V.E.1. Standards for Ultrasonic Inspection

Standard reference blocks are included as accessories with commercially available ultrasonic equipment used in nondestructive testing. These blocks are used for standardizing the ultrasonic equipment and for a reference in evaluating indications from the part being inspected. The types of standards available with the equipment depend upon the use for which the equipment is designed. The standards are made from carefully selected material which must meet predetermined standards of ultrasonic attenuation, grain size, and heat-treat condition.

Types of standard reference blocks include area/amplitude blocks (see Fig. 7.5), distance/amplitude blocks (see Fig. 7.6), ASTM basic set of blocks which includes both area/amplitude and distance/amplitude blocks, shear-wave testing blocks (see Fig. 7.7), and thickness-testing reference blocks (Fig. 7.8) which are used for calibrating resonance and pitch-catch types of thickness-measuring equipment.

The commercially available standard reference blocks fulfill many inspection needs. However, metallurgical and geometrical differences between the test block and the part to be inspected often require that special test blocks be fabricated from the same material as that to be inspected. Artificial defects which best simulate the types and sizes of defects to be detected are introduced into the test block.

Comparisons between the results of two different ultrasonic inspections carried out presumably under similar conditions may show differences. The reasons for these differences often may be traced to differences in the equipment rather than in the personnel operating the equipment. In cases where a base-line inspection is to be performed, it is good to file the characteristics of the inspection equipment with the test data. Such equipment data include make and model of the instrumentation, characteristics of the probe including directionality, sensitivity, and frequency response, and the coupling conditions. As an example, data for an immersion, pulse-type probe may be obtained in a small immersion-type, anechoic tank in which is located a rod for reflecting a signal. The probe is mounted to operate in a horizontal direction. The directionality of the probe is determined by recording the amplitude of a reflected pulse in each of several angular positions relative to the positions of the probe and reflector.

Additional information of importance includes an analysis of the frequency response of the probe and equipment. Often, the rated frequency of a commercial probe differs from the actual center frequency of the pulse. In order to compare test results with one another, it is essential that the equipment used in each test have identical characteristics.

Differences in transducers may be attributed to differences in piezoelectric characteristics, damping, and quality of faceplates. Equipment variations include bandwidth characteristics of the amplifiers and deterioration of components in the instrument.

V.E.2. Methods Used to Determine Flaw Size

Various methods are used for the determination of the actual flaw size during ultrasonic inspection. A few methods are presented here. Problems unique to specific applications of general interest will be discussed more fully in later sections. The factors that influence the nature of the indications include orientation, location, type of defect (void or inclusion), ratio of wavelength to dimensions of the defect, size of the defect compared with the dimensions of the ultrasonic beam, and geometry of the defect.

Defects may change the wave front of a beam on reflection and thus affect the amplitude of the defect indication. For example, a plane wave striking a large-plane reflecting surface will be reflected as a plane wave. However, a plane wave incident on a spherical surface is transformed into a spherical wave on reflection (or refraction). Diffraction also occurs, and if the beam is larger than the defect, the defect casts a shadow that may disappear within a distance which depends upon the wavelength and the dimensions of the defect.

It is possible to calculate theoretically the energy that might be reflected and the energy lost from an ultrasonic beam due to a reflector if the scattering cross section is known. The scattering cross section, therefore, may form the basis for a determination of defect size.

A laminar defect which lies in a known plane, which has a projected area in a plane that is normal to the direction of propagation, and which is large compared with the cross section of the beam provides a simple illustration. An example of this condition is large-area lack of bond between core and cladding in flat nuclear fuel elements. In this case, inspection by an immersion through-transmission method using a collimated beam can be used to map the extent of the flaw quite accurately. The accuracy depends upon how well the position at which the defect influences the signal amplitude can be determined.

The sizes of defects that are smaller than the cross section of the ultrasonic beam are more difficult to determine. Such defects may occur in an infinite number of configurations and sizes. They may be voids or they may contain inclusions which are capable of conducting energy away from the interface. An analysis of the energy reflected by defects of irregular shape can be simplified by the assumption that the defect approximates a simple geometry: cylinder, sphere, plane, or rectangular parallelpiped.

If the geometry of a void-type defect approximates a cylinder, the total power scattered by the cylinder may be calculated by (30):

$$P_s = \begin{cases} (6\pi^5 a^4/\lambda^3)I_o & (\lambda >> 2\pi a) \\ 2aI_o & (\lambda_1 \leq 2\pi a \leq \lambda_2) \\ 4aI_o & (\lambda << 2\pi a) \end{cases} \tag{7.10}$$

where I_o is the intensity of the wave incident on the defect (w/cm^2),

P_s is the total power scattered per unit length, (W) and

a is the radius of the cylinder (cm).

The limiting value of total scattered power for very short wavelengths is the power contained in a beam twice as wide as the cylinder.

For a spherical cavity in which $\lambda << 2\pi a$, an approximate value of the scattered power is

$$P_s = 2\pi a^2 I_o \tag{7.11}$$

where a is the radius of the sphere.

In terms of intensity, the ratio of transmitted average intensity (I_t) across a defect to that which would exist (I_i) were the defect not present and the corresponding voltage ratios (V_d/V_o) measured across the terminals of a receiving transducer in a through-transmission system are given by

$$\frac{I_t}{I_i} = \left(\frac{V_d}{V_o}\right)^2 = 1 - \left(\frac{D_1}{D_2}\right)^2 \tag{7.12}$$

for circular delaminations, or

$$\frac{I_t}{I_i} = \left(\frac{V_d}{V_o}\right)^2 = 1 - \frac{4\ A_1}{\pi D_2^{\ 2}} \tag{7.13}$$

for any geometry, where

D_1 is the diameter of the circular delamination,

D_2 is the diameter of the cross section of the beam, and

A_1 is the area of a delamination.

The corresponding equation for transmission past a cylindrical void for which the long-wavelength case applies is

$$\frac{I_t}{I_i} = \left(\frac{V_d}{V_o}\right)^2 = 1 - \frac{24\pi^4 a^4 D_2}{D_1^{\ 2}\lambda^3} \tag{7.14}$$

For the intermediate-wavelength case

$$\frac{I_t}{I_i} = \left(\frac{V_d}{V_o}\right)^2 = 1 - \frac{8a}{\pi D_2} \tag{7.15}$$

Theoretical and experimental intensity ratios and voltage ratios are given in Table 7.1, showing effects of defects in thin plates on receiver voltage in a through-transmission test in which collimators were used on both transducers. The agreement of experimental values with calculated values is typical of that found in several materials. If the flaw is an inclusion, the value of V_d/V_o may increase by an amount dependent upon the material of the inclusion and the manner in which it is coupled to the base material.

In pulse-echo inspection, sometimes a comparison of the echo indication from a defect with the amplitude of a back echo provides a more reliable indication of flaw size than does the defect echo alone. If the orientation is such that only a small amount of the energy reflected from the defect is returned to the receiver, the amplitude of the signal corresponding to the defect will be correspondingly small and give a false indication of the size of

TABLE 7.1

Effects of Defects in Thin Plates on Receiver Voltage

Type of defect	Size of defect, in.	Theoretical I_t/I_i [a]	Theoretical V_d/V_o [a]	Experimental V_d/V_o [a]	Remarks
Circular delamination	1/16 diam	0.75	0.866	0.80	
Circular delamination	3/32 diam	0.4375	0.662	0.65	
Circular delamination	1/16 diam	0.75	0.866	0.84	Defect opposite transmitter
Circular delamination	1/16 diam	0.75	0.866	0.82	Defect on transmitter side of plate
Rectangular void	1/16 x 1/8	0.364	0.603	0.60	
Long Cylindrical void[b]	0.010 diam	0.996	0.998	--	
Long Cylindrical void[b]	0.020 diam	0.934	0.966	--	
Long cylindrical void[b]	0.030 diam	0.668	0.818	--	Long-wavelength formula
Long Cylindrical void[b]	0.030 diam	0.694	0.832	--	Intermediate-wavelength formula
Long Cylindrical void[b]	0.040 diam	0.593	0.77	--	Intermediate-wavelength formula

TABLE 7.1 (con't)

Type of defect	Size of defect, in.	Theoretical I_t/I_i [a]	Theoretical V_d/V_o [a]	Experimental V_d/V_o [a]	Remarks
Short Cylindrical void	0.030 x 0.060	0.841	0.917	--	Long-wavelength formula
Short Cylindrical void	0.030 x 0.060	0.853	0.924	--	Intermediate-wavelength formula

[a]Transmitter collimator: 1/4 in.
I_t = transmitted intensity
I_i = incident intensity
Receiver collimator: 1/8 in.
V_d = receiver voltage measured as result of defect
V_o = receiver voltage measured as result of nondefective area of plate

[b]Cylinder longer than incident beam. Axis intersects axis of incident beam at right angles.

the defect. The amplitude of the back echo will decrease by an amount corresponding to the projected area of the defect.

Two characteristics of the ultrasonic beam are of importance in determining flaw size. These characteristics are (1) the effective beam form, i.e., the effective distribution of energy across the wavefront and (2) the effective bandwidth or the effective frequency content of the transmitted and received pulses.

In the far field, the energy distribution across the beam typically peaks at the center and drops off rapidly at the edges. Therefore, the maximum sensitivity coincides with location of the defect at the center of the beam. In the near field, the intensity varies according to near-field conditions.

The frequency effect relates to the relationship between the wavelength and the dimensions of the flaw. Equations (7.10) and (7.15) are examples of frequency effects. A defect in the path of a beam may be treated as a new source of ultrasonic energy for analytical purposes. Huygens' principle may be used to analyze the "radiation" pattern from the defect. Reflected energy corresponds to radiation of the same frequencies from a source of the size and location of the defect. Energy bypassing the defect corresponds to energy radiated from a transducer in the same location as the defect but with an area equal to the cross section of the beam at that location and the energy distribution in the plane corresponding to the exit side of the defect.

When ultrasound of short wavelengths impinges on a plane surface, energy is reflected in a plane wave. If the wavelength is large compared to the area of the surface, the surface simulates a point source and the reflected wave is spherical. If a plane wave impinges on a spherical surface, the reflected wave is spherical regardless of the wavelength; however, the amount of energy reflected is a function of the scattering cross section.

Since an ultrasonic pulse used in nondestructive testing

consists of a spectrum of frequencies, frequency analysis of pulse-echo indications from defects provides a possible means of determining flaw size. Calibration of the system is performed by using suitable standards.

In surface-wave inspection, the amplitude of waves reflected when the impinging beam is at normal incidence varies in proportion to the length of the defect if the depth is constant. For defects of constant width, the rate of increase of the amplitude of the indication decreases with depth to a limit which is approximately equal to the depth of the defect.

In shear-wave inspection by the angle beam method, the amplitude of echoes from cracks open to the surface is proportional to the depth of the cracks to depths of approximately two wavelengths (31). Interferences from reflections such as cylindrical waves from edges or corners prevent continued increase at a proportional rate as the cracks become deeper. However, depths of deep cracks may be determined geometrically. If the actual angle at which the beam is refracted within the testpiece is known, reflections corresponding to the top and bottom of the defect can be used to determine the approximate depth. The probe is moved to a position corresponding to the first indication of a defect (see Fig. 7.9). The transducer is then moved away from the defect area until the defect indication disappears. At this point, the transducer has moved a distance corresponding to the depth of the crack plus the effective beam width. The depth of the crack is

$$h = x \cos \theta \tag{7.16}$$

where x is the total distance that the transducer moved during the measurement less the effective beam width and θ is the refraction angle (32).

One method of determining flaw sizes and configurations combines B-scan and C-scan displays to present a three-dimensional view of the defect on an oscilloscope screen (33). A single transducer scans the testpiece in an immersion-test setup. The transducer oscillates so that a complete isometric view is obtained.

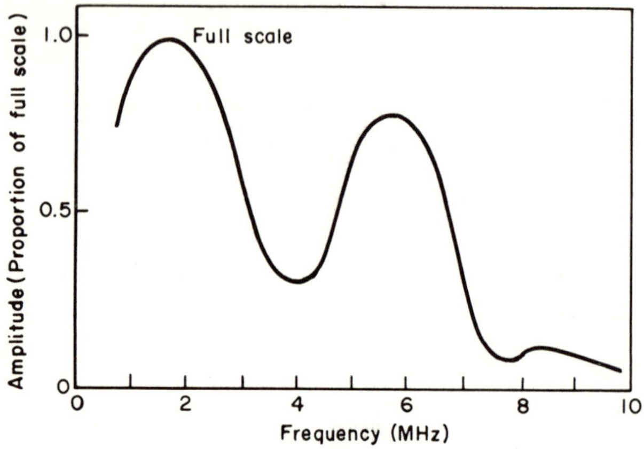

FIG. 7.1. Typical analysis of the spectrum of a nominal 5-MHz transducer excited by a direct current voltage spike.

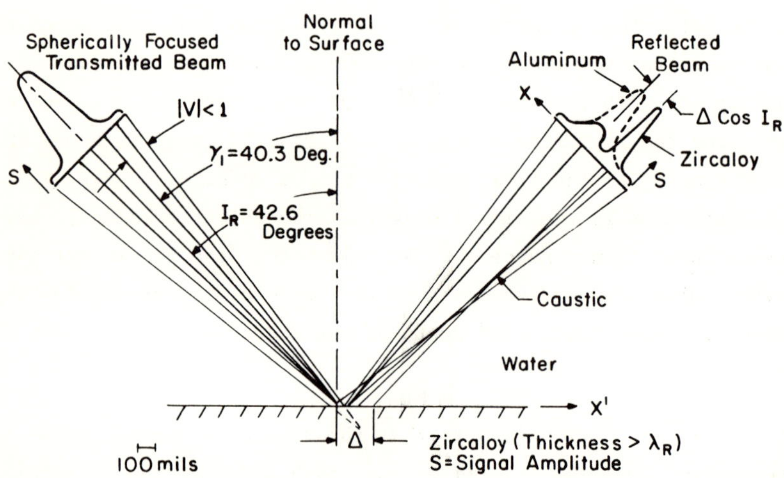

FIG. 7.2. Ray diagram illustrating the amplitude distribution and beam shift of a spherically focussed beam reflected from a liquid/solid interface at the Rayleigh critical angle.

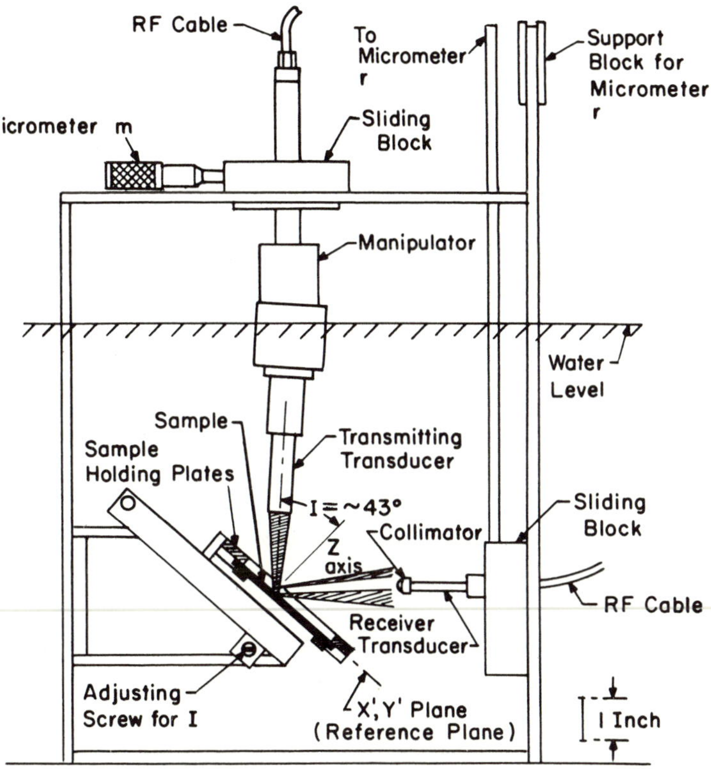

FIG. 7.3. Schematic of ultrasonic mechanical rig (Courtesy of Bettis Atomic Power Laboratory).

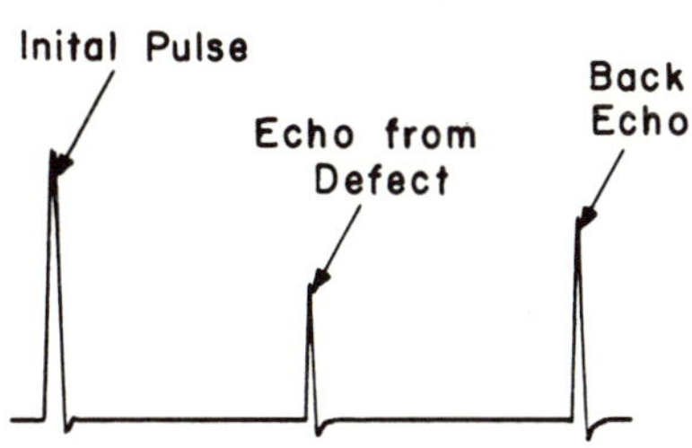

FIG. 7.4. A-scan presentation using pulse-echo method.

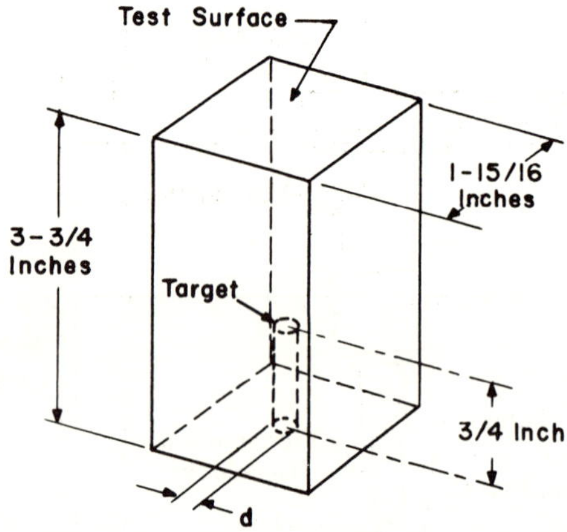

FIG. 7.5. Area/amplitude standard reference block.

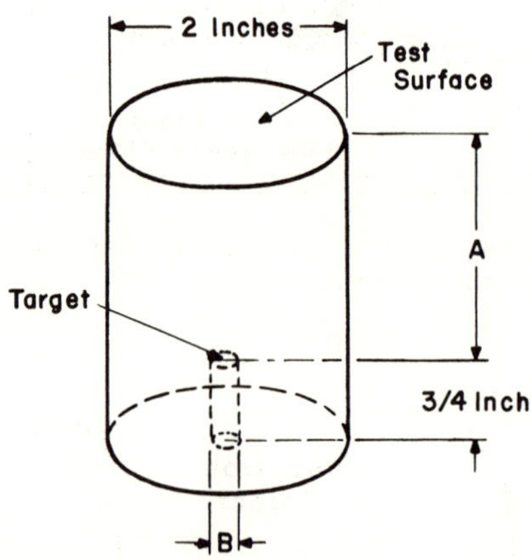

FIG. 7.6. Distance/amplitude standard reference block.

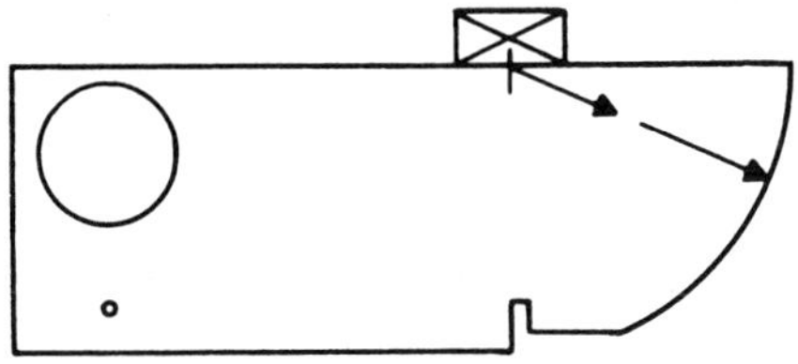

FIG. 7.7. Shear wave testing reference block.

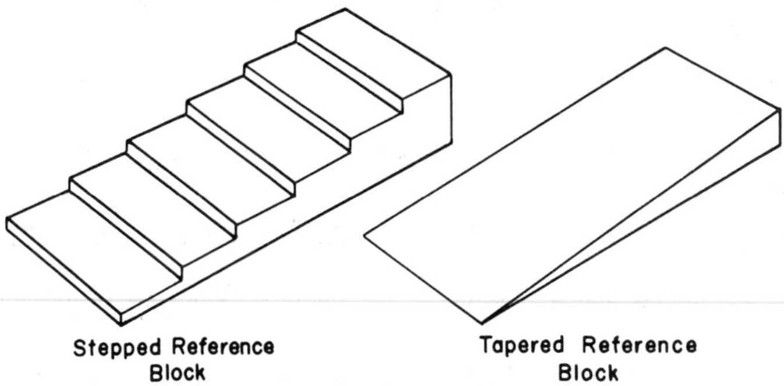

FIG. 7.8. Thickness testing reference blocks.

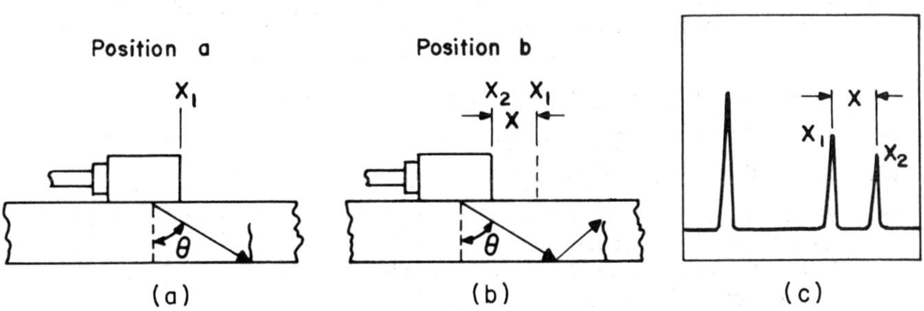

FIG. 7.9. Measuring crack depth by shear-wave technique.

REFERENCES

1. General Dynamics Convair Division, Nondestructive Testing--Ultrasonic Testing, Classroom Training Handbook, San Diego, California, 1967.
2. W. P. Mason and H. J. McSkimin, J. Acoust. Soc. Am., 19, 464-473 (1947).
3. W. P. Mason and H. J. McSkimin, J. Appl. Phys., 19, 940-950 (1948).
4. R. M. White, J. Acoust. Soc. Am., 30, 771-785 (1958).
5. N. G. Einspruch and R. Truell, J. Acoust. Soc. Am., 32, 214-220 (1960).
6. C. F. Ying and Rohn Truell, J. Acoust. Soc. Am., 32, 214-220 (1960).
7. C. F. Ying and Rohn Truell, J. Appl. Phys., 27, 1086-1097 (1956).
8. H. S. Hayre and D. E. Kaufman, J. Acoust. Soc. Am., 38, 599 (1965).
9. R. K. Moore and B. E. Parkins, J. Acoust. Soc. Am., 40, 170 (1966).
10. J. E. Burke and V. Twersky, J. Acoust. Soc. Am., 40, 883 (1966).
11. T. F. W. Embleton, J. Acoust. Soc. Am., 40, 667 (1966).
12. R. Botsco, Mater. Eval., 26(2), 21-26 (1968).
13. J. F. Moore, Mater. Eval., 25(2), 25-32 (1967).
14. I. R. Kraska and H. W. Kamm, Report No. AFML-TR-69-283, Air Force Materials Laboratory, Air Force Systems Command, Wright-Patterson Air Force Base, Ohio, August, 1970.
15. H. L. Godfrey, Combustion, pp. 19-20 (1965).
16. R. L. Martin, Instr. Control Systems, 43(12), 79-82 (1970).
17. J. R. Frederick, Mater. Eval., 28(2), 43-47 (1970).
18. B. H. Scofield, Technical Report No. AFML-TR-66-22, Air Force Materials Laboratory Research and Technology Division, Air Force Systems Command, Wright-Patterson Air Force Base, Ohio (November, 1966).
19. P. H. Hutton, Mater. Eval., 26(7), 125-129 (1968).
20. D. M. Romrell and L. R. Bunnell, Mater. Eval., 28(12), 267-270, 276 (1970).
21. R. Schroeer, R. Rowand, and H. Kamm, Mater. Eval., 28(11), 237-243 (1970).

22. O. R. Gericke, Technical Report No. AMRA-TR-66-38, U. S. Army Materials Research Agency, Watertown, Massachusetts, December, 1966, 10 pp.

23. O. R. Gericke, Ultrasonic Pulse-Echo Spectroscopy, Technical Report, Materials Testing Laboratory, Army Materials and Mechanics Research Center, Watertown, Massachusetts, June, 1968.

24. H. L. Whaley and K. V. Cook, Mater. Eval., 28(3), 61-66 (1970).

25. S. Serabian, Mater. Eval., 26(9), 173-179 (1968).

26. H. J. Willard, Report No. WAPD-TM-833, Bettis Atomic Power Laboratory, Pittsburgh, Pennsylvania, October, 1969. Available from Clearinghouse for Federal Scientific and Technical Information, National Bureau of Standards, U. S. Department of Commerce, Springfield, Virginia 22151.

27. D. O. Hunter, Ultrasonic Velocities and Critical-Angle-Method Changes in Irradiated A302-B and A542-B Steels, Battelle Memorial Institute, Pacific Northwest Laboratories, Richland, Washington, 99352, April, 1969.

28. R. A. Youshaw, C. H. Dyer, and E. L. Criscuolo, Final Report on Project 176 "Quality Assurance," to the Ship Structure Committee. U. S. Coast Guard Headquarters, Washington, D. C., October, 1968, 12 pp.

29. ASNT., Mater. Eval., 27(7), 17A-18A (1969).

30. P. M. Morse, Vibration and Sound, 2nd ed., McGraw-Hill, New York, 1948.

31. R. Frielinghaus, J. Krautkramer, and U. Schlengermann, Nondestructive Testing, 125-127 (1970).

32. D. Hagemaier, Proc. 1968 Symposium on the NDT of Welds and Materials Joining, Los Angeles, California, March 11-13, 1968.

33. L. J. Anderson, M. R. Wick, and G. J. Posakony, Technical Documentary Report No. ASD-TR-61-205, Part II, November, 1963.

Chapter 8

USE OF ULTRASONICS IN THE NONDESTRUCTIVE TESTING OF METALS

I. INTRODUCTION

Ultrasonic methods have been applied to nondestructive testing in numerous ways and are among the most important of the techniques used for inspecting metal products. Those applications presented in this chapter are representative only. For a more detailed approach to specific areas, the reader should consult the references at the end of this chapter. A book written by J. and H. Krautkramer (1) is devoted entirely to the applications of ultrasonics in nondestructive testing. The journal, Materials Evaluation, published by the American Society for Nondestructive Testing, is a good source of information on applications of ultrasonics in this field. Many codes include ultrasonic methods as being recommended or required for inspecting materials.

Ultrasonic inspection is safe, versatile, and economical.

Effective use of ultrasonic nondestructive testing requires skilled operators. It also requires an understanding of the theory of wave propagation and of the characteristics of available equipment on the part of the person responsible for prescribing the testing procedure. Many short courses are available to those interested in developing such an understanding and skill.

Presented in this chapter are common applications of ultrasonics in the inspection of metal products. The use of ultrasonics for evaluating nonmetallic products is discussed in Chap. 9.

II. INTERNAL STRUCTURE OF METALS

Measuring the propagation characteristics of ultrasonic energy is a means of determining internal (or metallurgical) structures. Evaluation may be based on attenuation characteristics of wave components at various frequencies, on the magnitude of echoes produced by internal structures, or on the velocity of sound in the material. A good reference standard is essential to a reliable interpretation of the data. The ultrasonic techniques have been used successfully in assessing the structure of heat-treated steel castings, the form and amount of graphite in iron castings, the tensile strength of materials, the hardening depth or quality of chilled iron rolls, and the grain size and orientation in uranium, brass, and other alloys.

In common methods of indicating average grain size, use is made of (1) decrease (attenuation) in amplitude of the first back-echo pulse, (2) decrease in amplitude of the received pulse in a through-transmission arrangement of transducers, or (3) rate of decrease of successive back echoes. The first two methods are most easily adapted to recording. Recording when using the third method requires at least two gates and a means of comparing the second back echo with the first. Standard, commercially available transducers are used for all three methods. They are useful in separating materials with grain sizes larger than an acceptable limit and at the same time respond to other factors that also affect the attenuation. As grain-size indicators, their main weakness lies in the fact that properties other than grain size contribute to attenuation. The bandwidth and center frequency of commercial transducers vary even between transducers supposedly operating at the same nominal frequencies. These variable characteristics of the transducers result in varying sensitivities to grain-size distribution. In recent years, a more accurate determination of grain size has been made possible by the development of methods of analyzing the frequency spectrum of ultrasonic pulses.

The relationship between attenuation (due to scattering at grain boundaries) and the ratio of wavelength to grain dimensions may vary among materials because of variations in anisotropy and elastic properties at interfaces, but generally, when the ratio between wavelength and grain size exceeds 30:1, little change in attenuation with change in wavelength occurs. When the ratio falls below 10:1, attenuation increases rapidly, reaching a maximum within the range 3:1 to 1:3.

Aisin (2) reports using the pulse-echo method to determine the macrostructure of 100 to 160-mm (4 to 6.4 in.)-diameter bars of ball-bearing steel.

Hafemeister (3) has used ultrasonic methods to detect grossly coarsened grain structure in finish-machined steel parts.

Dunegan (4) has developed the following equation for describing the attenuation of sound in uranium as a function of average grain diameter:

$$\alpha = Ae^{kD} \tag{8.1}$$

where α is the attenuation (db/μsec),

A is a frequency-dependent constant,

k is 1.17×10^3 db/μsec, and

D is the average grain diameter (in.).

Sharpe and Aveyard (5) demonstrate the facsimile recording of coarse grain areas in cast uranium bars using the ultrasonic pulse through-transmission technique. Ultrasonic transmission through coarse-grain regions is poor. Sections of the facsimile recording corresponding to these coarse-grain regions are shown as zero-intensity areas, thus producing a plot of the exact location of the coarse grains.

Nemkina, Radaev, and Zborovskii (6) evaluated the use of ultrasonic attenuation in determining the grain orientation of transformer steel. Using a pulse-echo method and a frequency of 5 MHz on samples 0.5 and 0.35 mm thick, they concluded that the position of the amplitude minimum of the reflected signal near the direction of rolling uniquely determines the orientation of the (001) axis and can be used to determine the deviations of the (001) orientation of the grains from the ideal orientation.

As early as 1941, Sokolov (7) reported a method of measuring the depth of hardening in steel by ultrasonic means. Yegorov (8,9) and Koppelman (10) also claim good accuracies using similar methods. These measurements are based upon the backscattering of ultrasonic pulses from grain boundaries using high-sensitivity detectors. Yegorov (8,9) uses a pitch-catch arrangement of transducers and determines the depth of hardening by the angle of incidence of the beam on the nontempered metal layer and the distance between transducers. Since the structure of the hardened layer differs from that of the nonhardened layer, reflections occur at the boundary between the two layers. Koppelman (10) uses a pulse-echo method to arrive at the statistical back scattering of ultrasonic waves (i.e., noise caused by scattering from the grain boundaries). By averaging over an area, he obtains sufficient accuracy for practical measurements. Measurements on hardened-steel rolls are done at frequencies ranging from 30 to 60 MHz.

Fenkner (11) has found a reproducible relationship between heat treatment and the velocity of sound in ball-bearing steel by which he is able to determine the amount of residual austenite in the steel. He measures the velocity of sound in the material by a pulse-echo method using an ultrasonic material tester with a calibrated transit-time base. The abrupt changes of elastic constants occurring with polymorphous transformations of solids form the basis of the measurements. Fenkner obtains the following formula for the retained austenite:

$$Y_g = RY_A + (1 - R)Y_M \tag{8.2}$$

where Y_g is the ultrasonically determined Young's modulus,

Y_A is the Young's modulus of pure austenite,

Y_M is the Young's modulus of pure martensite, and

R is the retained austenite.

Since Y_A cannot be measured because pure austenite is unstable at room temperature, Fenkner extrapolates its value from a retained-austenite value obtained by other methods. He claims an approximately linear relationship between retained austenite and longitudinal-wave velocity. The velocity decreases with increasing amounts of austenite, dropping from slightly over 5900 m/sec at 5% austenite to slightly under 5850 m/sec at 25% austenite.

Propagation characteristics of ultrasonic energy are influenced even by factors operating on a microscopic scale. Merkulov and others (12) claim to have shown good agreement in the determination of trace impurities in metals by ultrasonic means and by other means. The total number of impurity atoms determines the degree of absorption of ultrasound and the frequency at which maximum absorption occurs.

II.A. MATERIAL EVALUATION BASED ON VELOCITY AND ATTENUATION OF ULTRASOUND

In general, an increase in the attenuation or a decrease in the velocity of ultrasound in a material is an indication of possible degradation or loss of strength of the material. Measurements of these properties can be used to detect (1) changes in the properties of materials that occur during such processes as nuclear irradiation, deformation, stress cycling and fatigue, creep, temperature fluctuations, pressure fluctuations; (2) changes in purity or changes in state (such as aging, hardening, formation of precipitates; or (3) phase changes. These conditions influence the propagation of all modes. The choice of method depends upon which one is most affected by the condition to be monitored and the ability to implement the technique. When a study involves both ultrasonic attenuation and velocity measurements, these should be performed concurrently.

The quality of certain geometrical structures can be ascertained by resonating the parts. When cracks or other defects form in a vibrating part, the resonant frequency drops and frictional losses at the new mating surfaces cause an increase in the damping factor. Resonance methods have been used to inspect grinding wheels, crankshafts, honeycomb panels, laminated materials, and corrugated materials. Types of defects located by this method include lack of bond, cracks, crushed cores, and voids.

Attenuation measurements have been used to determine average grain size in metals. Grain size is a determining factor in the size of a defect that can be detected ultrasonically.

Measurement of attenuation also is useful for basic studies of the properties of materials. Bratina et al. (13) explored the use of the attenuation of ultrasonic energy in the megahertz range associated with vibration of free dislocation segments to study mechanical characteristics of metals over a wide temperature range (-150 to +150°C). Attenuation associated with vibration of free dislocation segments is called dislocation damping. The studies of these investigators included the strain aging and ductile-brittle transition in iron and carbon steels. A measure of the depinning of dislocations from interstitial atoms was indicated by the amount of dislocation damping caused when an external load was applied to the specimen.

Tietz (14) used ultrasonic loss measurements at 4 MHz to determine the yield point of materials during elastic-plastic deformation. Both shear and longitudinal waves propagating in a direction normal to the applied tensile stress were used during these studies. An abrupt increase in the value of the loss coefficient for structural steel occurs at the yield point. Tietz attributes the increased attenuation to formation and action of dislocations which are formed in certain grains at the upper yield point, are torn from the cloud of foreign atoms by the critical stress, and pile up at grain boundaries to form stress fields in which the separation of the dislocations in adjacent grains occurs.

II.B. HARDNESS MEASUREMENTS

Equipment is available commercially for measuring the surface hardness of a wide range of materials (15-18). The hardness can be registered directly on a dial or the process can be automated in a go-no-go form on a production line. It is claimed that the equipment can be calibrated in Rockwell units in the range of 20 to 70 with an accuracy of ± 1 unit. This technique produces an instantaneous electronic output that is a function of the contact area of the indenter. Commercial equipment is available in a portable version and in industrial versions offering variable loads. A laboratory microhardness tester also is available.

The ultrasonic hardness tester uses a vickers indenter with a load of either 150 or 675 g. For best results, the surface finish should not exceed 50 μ rms, but by averaging, surfaces of 125 μ rms can be measured. The depth of penetration is only a few microns and, although primarily designed for steel, the instrument can be calibrated for most other materials. The advantages of the ultrasonic test are speed, ease of automation, accuracy, and nondestructiveness. In addition to the Vickers diamond indenter, the equipment used in the ultrasonic test includes a piezoelectric pickup, an exciter coil, and a magnetostrictive rod.

To measure hardness, a vibrating rod is applied lightly to the surface. The rod is mounted on a heavy mass at one end. The other end is free and supports the Vickers diamond indenter. Under no-load, the resonant rod is a quarter-wavelength long. As the load is applied and the area of the indentation on the indenter increases, the impedance of the material being tested causes the point of contact to move toward a nodal condition, thus shortening the wavelength. The resonant frequency is forced upward by an amount which depends inversely upon the hardness of the material. The accuracy of the measurement is dependent upon having both the mass to which the rod is permanently clamped and that of the load to appear as infinite masses.

II.C. EVALUATION OF SINTERED PRODUCTS

The nondestructive testing of powdered metals is in many respects similar to nondestructive testing of concrete. The velocity of sound in a material is a function of the elastic moduli, the density, and Poisson's ratio. In sintered materials the elastic moduli are related to the density of compaction of the materials. The moduli increase at a greater rate than does the density of the material and, therefore, the velocity of sound increases with increased density. Measurements have been made to correlate the velocity of sound in sintered products with density.

Brockelman (19,20) claims a linear relationship between the tensile strength of powder-iron compacts and both the resonances of specimens and the velocity of sound in them. The strength of critical regions in complex shapes can be evaluated by using a narrow ultrasonic beam. The velocity of sound provides a better nondestructive test of tensile strength than of the sintered density. Both the velocity of sound and the tensile strength are influenced in the same way by pore size, shape, and distribution.

Abbe and Korytoski (21) use an immersion through-transmission technique for evaluating powder-metal materials. Information is stored on a magnetic tape recorder and the data are compared with data taken from a standard test specimen. A technique similar to this has been used to test brake friction disks in which the material is evaluated on the basis of attenuation of a signal through the disk. Poor bonding between individual particles within the powdered-metal friction layers increases the scattering from such particles and thus contributes to attenuation. Techniques have been developed for semiautomatic inspection on production lines.

II.D. ELASTIC AND ANELASTIC ASYMMETRY OF METALS AND THE ACOUSTIC MEASUREMENT OF RESIDUAL STRESS

Both elastic and anelastic characteristics of materials are nonlinear at relatively low stress amplitudes (22). Since these are conditions which affect the propagation of ultrasonic waves, it is natural to look to ultrasonics as a means of indicating stress levels in materials--particularly the residual stress.

The velocity of a transverse wave propagating in a direction which is perpendicular to the direction of stress depends upon the direction of its polarization relative to that of the stress. The velocity of ultrasonic waves polarized perpendicular to the stress direction differs from the velocity of those polarized parallel to the stress direction by a quantity that increases linearly with the stress (23).

Several investigators have studied the effects of stress-induced anisotropy on the propagation of ultrasonic waves. Most of the proposed techniques are in the development stage and have been used primarily in laboratories. Change in velocity with change in stress in metals is small and many other factors also affect the velocity by amounts that may obscure stress-induced changes. Some of these factors are grain orientation, variations in mechanical treatment, and varying concentrations of alloying elements. In the laboratory, these factors can be identified, but in the field, identification may not be possible.

Conventional pulse-echo methods of indicating velocity changes in metals due to stress usually are not practical. Various investigators have tried different techniques of indicating velocity changes due to stress. Benson (24) and others have investigated the birefringent effect of stress on polarized shear waves. By orienting the wave at 45° to the stress orientation, the ultrasonic wave is broken down into two components. One component is polarized parallel to the direction of the stress and the other is polarized normal to the direction of stress. Since these two components propagate at different velocities, the wave undergoes a change in polarization as it propagates through the medium. Benson used two identical shear-wave transducers aligned at 45° to the direction of stress on opposite and parallel sides of the specimen. The amplitude of the first received signal was plotted as a function of applied stress. As the stress increased, the phase relationships between the two wave components shifted accordingly. As a result, the received-signal amplitude went through a series of maxima and minima, maxima occurring when the velocity difference produced a phase difference of $2n\pi$, where n is an integer (including 0 when the medium is isotropic and the components are in phase).

An objection to this method of measuring residual stress is that it gives an integrated average and, except in structures of certain geometries, the algebraic sum of the residual stresses through a test section is likely to be zero. Thus, on ideal rectangular specimens, the polarized-beam method of Benson could measure applied stress; however, its use in measuring local residual stresses is limited.

Benson and others have investigated the effects of stresses on the velocity of surface waves to detect residual stresses in the surfaces of materials. Benson (25) used a phase-comparator method.

McGonagle and Yun (26) refer to the "sing-around" method of measuring surface-wave velocity as a possible means of measuring surface residual stresses. (In the sing-around method, each received pulse is used to trigger the succeeding ultrasonic pulse. Velocity is determined as a function of pulse repetition rate.)

Rollins et al. (27) attempted to develop a technique of measuring residual stresses at any location within a material by analyzing the interaction of two intersecting plane waves. Such an interaction results, in theory, in a scattered wave with a frequency corresponding either to the sum or to the difference between the frequencies of the intersecting waves. Although Rollins apparently was successful in generating such waves, the principle has not been successfully applied to measurement of local stresses.

II.E. FATIGUE IN METALS

Ultrasonics has been used to study and detect fatigue in metals. High-intensity ultrasonic waves often induce fatigue cracks in materials, and this phenomenon has sometimes been applied to accelerate fatigue tests.

Nondestructive testing methods of monitoring or detecting the presence of fatigue in metals are based on either the attenuation of the amplitude of the waves or the reflection of energy from the crack as a discontinuity in the material. The methods have been applied successfully to follow the growth of cracks on specimens in fatigue-testing machines. In other studies attempts have been made to detect the initiation of cracks and to predict the remaining life of parts.

Lamb waves and surface waves are especially sensitive to the development of fatigue cracks. The effect on surface waves can be visualized by the fact that the effective depth of penetration of such waves is approximately 1 wavelength. For example, the velocity of a Rayleigh wave in steel is approximately 0.11×10^6 in./sec. The effective depth of penetration of a surface wave at a frequency of 5 MHz in steel is 0.022 in. The wave is influenced most by defects nearest the surface. A 5-MHz Rayleigh wave in steel can be used to detect fatigue cracks less than 0.002 in. in depth.

According to Rasmussen (28), the first indication of a change in the properties of the material under dynamic stress is given by the formation of slip lines which may be observed using electron microscopy. These slip lines may form as early as at 1% of the fatigue life of the part. Fatigue fracture begins as microcracks forming along active slip planes at an early stage. In some cases, extrusions, intrusions, or pits are formed in the surface at an early stage. At about 60 to 70% of the fatigue life, the microcracks start joining to form larger cracks. Some of these growing cracks cause the final failure. Fatigue life of a part depends upon the manner in which the cracks form under fatigue-stress conditions. Prediction of fatigue life, then, is based upon locating points on a characteristic curve corresponding to the developing cracks under known conditions of cyclic stress. Rasmussen claims to have detected fatigue damage in the form of microcracks and other discontinuities mentioned previously in 2024ST3 aluminum alloy plate which was electropolished at 39-44% of the fatigue life during dynamic loading. The primary variable which influences how early detection can occur is the surface roughness of the specimen. Rasmussen used surface waves and claims very reproducible results, and claims further to be able to extrapolate from the given indications when total fatigue failure would occur. The method was based upon measuring the attenuation of surface waves due to the formation of the microcracks.

Critical-angle tests have been used to detect fatigue in metals (29). These tests involve the use of surface waves generated when the transmitted wave is incident on a test specimen at or near the Rayleigh critical angle. The results were similar to those of Rasmussen.

Klima and Freche (30) have applied a standard longitudinal-mode pulse-echo technique to detect and measure the growth of fatigue cracks in notched cylindrical specimens of 2014-T6 aluminum, 5Al-2.5Sn-titanium, a maraging steel, and a cobalt-base alloy, L-605.

These investigators were able to detect cracks, which generally extended around the specimen circumference, by the time they reached a depth of 0.0005-0.004 in. Initially, detectable cracks were formed within approximately 10-40% of life-to-fracture, depending on material and cyclic stress.

In a similar method using 10-MHz shear waves, Clark and Ceschini (31) move the transducer along the surface of a specimen relative to a crack tip so that the crack length can be interpreted in terms of transducer location.

Norris et al. (32) combined surface-wave and longitudinal-mode techniques to study fatigue-crack initiation and growth in low-cycle fatigue testing. The surface waves were used to determine crack initiation and longitudinal waves for monitoring crack propagation.

The attenuation of through-transmission ultrasonic energy has been applied by Cole (33) to determine the location and criticality of fatigue-induced damage in filament-wound fiber-glass cylinders. The application is the determination of residual life remaining in composite, deep submergence, multidive vehicles.

The adoption of an ultrasonic monitoring technique to estimate fatigue-limited serviceability of aircraft structures has been proposed by Zirinsky (34). The proposed technique would involve the use of a single instrument to monitor several permanently-mounted shear-wave transducers which can be manually operated or used with a recorder. Although the method proposed is based on sound principles under ideal conditions, there appears to be no extensive use of the technique in the field.

III. INSPECTION OF BASIC STRUCTURES AND PRODUCTS

Many structures are readily adaptable to ultrasonic inspection. Shafts and bar stock of simple geometries with flat, parallel ends may be easily inspected using longitudinal modes. If the material has low attenuation (for example aluminum or low-alloy steel with small grain size), fairly accurate and rapid inspection can be performed. Wall effects, which include mode conversion as the beam spreads and impinges on the walls, will cause trailing signals of noise in long, relatively slender bars. However, it usually is po possible to isolate the back-echo and defect indications from these noise signals. Bar-type structures typically inspected by ultrasonics include ordinary steel round stock, engine crankshafts, railroad axles, large bolts, steel billets used in the fabrication of landing-gear pistons (AISI 4340 steel), cold-extruded transmission shafts (SAE 4027 steel) and axle shafts (SAE 1052 steel) with hot-forged flanges, generator rotors and armature shafts, rolls of cast and forged Cr-Mo steel alloys used in rolling mills (inspected for macrodefects), reconditioned or built-up drill bits, and steel trunnions for steel casting ladles.

Pulse-echo longitudinal- and shear-wave techniques have been used to detect defects in high-pressure vessels, high-pressure gas cylinders, riveted drums, and many other similar structures.

Surface-wave inspection has been useful in locating defects such as fatigue cracks in helicopter and propeller blades, turbine and compressor blades, and turbine and compressor disks. The advantage of surface waves over other types in such applications is that they follow the contour of the pieces without loss due to direction changes.

III.A. WELDS

Ultrasonics is one of the most widely used techniques for the nondestructive inspection of welds, and consequently much has been written on the subject. The most commonly used method is angle beam testing in which shear waves are transmitted through the part as shown in Fig. 8.1. The shear wave is generated by mode conversion from longitudinal waves incident on the surface of the test material. The longitudinal wave is eliminated from the test material by using an angle of incidence which exceeds the critical angle for total reflection.

Methods of inspecting T-sections using longitudinal waves are illustrated in Fig. 8.2.

Any of the three common methods of coupling energy to the part are used in weld inspection: (1) direct contact, with the transducer mounted on a plastic wedge; (2) immersion; and (3) gap, using a liquid stream, liquid-filled wheel, or liquid-filled boot.

The proper design of any ultrasonic inspection method takes into account the possible generation of unwanted wave modes which cause confusion by the fact that they propagate at different velocities. Figure 8.3 illustrates what can occur when using the same ultrasonic beam dimensions to inspect welds in tubing of two different sizes. In the large-diameter tubing, the variation in the angle of incidence across the beam is small, and the shear wave that is generated is close to being similar to that which would occur if the tube were a flat plate. In the small-diameter tubing, the incidence angle varies from front to back by a large amount. As a result, longitudinal waves, shear waves, and surface waves, and sometimes Lamb waves may travel in the material at the same time and in both directions around the circumference. A thinner beam in the circumferential direction is necessary to eliminate this problem.

Typical weld defects include cracks from various causes (crater, hot cracking, cold cracking, etc.), incomplete penetration, pickup, blowholes, porosity, slag inclusions, lack of fusion, undercutting, and pipes.

Crater cracks are formed while a weld cools. The crater solidifies from the outside toward the center, a condition which favors the production of shrinkage cracks.

Hot cracking refers to intergranular cracks which form as a weld metal cools and solidifies. When these cracks are opened at the surface of the metal, oxidation occurs on the walls of the crack because of the temperature of the metal and exposure to the atmosphere.

Cold cracking refers to cracks that form after freezing has been completed. These cracks usually cut across grains. There are many causes of this type of defect--all of which can be related to high stress concentrations and a brittle characteristic of the material at the site of the fault.

Penetration is a term used to refer to any of several conditions. In fillet welds it may refer to fusing completely through a layer of mill scale on the surface of the welded parts. It may also refer to the depth below the surface of the base metal to which the fused zone penetrates. In a butt joint, it may refer to the depth of the joint to which the weld metal is able to flow.

Pickup refers to the intermixing of a filler rod and base metal where these materials are of different composition.

Blowholes are rounded gas bubbles with smooth, bright surfaces. They are caused by gases either being released from solution in the base metal or being produced by fluxes which dissociate at high temperatures. The gas bubbles are prevented from rising to the surface by the slowing effect of the viscosity of the melt and by the cooling rate of the weld metal.

Porosity is related to the formation of blowholes. Gas bubbles liberated during crystallization of the weld metal are entrapped by the growing crystals which results in clusters of pores.

Slag inclusions are nonmetallic materials other than graphite which are imbedded in a metal. They may originate in slag or flux on the surface of the metal or they may originate in the metal as products of a deoxidation reaction.

Lack of fusion refers to a lack of coalescence between filler metal and base metal, or of base metal only, during welding.

Undercutting is illustrated in Fig. 8.4. This refers to a groove melted into the base metal adjacent to the toe of a weld which is left unfilled by weld metal.

Pipe may occur in fillet welds as a result of solidification starting at both sides of the weld. As freezing progresses, the solid, being more dense than the liquid, occupies less space than it did when it was liquid and the level of the remaining liquid falls. Thus, when solidification is complete, there remains a deep cavity, or pipe.

Sometimes the type of the defect can be ascertained by a knowledge of the materials being inspected and the nature of the ultrasonic indication. More often, the identification of defect types requires a knowledge of the location of the defects within the weld since certain types may occur only in certain parts of the weld.

One may understand from the foregoing the importance of using good test equipment, good standards for calibrating the equipment and assessing the nature of defects, and adequate training of the inspector.

A complete discussion of weld inspection by ultrasonic means would be a nebulous task. Those interested in more detailed descriptions of procedures and problems are directed to some excellent literature on the subject.

Ostrofsky (35) discusses the control of beam spread, calibration procedures, weld-inspection methods, and characterization of defects and difficulties in interpretation. He also includes a table of helpful information for identifying weld defects from ultrasonic indications.

An excellent treatise on the fundamentals of weld inspection has been presented by Hagemaier (36). He discusses various aspects of instrument calibration and transducer characteristics as well as factors involved in the test setup, identification of defect location, and interpretation of test results.

Others, Harris-Maddox (37) for example, provide useful discussion and tables on the identification of weld defects.

The training handbooks issued by General Dynamics (38) are of considerable value to those interested in developing skill in ultrasonic inspection of welds.

Discussions of methods of determining the dimensions and locations of flaws in welds can be found in Refs. 36-45, listed at the end of this chapter.

Weld beads and certain metallurgical characteristics of some types of weld metals may cause such severe scattering of ultrasonic energy that effective inspection for internal weld-metal defects is precluded. Grinding weld beads flush with the base metal can eliminate scattering from this source, but scattering from internal structures, such as large grain boundaries, makes other inspection methods necessary.

Specific areas of application of ultrasonic inspection of welds are pipe and tubing welds, welds in rocket-engine hardware (aluminum, steel, and stainless alloys), structural-steel welds, ship and submarine structures, pressure vessels, and end-cap welds of nuclear containment capsules. Hagemaier and Posakony (46) describe the use of a 25-MHz transducer and high-resolution equipment to inspect welded couplers in 1/4- to 1-in.-OD hydraulic lines. They found ultrasonics to be the only feasible method of inspecting melt-through fusion welds joining the sleeve or adapter to the tube.

Center and Roches (47) propose combining ultrasonics and radiography for inspecting heavy-walled pressure-vessel weldments, an area in which, until recently, there has been a reluctance to use ultrasonics. Their tests show that ultrasonics could make a valuable contribution, particularly to the detection of planar-type defects such as cracks, lack of fusion, and lack of penetration. Regarding lack of fusion in any weld, it is possible to have such defects under so much compression that little energy is reflected from the defect and it goes undetected by ultrasonics.

Lamb waves are often used for the inspection of welds in thin-walled tubing (48). Stainless steel or other thin-walled tubing 10-70 mm in diameter can be inspected and recorded at the rate of 3 m/min.

Automatic and semiautomatic systems are being used for ultrasonic nondestructive testing of welds (49-52). The problem in the development of such systems is to incorporate or simulate the versatility of manual operation. Certain types of welds in which the defects are oriented along a known direction or plane are easily adapted to automated ultrasonic tests.

Methods of inspecting spot welds include measuring the transmission of an ultrasonic beam through the weld as it is being formed (53), or observing differences in the pattern of multiple echoes from different welds (54,55). To facilitate inspection of the weld as it is being formed, ultrasonic transducers are built into the electrodes.

The orientation of a defect with respect to the direction of propagation of the ultrasonic beam determines whether the reflected pulse will return to the receiver. Cross and his co-workers (56) have developed a method of detecting weld defects regardless of orientation. This method, called the Delta method, is illustrated in Fig. 8.5. The transmitter produces an ultrasonic wave at an angle to the surface of the base metal. The energy propagates into the weld metal and strikes a discontinuity. If the defect has a surface oriented so that energy is reflected upward, the receiver, which is located directly over the weld, receives the reflected pulse. If the defect is oriented normal to the surface, a surface wave is generated by mode conversion on the surface of the defect. The wave propagates downward, converts into a longitudinal wave at the edge of the defect, and reflects from the bottom. The receiver then receives "reradiated" energy from the defect. Another orientation may reflect energy toward the underside of the weld where it is reflected upward and into the receiver.

III.B. TUBES, PIPES, AND SHELLS

Ultrasonic inspection of tubes, pipes, and shells is closely related to and may include weld inspection, but it also includes the location of defects in any part of the base metal. As illustrated in the previous section (Fig. 8.3), the geometry of the test specimen enters into the inspection-system design. Not only is it important to control the relationship of the beam dimensions to the tube diameter but it is also important to maintain a fixed alignment between the transducers and the surface of the item being tested. A slight shift in the position of the tubing axis may result in a large shift in the angle of incidence of the ultrasonic beam if the transducers are not moved in synchronism with the tube.

Most wave modes used in other types of ultrasonic inspection are used for inspecting pipe and tubing. Shear waves are used most often, but longitudinal waves for thick-wall tubing and Lamb waves for thin-wall tubing may be used occasionally.

Tubing as small as 0.2-in.OD with walls as thin as 0.008 in. have been inspected satisfactorily with sensitivities to defects of 3% of the wall thickness.

Sometimes pipe and tubing are inspected manually using direct contact methods and contoured plastic shoes. Production speeds require automated systems. Coupling in automated systems is done by immersion, gap, sliding boot, or wheel filled with liquid. The positions of the transducers relative to the tubing under inspection are rigidly maintained. For external application, the control of the transducer position may be maintained by guides on the transducer holder which may consist of guide sleeves matching and concentric with the tubing or of wheels which support the transducer carriage. Either the lateral motion of the tubing must be completely restrained or the transducer carriage should be free floating to follow the motion of the tubing as it passes through the inspection system.

The ultrasonic beam is introduced into the tubing in such a manner as to minimize the number of wave modes present in the material. The critical angle for longitudinal waves in steel corresponds to a shear-wave angle of refraction which is slightly less than 36°. Therefore, to eliminate the longitudinal wave from a shear-wave test in steel requires a shear-wave refraction angle greater than 36°. With shear mode inspection, the angle of refraction determines the wall thickness that can be inspected. For instance, the maximum wall thickness that can be inspected using a 45° shear wave is 20% of the ID.

The ultrasonic beam may be collimated or focused to minimize the spread of the incidence angle. Focusing to a "point" provides the greater concentration of energy and, therefore, greater sensitivity to defects. However, a "line" focus is more practical for production-line speeds because it covers more of the tubing per rotation of the tube. True point and line focusing are impossible since the wavelengths of the ultrasound are finite. By making the circumferential dimension of the focus on the surface of the test material small compared with the diameter, the incidence angles vary little from those that would occur if the surface were a flat plate.

There are many approaches to inspecting pipe and tubing ultrasonically. These may involve only one transducer or a number of transducers. In one arrangement, one transducer is used to transmit a wave in a circumferential direction and another to transmit a wave in the axial direction. This arrangement increases the possibility of detecting defects regardless of their orientation. Additional protection is provided by using a third probe which also transmits in a circumferential direction but opposite to that of the first probe, so that waves are propagated in both directions simultaneously around the tube. Thus, certain types of defects such as laptype seams, that might be missed in one direction may be detected in the other. Figure 8.6 illustrates several transducer arrangements.

When an angled shear wave is used to inspect pipe or tubing in which the OD and the ID are not concentric, false echoes may be received when the wave propagates in the direction of decreasing thickness. The angle of incidence decreases for each reflection from the surfaces of the tube until after a few reflections, at which point the wave reverses itself and returns to the point of entry. This type of reflection can be distinguished from a true defect by the fact that as the tubing rotates, the indication moves across the screen as does a defect but it dies out as it approaches the transducer, whereas the true-defect indication remains. For automatic testing, the gate can be set so that the false echo disappears before approaching the gated area. Another method of reducing or eliminating such false indications is to cause the beam to propagate in a helical pattern around the tube.

As with all ultrasonic-inspection applications, suitable standards are essential for adjusting and calibrating the equipment. Standards are produced usually from the same material as that from which the tubing to be inspected is made. Since the geometry of the test specimen is an important factor in determining the inspection modes, the standard may be made from a piece of tubing taken from the production lot. This tubing is then carefully inspected for defects. After thorough inspection assures that the tubing to be used for a standard is free of defects, artificial defects are punched, drilled, or electromachined into the tubing to be used for calibration purposes.

The echo method of inspecting tubing is more sensitive to defects than the transmit-receive method.

Ultrasonic inspection of tubing often involves rotating the tubing in order to obtain total inspection. In some cases, it is practically impossible to rotate the tubing. The Krautkramer Company has developed a method which involves rotating the ultrasonic transducers about the tubing instead of rotating the tubing (57). Rotation is at speeds up to 3000 rpm. At least three probes are used, two of which scan in a circumferential direction in opposite directions to detect longitudinal defects and the third scans in an axial direction to detect circumferential defects. Electrical connection between the test probes and the electronic components is by inductive coupling. Pulse repetition rate is 3000/sec. Coupling is provided through water pumped under pressure into the rotary test block. Alignment is maintained by guides and rollers. This equipment is reportedly capable of increasing the inspection speed by a factor of 10 over previous methods.

The speed at which a tube can be inspected by pulse methods depends upon the effective dimensions of the ultrasonic beam and the pulse repetition rate. The defect should be interrogated by a sufficient number of individual pulses to provide the maximum signal possible from that defect. Speeds as high as 200 ft/min have been claimed for the Krautkramer equipment.

As mentioned previously, controlling the orientation of the ultrasonic beam with the surface is important in the proper ultrasonic inspection of tubing. Obviously, tapered tubing presents a problem in controlling the angle of incidence due to the varying diameters. McClung and Cook (58) have developed a method of inspecting tapered tubing of 0.009-in. and 0.050-in. wall thicknesses used in nuclear heat exchangers. They claim reproducible sensitivity to discontinuities only 0.0005 in. in depth and 0.125 in. long. Their method involves scanning the tube in an immersion tank by rotating the tubing and translating the transducers in an axial direction. The ultrasonic beam is transmitted circumferentially. Alignment of the transducers with the tubing surface is maintained constant by rotating the tubes in Teflon bushings machined to fit the maximum diameter of the tubing and located on either side of the transducers. V-blocks mounted alongside each bushing apply mechanical pressure to the tubing and maintain a constant angle of incidence.

Inspection of silver-brazed pipe joints has been accomplished using an automated pulse-echo system (51). A single probe on a plastic delay line is caused to rotate about the circumference of the brazed section. Longitudinal waves at 5 MHz are transmitted into the joint. If the bond is good, the energy reflects from the ID of the pipe. If the bond is poor, energy is reflected from the ID of the fitting. The acoustic impedances of the fitting, the silver braze, and the pipe are sufficiently similar that little energy is reflected from the bonded interface.

Measurement of thickness is also an important application of ultrasonics in tubing inspection. At one time, boiler tubes were inspected in situ by manual contact methods from the outside. A more recent development makes use of an immersion method to inspect boiler tubes for corrosion from within (59). The inspection unit is carried by six wheels mounted on retractable legs that permit the assembly to pass through small IDs and to expand into a larger ID. The mechanism maintains the probe to produce normal incidence. It can be rotated to inspect the circumference at any desired location. The electrical cables to the transducer are located inside a 40-ft-long rubber-coated flexible shaft which is used to manipulate the probes. A pulse-echo technique is used.

The method is not effective in areas having a substantial amount of scale. Nor is it suited to finding localized pitting since this would require full coverage of an area, which is time consuming.

Resonance measurements of thickness are possible in tubing if the surfaces are regular and parallel. The transducer, however, should be contoured to match the surface of the tubing.

The dual-probe, pulse method of measuring thickness also is becoming widely accepted.

The ultrasonic micrometer, based upon the interferometric cancellation of an ultrasonic wave in a quarter-wave thickness and minimum loss in a half-wave thickness, has been used to measure wall-thickness variations of 2×10^{-5} in. in 0.015-in.-wall tubing (60).

Chretien and Huard discuss a Lamb-wave technique of inspecting tubes intended for use as fuel-element cladding (61). The method is based on differences in the damping of the Lamb waves at the air-metal and the metal-water interfaces which allow separate recordings to distinguish between internal and external defects. The instrument employs a variable gain along the time base. It has been used to inspect zirconium-copper cladding of thicknesses in the range 0.7 to 1.2 mm. It has also been used for stainless steel tubes 6.3 and 0.35 mm thick.

Typical applications of ultrasonic inspection of tubing and pipe include the inspection of seamless tubes; arc-welded pipe; resistance welded pipe; high-pressure couplings and tubings; Zircaloy-2 cladding tubes; thin-walled, aluminum-clad, uranium-aluminum tubes; aluminum tubes; stainless steel tubes; welded pipes as large as 42 in. or more; and many others. Automatic systems capable of inspecting welds in pipes at speeds exceeding 300 fpm are possible.

III.C. PLATES AND STRIP

Conventional ultrasonic methods are used to inspect plates and strips for defects and for thickness variations. The geometry of plates provides a fairly constant surface for maintaining a constant angle of incidence and this is an advantage to the inspector.

Inspection of plates and strips may be divided into methods applicable to thin sheets and those applicable to thick plates. Surface waves and shear waves are used to inspect thick plates for defects such as cracks or inclusions at positions remote from the transducer. Longitudinal waves are useful for detecting defects such as laminations and inclusions in a plane parallel to the major surfaces of the plates. If the conditions of the edges of the plate permit, longitudinal waves also can detect cracks normal to the major surfaces.

Lamb-wave techniques are widely used to inspect thin plates. There are numerous discrete values of velocities of Lamb waves in a plate which depend upon the plate thickness and test frequency.

Lamb waves affect the entire thickness of the plate and travel parallel to its surface. In accordance with Snell's law, a particular Lamb wave mode will be excited most easily at discrete values of incidence angle corresponding to the velocity of the particular wave mode. The velocity of Lamb waves, in addition to being a function of the product of the frequency times the plate thickness, is also a function of the elastic properties and density of the materials of the plate as are all other types of ultrasonic waves. Figure 8.7 shows the incident angle required to produce given Lamb-wave modes in aluminum sheets in water (62).

Techniques of inspecting plates and sheets vary from manually scanning the materials with direct contact probes to fairly rapid automatic systems. Manual inspection with longitudinal waves is slow and impractical for production use except, perhaps, where the sides of the plates are suitable for coupling.

Thickness measurements of plates are made using longitudinal waves and either resonance or pulse methods. The probes must scan the entire surface of the plate for complete coverage, if this is desired. To make such an inspection less time consuming and more adaptable to production speeds, complete scanning can be accomplished by employing an array or assembly of transducers which extends the width of the area covered by a given movement of the apparatus. The transducers are "sampled" by switching from one to the other and correlating the output signal with the position of the probe sampled. Each of the probes must have its own associated electronic system for generating and detecting the signal in order that they may be calibrated to similar references.

Harris (63) describes one multiple unit device for measuring the thickness of sheets of metal moving at 43 in./sec. The system uses the resonance principle to cover the thickness range 0.010 to 0.100 in. Five identical transducers are water coupled to the sheet of metal. Each transducer operates through its own electronic system. The system provides both visual displays and permanent recordings.

An automatic system described by Silber and Ganglbauer (64) is used to test slabs in the thickness range 80 to 250 mm (approximately 3 to 10 in.) at a speed of 6 m/min. The transducers operate through a 13% chromium steel wear plate.

A manually operated instrument for inspecting plates ultrasonically using multiple transducers has been developed by Homer Research Institute, Bethlehem Steel Corporation (65). The inspection device is mounted on a cart which is equipped with push-type handles so that a man can manipulate the unit in an erect position. The transducers operate through a water column and generate longitudinal waves in the thickness direction in the plate. The transducers are located in direct contact with one another to provide a continuous inspection the full width of the probe assembly. Each transducer section is separately calibrated so that all sections are identically sensitive to defects.

A wheel type of inspection unit has been used to continuously inspect 2.25-in.-wide by 0.050-in.-thick 1050 steel strip for laminar defects (66). The liquid-filled wheel contains two transducers transmitting Lamb waves in a cross pattern which provides 100% coverage of the entire width of the strip. The direction of propagation is opposite the direction in which the strip moves through the test area. The strip is uncoiled and passed through a straightener to eliminate a crown, or curvature, perpendicular to the length of the strip before it passes under the wheel. The wheel is oil coupled to the strip. The testing speed is limited to 40 fpm by the straightener.

A method that has been used to eliminate the need to readjust the incident angle when continuously inspecting strip when the thickness of the strip changes involves the use of a wide angle of spreading of the incident wave as it propagates from the source transducer (67). With this type of probe, the proper angle of incidence for a specific Lamb-wave mode is always available.

Lamb waves, shear waves, and surface waves are used for automatic inspection of wide strip. In most cases, immersion testing is not practical, and coupling of the energy to the test strip is best accomplished by water jets of bubblers which produce a water film between the transducer and the strip. Grasshoff and Tobolski (68) have used Lamb waves in a pulse-echo manner to inspect strip and sheet up to 8 mm thick and 600 mm wide. Their apparatus is used in the pickling line between the scale breaker and the pickling bath and has good sensitivity to shrinkage cavities and other defects.

Figure 8.8 illustrates a side-looking ultrasonic wheel for generating 45-degree shear waves in a steel plate to be inspected. The wheel contains a 1/2 by 1-in. 2.25-MHz transducer. Table 8.1 shows the effectiveness of the transducer in detecting defects in 0.083-in. thick steel strip at various speeds. The strip was 12 in. wide and contained artificially milled defects 0.005, 0.010, and 0.025 in. deep, each 1 in. long. The wheel was located at one edge of the strip and water coupled to the strip. The pulse repetition rate was 500/sec. The gate was adjusted to correspond to the width of the strip from slightly past the edge of the wheel to slightly short of the opposite edge (to prevent false alarms from back echoes).

TABLE 8.1

Amplitude of Indications From 1-In.-Long Defects in 1055 Steel Strip Using Ultrasonic Wheel and 2.25-MHz Shear Wave

(Defects were on the same side of the strip as the transducer.)

Plate Speed, fpm	0.005-in. defect	0.010-in. defect	0.025-in. defect
7.0	12	22	Off scale
26.4	5	17	Off scale
115.0	4	14	Off scale
194.0	3	7	Off scale
317	-	5	35

Figure 8.9 illustrates another inspection method in which a collimator with a 1/8-in. by 1/2-in. aperture terminating in a

liquid-filled latex boot is used. The assembly into which the transducer and collimator were mounted could be angularly adjusted to optimize the incidence conditions. Strip-chart recordings obtained by using the transducer and collimator shown in Fig. 8.9 to inspect the same material used to obtain the data of Table 8.1 are given in Figs. 8.10 and 8.11. In Fig. 8.10 the defects are on the same side of the plate as the transducer and in Fig. 8.11 they are on the opposite sides of the plate.

The same factors affect the attenuation of Lamb waves used for inspecting plate and strip that affect the attenuation of other modes. Curvature in sheet materials has no appreciable effect. The frequency spectra of Lamb waves influence the attenuation by factors related to the dependence of the velocity of such waves on the product of frequency and thickness.

Isono, Ueno, and Udagawa (69) claim to have discovered an extraordinary attenuation of Lamb waves in mild-steel sheets up to 500°C which they attribute to damping associated with either carbon or nitrogen, or both, dissolved in the steel. They claim that measuring the group velocity of Lamb waves is a suitable means of obtaining a measure of the internal friction associated with the concentration of solute carbon and nitrogen atoms.

III.D. FORGINGS

Materials suitable for forging are usually suitable for ultrasonic inspection both before and after forming. For this reason, ultrasonic nondestructive testing has been used successfully for inspecting forgings such as large turbine-generator rotors, aluminum-alloy die forgings, and many others.

Typical defects found in forgings and for which ultrasonic inspection is recommended are bursts, cracks, hydrogen flakes, inclusions, laminations, laps and seams, and unfused porosity (38).

Forging bursts are ruptures, usually running parallel with the grain, which can occur either at the surface or internally. They are attributed to excessive working or movement of the metal or to forging at too low a temperature. Bursts vary in size and are often large. Ultrasonic pulse-echo methods are perhaps the best means of detecting bursts.

Hydrogen flakes are very thin internal fissures with bright silvery surfaces usually found in heavy steel forgings, billets, and bars. They usually run parallel with the grain. They are caused by high localized stresses caused by hydrogen coming out of solution as hot-worked material cools. Ultrasonic immersion or contact testing methods are used to detect hydrogen flakes. Flakes usually present a number of scattering (or reflecting) sites that cause a loss of back echo and considerable "hash" on the oscilloscope screen with the A-scan presentation.

Inclusions may consist of many types, but they may be categorized into two groups: (1) nonmetallic, tightly adhering particles which are elongated during processing parallel to the flow lines; and (2) nonplastic materials which appear as randomly oriented masses. Inclusions appear as definite interfaces within the metal; however, the amplitude of reflections from the interfaces depends upon the coupling at the interface and the material of the inclusions. The maximum signal corresponding to echoes from the interfaces occurs when the beam is incident at 90° to the reflecting surface. For this reason, the preferred direction of the beam during ultrasonic testing usually is normal to the direction of grain flow. Pulse-echo methods generally are used for detecting inclusions coupling by either contact or immersion methods. Indications may range from loss of back echo to numerous echoes corresponding to scattered inclusions.

Laminations are impurities in metals that have become very thin during processing. They form layers parallel to the work surface and alternate between layers of the metal. The choice of ultrasonic inspection method and wave mode depends upon the dimensions of the material to be inspected.

Laps are characterized by shallow wavy lines on the surface of a forging, which usually result from improper forging operation. Corners may be folded over or excess material may be smeared and pressed into the forging without fusing.

Seams are similar to laps but are caused by defects introduced into the material during a prior operation. They are elongated fissures parallel with the direction of forging, and are often deep and sometimes very tight.

Ultrasonic testing generally is used to evaluate wrought material prior to forming. The surface-wave technique is used to locate and determine the dimensions of laps and seams. Laps and seams produce clear, definite indications of the interfaces on an A-scan presentation.

Unfused porosity consists of very thin fissures aligned with the grain flow existing in wrought aluminum forgings. It is the result of porosity existing in the cast ingot which is flattened without fusion during the forging operation. Pulse-echo methods are capable of locating and establishing the extent of these defects if the geometry of the part permits testing in more than one direction.

III.E. BEARINGS AND BEARING MATERIALS

Conventional methods of ultrasonic inspection have been used for detecting defects in bearings and bearing races. Ball bearings as small as 0.186 in. in diameter have been inspected successfully for cracks. Brown and Dunegan (70) used a focused, 15-MHz, lithium sulfate transducer in conjunction with a collimator with a terminating inside diameter of 0.157 in. to inspect ball bearings. The end of the collimator was dished out to accept the bearings and to hold them in the proper position for testing. Water was pumped through the collimator to provide coupling. The focal region was adjusted to the bearing surface to minimize mode conversion. The appearance of a crack in the bearing produced a reflection, and when the defect was properly oriented with respect to the incident beam, the defect indication on the oscilloscope screen appeared between those corresponding to the front reflection and the back reflection. If the crack is large enough it will completely eliminate the true back reflection although reflections from the surface returning to the transducer by reflection from the defect may give the appearance of a back echo. The existence of a crack, however, is easily ascertained by observing the effects on amplitude and position of indications coresponding to internal reflections when the orientation of the ball with respect to the incident beam is varied.

Accurate measurement of the velocity of sound provides a means to detect polymorphous transformations of solids by the corresponding changes in elastic constants accompanying these changes. Equipment providing a capability of measuring velocity of sound in metals is available commercially. Fenkner (11) has used such equipment to determine a correlation between the velocity of sound and the amount of residual austenite in roller-bearing steel. Fenkner's data indicate a linear drop in velocity with increased retained austenite in chromium-steel bearing alloy. Between 4 and 28% retained austenite, the velocity drops from approximately 5905 m/sec to approximately 5840 m/sec.

Through-transmission and pulse-echo methods are effective in locating unbonded regions between the liner and the base metal of bearings, as well as inclusions and other defects within these materials (71). Diesel-engine bearings and other large bearings are especially adaptable to ultrasonic inspection.

Lamb waves have been used to detect defects that may lead to fatigue failure in aircraft bearing races, employing a spherically focused transmitter and receiver. The test reveals lack of bond between lining and base material, cracks, and inclusions.

III.F. CASTINGS

The metallurgical and physical characteristics of many types of castings preclude their inspection by conventional ultrasonic means, precluding viz., primarily rough surfaces, large grain size, and unfavorable geometries. In the present context, conventional means refers to the typical pulse-echo and through-transmission techniques involving either immersion or direct contact coupling to the part and generally testing at frequencies exceeding 1 MHz. Not all castings are excluded from ultrasonic testing as attested by the number of such applications presently being used. Certain types of castings can be inspected by the conventional ultrasonic means and many of those that cannot be inspected by the high-frequency ultrasonic methods may be inspected at lower frequencies for certain factors that affect quality.

In general, ordinary cast iron parts are not inspected ultrasonically, except for those with fairly simple geometries which are inspected by resonance methods. Low-frequency resonance methods can be used to detect cracks, voids, porosity, mottle, shrink, and inclusions in both cast iron and cast steel parts if the extent and location of such defects are of such a nature that they affect the resonance characteristics. For example, voids at high-stress locations within a resonant structure lower the resonance frequency of the part, while similar voids located at a stress node may have little effect on the resonance.

Various resonances are associated with complex parts. If these resonances can be identified for a standard good part, it is possible to evaluate similar parts for defects by shifts in the points of resonance if the part can be excited through a range of frequencies.

An example of inspection of cast iron that can be accomplished ultrasonically is that of cast iron rolls. Sasaki and Ono (72) have employed immersion, pulse-echo ultrasonics to estimate the thickness of the hardened layer. They found that the most suitable frequency for this measurement was always less than 3 MHz. To detect internal defects, Sasaki and Ono used frequencies of 0.4 and 0.75 MHz.

Typical types of defects that can be located in castings and billets, but only if the conditions are favorable, are shrinkage defects, hot tears, cracks, gas pockets and porosity, pipe, and coarse nonmetallic inclusions such as slag and sand. Typical castings that have been inspected successfully ultrasonically include cast iron brake drums, cast iron rolls (as mentioned previously), cast iron crank shafts containing globoidal graphite, cast aluminum and aluminum alloys, cast piston rings, and ductile-iron parts.

The quality of cast-iron products is often related to the velocity of sound or to the relative attenuation through the material. These parameters are affected by the size, shape, and distribution of graphite and alloy constituents, by the microstructure, and by the presence of defects.

Figures 8.12 through 8.20 illustrate various steps that have been used successfully by the author to inspect ductile-iron industrial wheels. These figures illustrate typical defects, methods of locating and identifying these defects, and the types of indications observed on an A-scan presentation. Both contact and immersion pulse-echo inspection were used; however, the latter method is preferred because of the roughness and irregularity of the surface. Both longitudinal- and shear-wave methods were necessary to obtain the required coverage using both 2.25 and 5.0 MHz nominal test frequencies.

III.G. RAILS

The railroads provided one of the earliest commercial uses of ultrasonics in nondestructive testing. Both resonance (73) and pulse-echo methods (74) were used to inspect rails.

The resonance type of rail-flaw detector consisted of a piezoelectric quartz-crystal type of transducer mounted in the end of a cane. The operator carried the oscillator and a battery pack on his back and inspected the rails in a walking position. The transducer signal was coupled to the rail through a film of oil. The electrical signal to the transducer was frequency modulated. The modulation width included several harmonics of the resonant frequency of the rail. The operator would hear a tone through a set of earphones at a frequency equal to the product of the number of resonance peaks contained in one sweep times the number of sweeps per second. The resonance frequency of a shallower depth corresponding to the distance to a bolt hole or to a crack which lay parallel to the surface would be higher than that of sound rail. Such a discontinuity would allow fewer harmonic peaks within a sweep and the audible tone generated would drop accordingly. If the crack was not parallel with the running surface, the wave would be reflected from the crystal and the audible tone would disappear, except for noise associated with the electronics of the system.

The pulse-echo equipment was mounted on a small test car. The operator worked from a low hanging seat at the rear of the car. The signal was coupled into the rail through an oil film. Most of the inspection was done using a 45° shear wave to detect cracks of various orientations to the surface. Occasionally, a longitudinal wave normal to the surface was used to aid in interpreting certain types of defects. Typical A-scan indications were obtained on an oscilloscope.

Gowan (75) has devised a method of inspecting railroad rails automatically at high speed using the pulse-echo technique. Differentially oriented beams of ultrasound simultaneously probe different regions of the rail. The characteristics of the targets are determined by counting the returning echoes. Logic systems are incorporated so that response characteristics are adjusted to correlate with the types of target being inspected, leading to an automatic decision as to the type of target and whether it is a defect.

III.H. WIRES

Ultrasonic inspection of wires has received some attention for several years, but to date its use in industry is limited, if not nonexistent. Techniques of generating wire waves include (1) use of the magnetostriction characteristics of (steel and nickel) wire, (2) setting up flexural vibrations by drawing the wire across a metallic shoe, (3) drawing the wire through a mating groove in the shoe of an angle probe used conventionally to generate pulses of shear or surface waves and coupling through a thin film of oil, and (4) passing the wire through an immersion bath for pulse-echo or transmit-receive testing.

The magnetostrictive technique has the advantage that energy is coupled to the wire electromagnetically so that the driving coil does not touch the wire. The decrease in depth of flux penetration with increasing frequencies limits the method to inspection at low frequencies. Wires 0.080 in. in diameter and larger can be inspected by this means at approximately 12 ft/sec. Defects can be detected at depths of 0.002 in. or greater.

Wires may sustain torsional (shear) waves, longitudinal waves, or flexural waves. Like Lamb waves, wire waves may be produced in many modes and the velocities of propagation of these modes depend upon the mode and frequency. Therefore, ultrasonic pulses in wires become somewhat broadened as a result of velocity dispersion of the various frequencies contained in the pulse.

III.I. RIVETS

The geometry of rivets is amenable to ultrasonic testing. Wiseman and Garcia (76) have been successful in detecting corrosion at the roots of the flush and formed heads of aluminum rivets in a riveted aluminum diesel cruiser. The rivets were flush-head 2S aluminum, 3/16 in. in diameter and 1.2 in. long. By using a 2.25-MHz, 1/4 in.-diameter longitudinal-wave transducer, a pulse repetition rate of 13.5 kHz, and a Polaroid camera, these investigators were able to detect hairline cracks as well as larger defects extending over 20% of the total cross-sectional area of the rivet.

Inspection of smaller rivets and greater defect sensitivity should be possible, especially in materials with as good ultrasonic transmission characteristics as those of aluminum. Higher frequencies would increase resolution and defect sensitivity.

IV. INSPECTION OF HOT METALS

IV.A. HOT STEEL

Ultrasonic inspection of materials has been limited to temperatures only slightly above room temperature by the low Curie points of commonly used transducer materials. Beaujard et al. (77) have used water for the dual purpose of cooling transducers employed for inspecting hot steel and of coupling the energy into the steel. A stream of water is forced between the transducer and the hot surface and the area beneath the transducer is cooled locally below the boiling point. Either longitudinal or shear waves may be coupled momentarily into the part by this method. It is claimed that steel slabs and billets have been inspected at temperatures as high as 1100°C. The method has been recommended primarily for use with soft or extra soft steels at high temperatures; however, one might do well to view this application with considerable caution.

A typical application of ultrasonic testing of hot steel is in determining the extent of pipe in billets prior to cropping. Atkins (78) proposed to locate transducers inside rolls which could be pressed against the surfaces of billets. Oil under pressure was used as a couplant, lubricant, and coolant.

In addition to detecting pipe, Lynnworth and co-workers (79) have used both longitudinal- and shear-wave ultrasonics to measure thickness and temperature during the continuous casting of steel. Average temperature along a chosen path is determined by measuring the longitudinal- and shear-wave velocities. The temperature dependence of these velocities was predetermined in similar materials.

The velocity of sound has been measured at elevated temperatures to determine the physical properties of hot metals, including shear and Young's moduli and Poisson's ratio.

The high Curie point of lithium niobate (1210°C) makes it potentially useful for high-temperature studies. Care must be taken to prevent thermal shock to the piezoelectric element.

IV.B. FOLLOWING THE LIQUID-SOLID INTERFACE DURING COOLING OF INGOTS

The velocity of sound and, therefore, the acoustic impedance of a material differs considerably between the solid and the liquid state. Ultrasonic energy is reflected at a liquid-solid boundary. Ulrasonic pulse-echo methods have been used to monitor the location of solid-liquid interfaces during the freezing of various materials. The accuracy of the measurement is affected by the nature of the freezing interface. Dendritic formations or other roughness characteristics cause excessive scattering and a corresponding loss of signal. A distinct interfacial surface may not be formed, so that accurate measurements are impossible.

Bailey and Dule (80) have used miniature probes with pulse-echo methods to follow the motion of the solid-liquid interface during the freezing of water, where the heat transfer is unidirectional. Kurz and Lux (81) have used ultrasonic-reflection methods to study the freezing of Wood's metal (melting point 80°C), a Pb-Sn alloy containing 20% tin, and an alloyed steel solidifying between 1530 and 1430°C. An advantage of the ultrasonic method of monitoring the interface is that no foreign substances are introduced into the system. The interfaces between the liquid and solid phases of these materials are rough and cause severe scattering of reflected energy.

Kurz and Lux (82) report that casting conditions in the continuous casting of steel are more easily controlled by continuous measurement of the thickness of the solidified portion. In actual tests, they found that reflections are ideal only during the initial process of solidification. As the thickness of the solidified portion increases, dendritic formation inhibits return of reflected energy and the measurement becomes less reliable. However, they claim that the method is suitable for locating the solid-liquid boundary in steel during solification and remelting, and for determining the depth of the first "bubble-ring" during casting of rimmed steel.

V. DETERMINATION OF BOND INTEGRITY

Bonded materials may be inspected for either cohesive strength or adhesion by ultrasonic means, with certain limitations. There is no certain way of determining the actual strength of adhesion. However, if the bonding material is sufficiently matched acoustically to the base material, as with silver brazing of steel to steel, nonbonded areas can be detected by the size of echo received from the interface, by the amplitude of an indication corresponding to the depth of the bond line from the surface obtained by a resonance-type thickness gauge, and by the reduced signal received in a through-transmission test.

Lamb-waves have been used to detect nonbonds between laminations when at least one is an outer layer. The incidence angle of the transmitted wave is set to generate a Lamb wave of a specific mode corresponding to the material and thickness of the outer layer. The Lamb wave will be generated only when the interface is not damped by a bond to the adjacent surface.

The sensitivity of techniques based upon through-transmission, pulse-echo, or resonance methods to nonbond defects depends upon the relationship between the wavelength of the ultrasound used and the area of the defect. The sensitivity of through-transmission and resonance methods is also a function of the ratio of beam diameter to defect size. Sensitivity decreases with decreasing frequency or increasing wavelength.

Ultrasonic methods have been used to examine multiple thin-layered diffusion-bonded composites for nonbonds and interface locations and for inspection of brazed and diffusion-bonded honeycombs.

Interference phenomena occurring in thin plates may affect the amplitude of received signal during the inspection for delaminations in thin sections by the through-transmission method. The effects can be minimized by tilting the specimen a small amount with respect to the axis of the incident beam or by avoiding frequencies at which destructive interference may occur. When an acoustic plane wave of finite cross section traveling in a medium encounters an extremely thin flexible plate in which the wavelength is very large compared with the thickness, the plate flexes with the medium. Very little energy is lost in the flexural wave thus produced in the plate, and the pressures on the two faces of the plate are approximately equal in magnitude. As the plate thickness increases, multiple reflections from the interfaces interfere with the primary transmitted wave. The extent of interference depends upon the ratio of thickness of the plate to wavelength, the coupling to the surrounding medium, the pulse length and shape (when pulses are used), and the internal-absorption losses. When the ratio of thickness to wavelength is an odd multiple of quarter wavelengths, destructive interference occurs within the plate which may completely prevent transmission through the plate. Resonance occurs when the plate thickness is a multiple of half-wavelengths thick and the energy is transferred through the plate with minimum loss. Material thickness is much less important when the specimen is several wavelengths long.

If a nonbond region in a thin section is large compared with the cross section of the incident beam in through-transmission inspection, a barrier to transmission results and no signal is received by the receiving transducer. When the collimated beam is larger than the defect, the total energy transferred past the defect is

$$E_t = a_2^{-2}(a_2^2 - a_1^2)E_i \tag{8.3}$$

where E_i is the total energy that would be transferred were the defect not present, and

a_2 is the radius of a collimated beam concentric with a delamination of radius a_1.

Cohesive strength of adhesive bonds is related to the acoustic impedance of the adhesive. The response of a joint to ultrasonic excitation is primarily determined by the properties of the adhesive layer beneath the transducer. Viscoelastic properties, porosity, and mass are the primary factors that determine the cohesive strength of an adhesive. These same factors affect the velocity of sound and density, i.e., acoustic impedance. Inspection of adhesive bonds is based on this effect on acoustic impedance. The load that the adhesive imposes on the top-face, or lamination, influences the resonance frequencies of the layer. Resonance-type instruments, which have been calibrated against a free plate of the same material that is used for the top lamination, are employed for the inspection. Nonbonds, voids in the glue line, and degree of cure are determined by the shift in resonance indication.

Smith and Cagle (83) have been able to correlate bond strength determined by a tedious procedure of destructive measurements with ultrasonic resonance measurements.

Nonbonds are often assessed by means of pulse-echo methods, by comparing the echo from the bondline with that of the back echo.

VI. THICKNESS MEASUREMENTS

The thickness of materials may be determined by the time of travel of an ultrasonic pulse between opposing surfaces or by the thickness resonance frequencies. Portable instruments of both types are commercially available.

The principle of a common pulse type is illustrated in Fig. 8.21. In Fig. 8.21(a) the transmitting and the receiving piezoelectric elements are separated from the surface of the specimen to be measured by a distance which allows the electronic system to recover from the effects of the initial pulse before the received pulse returns. The triggering of the oscilloscope is delayed by a corresponding amount of time so that the left side of the oscilloscope screen corresponds to the top surface of the specimen. Little error is caused by the angle between the incident wave and the reflected wave from the back surface if the specimen is thick compared with the spacing between transducer elements. Errors in measurement in thinner specimens may be caused by multiple reflections within the specimen. Most commercially available instruments of this type are suitable for measuring thicknesses down to 0.050 in.; however, instruments are available that can measure thicknesses down to 0.010 in.

A combination of pulse and resonance methods has been used to produce an "ultrasonic micrometer" (84). An oscillator produces long pulses of carrier driving an ultrasonic transducer. The transducer is coupled to the material to be measured through a water column. The frequency of the carrier is adjusted so that the electrical output signal of the transducer exhibits a null in the center of the pulse, which is an indication that the sheet is very nearly in half-wave resonance or resonating at one of the overtones. If the acoustic-impedance mismatch between the sheet to be measured and the water column is sufficiently great and if the beam width and frequency are chosen properly, the receiving transducer will exhibit a null output. The null can be made to occur close enough to a half-wave dilatational resonance to be used for measuring thicknesses.

Arave (85) reports the use of an ultrasonic system to plot the thickness profile of a fuel-element coolant channel. Two 8-MHz lead zirconate-titanate piezoelectric crystals are mounted one on each wall of the channel. One transducer emits a signal which is detected by the second transducer and the received signal is used to trigger a time-to-analog converter that is connected to the Y-axis of a recorder. Longitudinal distance in the channel is converted to an analog output to drive the X-axis of the same recorder. The claimed accuracy is 0.0005 in., and the resolution is 0.0001 in. with a range of 0.050 to 0.175 in.

A method of inspecting copper, silver, brass, and bronze platings on steel (86) has as its basis the relationship between the thickness of the metallic layers and the phase velocity of ultrasonic waves. The method is sensitive both to defects in the metal and plate and to the size of the particles in the plate. Interference patterns developed as the result of the difference in velocities of sound between the plating and the substrate are used to assess the character of the metal/plating interface.

The pitch-catch type of thickness-measuring device can be obtained with digital readout. Such instruments are capable of measuring thicknesses of metal within the range 0.100 to 10 in. with an accuracy of 0.01 in.

Both continuous-wave and pulse-type ultrasonic micrometers are used in nuclear applications to measure wall thickness. Corrosion inside forced-circulation piping in reactor pressure systems is one such application.

Resonance methods have been used to attempt to assess the degree of corrosion in harbor structural elements located underwater. The corrosion rate of unprotected steel wharf and pier pilings and bulkheads in harbors has been monitored by means of ultrasonic probes.

Typical thickness measurements made by ultrasonic means, in addition to those mentioned previously, include measurement of thickness of hollow helicopter shafts, detection of corrosion in aircraft fuel tanks, corrosion and erosion in pipelines and closed vessels, pressure vessels in particular, determination of the thickness of ship steelwork, measurement of wall thickness in diesel cylinder liners, and testing for corrosion and wall-thickness reduction in petroleum refineries.

Another thickness-measurement technique is based upon the propagation of Lamb waves in thin plates (87). Pulsed ultrasonic waves having a wide frequency spectrum are caused to travel through a coupling medium and to strike the test sample at a predetermined angle of incidence. Lamb-wave modes are excited within the test sample by discrete frequencies within the incident ultrasonic waves. The types of Lamb-wave modes generated depend upon the thickness of the plate and the velocity of the ultrasonic waves along the surface of the test sample. The excited Lamb waves radiate pulsed ultrasonic waves from the plate at an angle of exit equal to the incidence angle. The waves radiated contain only the discrete frequencies of the Lamb waves and detection of these discrete frequencies gives a measurement of the thickness of the metal plate.

Stress corrosion cannot be detected by means of thickness measurements. However, cracks formed by stress corrosion can be detected by shear-wave techniques and the degree of damage can be assessed by the magnitude of echoes obtained from the cracks or by the attenuation of surface waves propagating through stress-corroded regions.

VII. IN-SERVICE INSPECTION OF NUCLEAR REACTORS

Ultrasonic methods of nondestructive inspection are used widely in the nuclear-reactor field, where a real concern for safety exists. Inspection procedures are applied at many stages of reactor development--when materials are selected, while the components are being fabricated, and after the elements have been in service.

Ultrasonic methods are used to indicate size and homogeneity of grains by attenuation; to measure cladding thickness; to locate lack of bond between cladding and core; to detect pipe, shrink porosity, inclusions, and micro-inclusions in uranium-alloy castings; to inspect welds for wall penetration, porosity, laminar voids, unfused joint interfaces, cavities at weld root, intergranular cracking and cracks; to locate wall and core defects; and to indicate density and porosity of metallic-uranium core. Density and porosity are determined by ultrasonic velocity and attenuation measurements.

Ultrasonic methods of inspecting reactor materials and components include resonance methods for thickness measurements (sometimes), through-transmission methods, and reflection methods. The latter two usually are pulse methods.

In addition, ultrasonics may be used to indicate the position of control rods in a nuclear reactor, to indicate the steam-water interface in a boiling-water reactor, and to detect boiling in pressurized-water systems.

The immersion method of coupling is commonly used for ultrasonic inspection of fuel elements. Automatic scanning and recording devices have been developed. The system design may vary between laboratories, but in general, a C-scan recorder is included for recording data. The position of the recording head is mechanically or electrically synchronized with the position of the probes relative to the fuel element. Plate-type elements are scanned linearly, and cylindrical elements are scanned in a helical pattern.

Years of development and experience have produced reliable inspection methods for use during fabrication of certain components used in nuclear power plants. However, after a reactor goes into service, several of these methods are no longer useful. Access to many areas of the reactor is difficult. Radiation levels prevent an individual's remaining in an area for an extended length of time, and thus remote methods of inspection are preferred. Present methods of in-service inspection require a complete shutdown of the reactor until the required inspection can be completed.

The ASME Code for in-service inspection calls for volumetric examination of a large percentage of the components of nuclear power plants. This means that a penetrating form of energy is required. Only two nondestructive methods are capable of practical examination of thick steel structures. These are radiography and ultrasonics. Radiography can be used before the plant is placed into service. However, after the reactor goes into service and becomes radioactive, ultrasonics is about the only method that can be used where volumetric inspection is required.

Automatically operating ultrasonic systems can be developed for remote inspection. The waves are insensitive to gamma rays (although the materials of the ultrasonic system might be affected). The problem is not exactly and completely solved, however, because the materials used in certain components of the reactor cause severe scattering of ultrasonic waves generated by conventional inspection equipment. Problems exist in the use of ultrasonics even when the energy is coupled through a somewhat ideal surface material such as the outer shell of the reactor vessel. Undulations at the cladding/shell interface cause serious scattering and detecting defects in clad materials requires a systematic comparison between initial base-line data and later inspection data. This process of data comparison with the transmit-receive method of inspection is susceptible to the introduction of a number of errors. Slight changes in incidence angle, coupling, or indexing between inspections could result in errors unless their effects are detected and adaptively corrected in an accurate fashion.

In tests during fabrication of certain reactor vessels (88), inspection has been performed through cladding at normal inspection frequencies of 2.25 MHz. The cladding was weld-deposited stainless steel, and longitudinal waves at 2.25 MHz might be expected to penetrate the cladding satisfactorily. The cladding was inspected for quality of fusion to the base metal, using a wheel-type search unit.

Shear waves at 2.25 MHz, using a standard 45-degree angle probe, have revealed cracks during inspection from the cladding side. However, the wavelength of shear waves in steel is only approximately half that of longitudinal waves at the same frequency. Scattering is so excessive that the gain of the instrument must be increased by an order of magnitude above what might be considered normal; i.e., the gain setting must be increased to at least 10 times the calibration sensitivity. This practice would not be acceptable in a remote, automatic device and more reliable means of inspection should be developed.

Certain components of the reactor system are especially difficult to inspect ultrasonically. For instance, cast stainless steel pump castings and valve bodies consist of large grain structures that are impervious to ultrasonic energy of conventional nondestructive testing methods.

In spite of the difficulties, the development of nuclear power plants for supplying electrical power to the public is proceeding, and inspection methods must be improved to meet the demands of the codes. Certain areas can be inspected by present methods if means of access for the inspection devices can be provided. Perhaps means of controlling grain size can be incorporated during fabrication to provide easier transmission of ultrasonic energy in those components that presently are uninspectable. If this is impractical, then lower frequency methods based upon changes in propagation characteristics indicative of developing damage may be adequate. The latter approach probably would not give the resolution and sensitivity to defects that the conventional methods would provide, assuming that conventional methods were practical.

Methods of remote inspection already are under consideration and development. Permanent tracks located in appropriate regions of the reactor, such as in the space between the reactor vessel and the surrounding structure, have been proposed (89). The tracks would guide a device fitted with such items as a closed-circuit television camera, ultrasonic inspection equipment, and other tools for use on the reactor vessel.

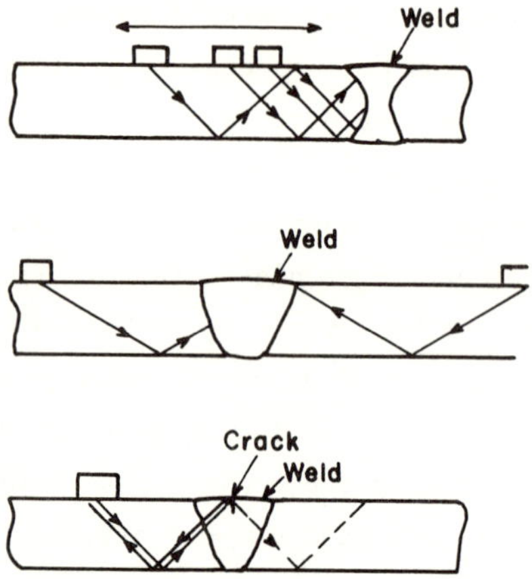

FIG. 8.1. Angle-beam method of weld inspection by ultrasonics.

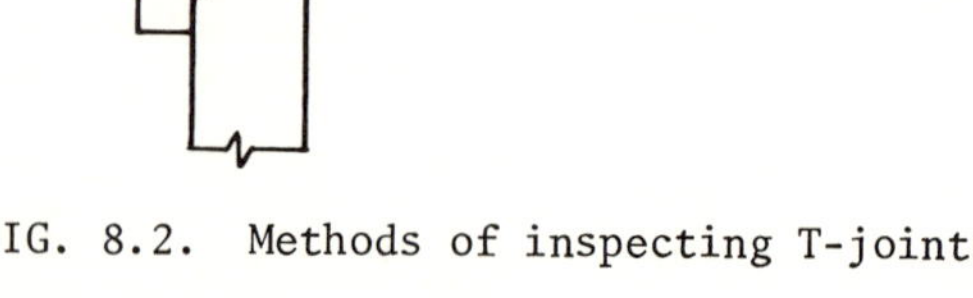

FIG. 8.2. Methods of inspecting T-joints.

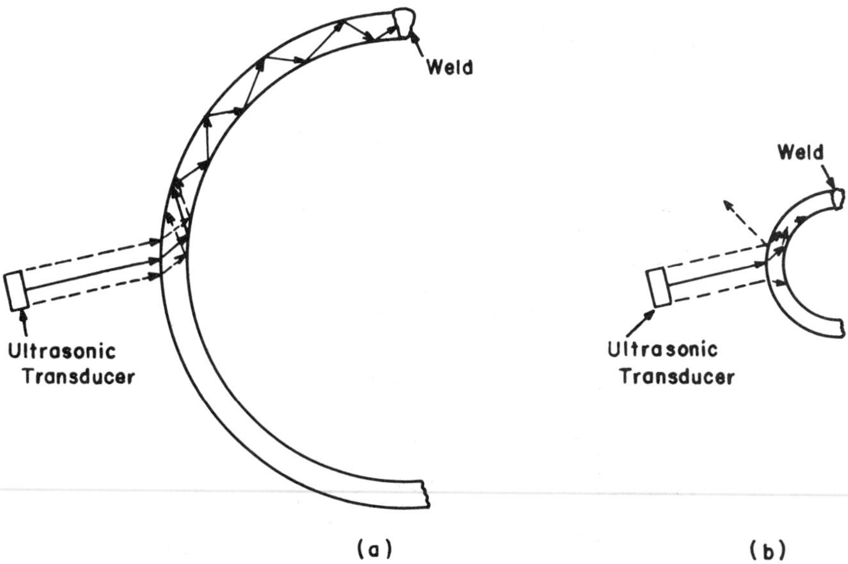

FIG. 8.3. Problem of applying same wide beam to large and small diameter tubes.

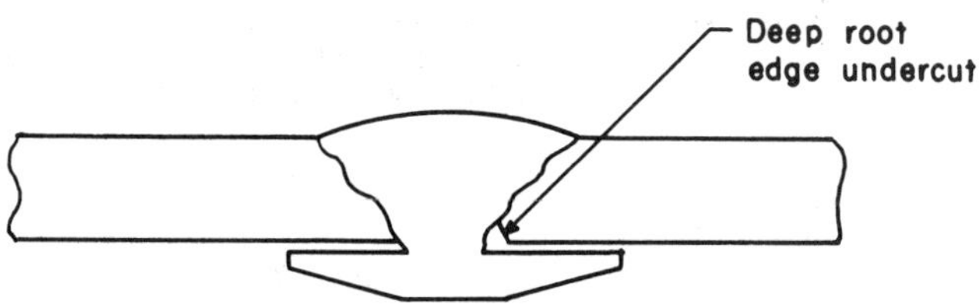

FIG. 8.4. Deep root edge undercut sometimes encountered in backing type of pipe weld.

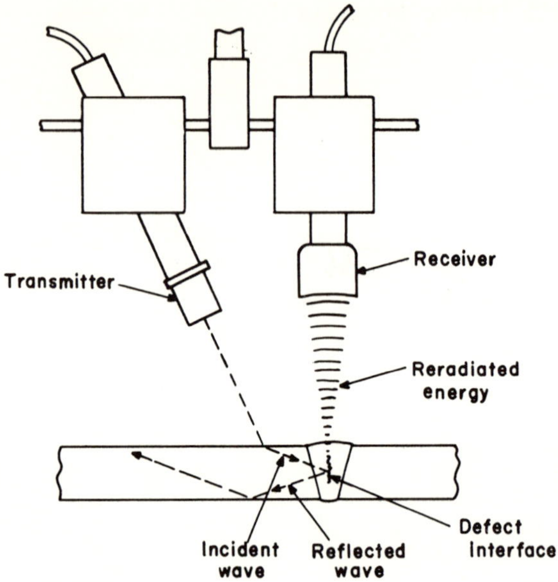

FIG. 8.5. Delta configuration technique.

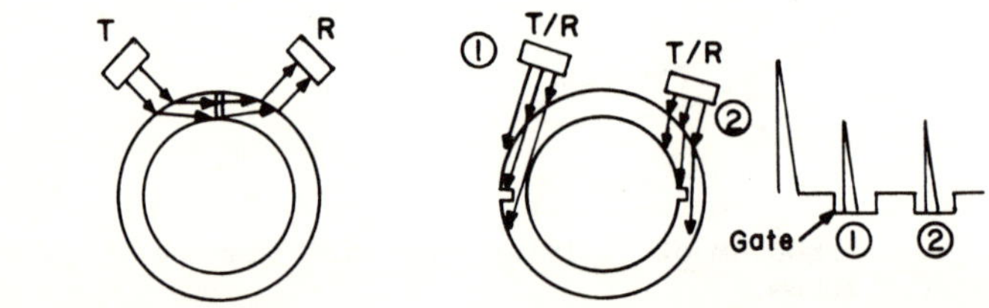

(a) Transmit/receive method of inspecting electric resistance-welded tubing, etc.

(b) Two probe method of distinguishing internal and external flaws. Probe 1 is adjusted to obtain maximum echo from external defect. Probe 2 is adjusted to obtain maximum echo from internal defects.

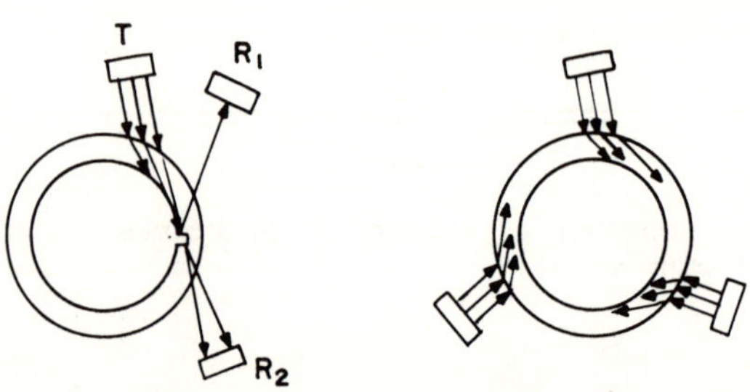

(c) Three probe echo and transmit/receive method.

(d) Multiple probe method of covering entire circumference of tubing without rotation.

FIG. 8.6. Examples of transducer arrangements for inspecting pipe and tubing.

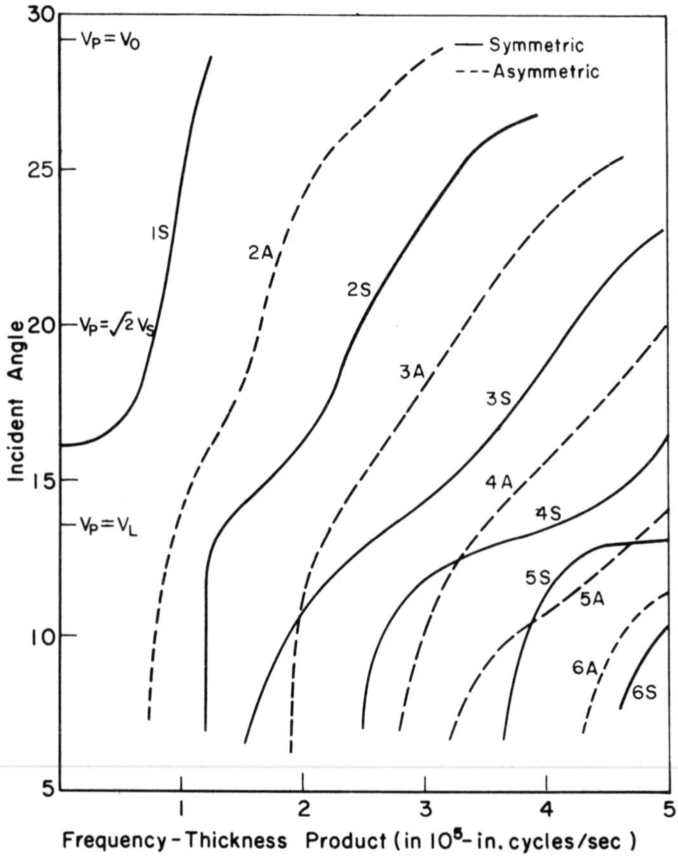

FIG. 8.7. Incident angle required to produce various Lamb-wave modes in aluminum in water.

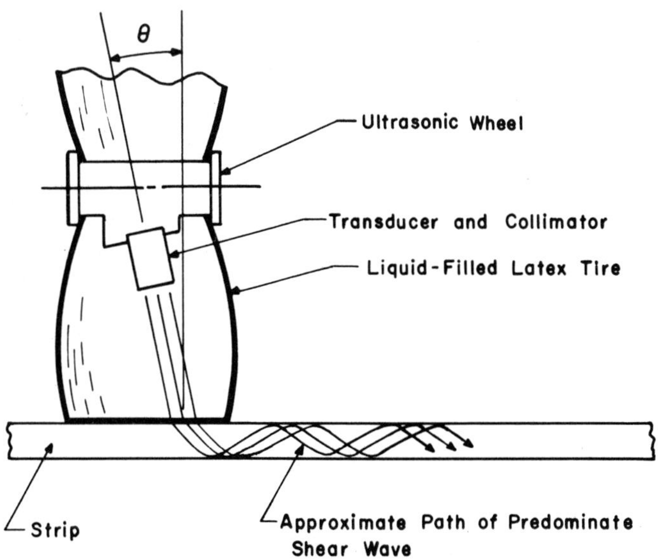

FIG. 8.8. Ultrasonic wheel assembly used for inspecting plates and pipes.

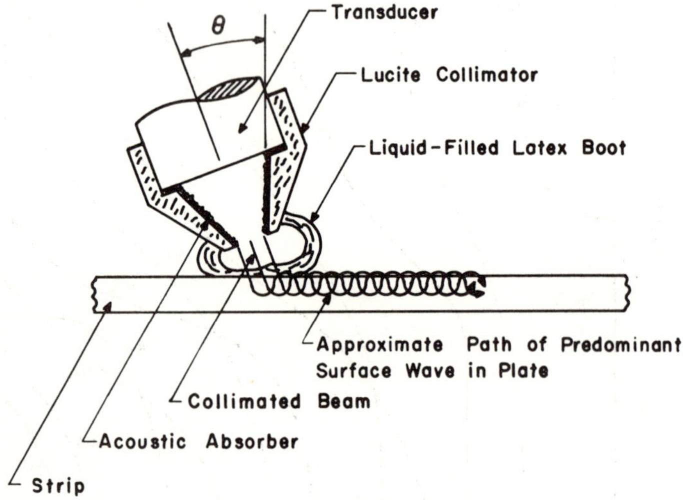

FIG. 8.9. Ultrasonic rectangular collimator assembly and sliding boot used for inspecting plates and pipes.

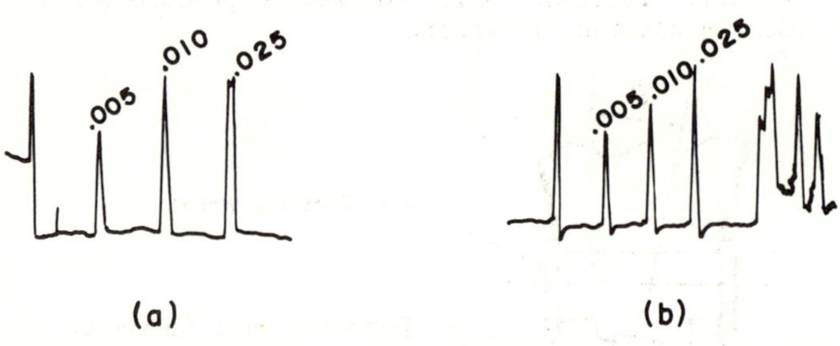

FIG. 8.10. Indications of defects obtained with collimated ultrasonic probe at speed of (a) 23.2 fpm and (b) 309 fpm. Defects and transducer on same side of plate. Incidence angle = 18 degrees. Plate 3A (1055 steel). Plate thickness is 0.083 in. f = 5 MHz.

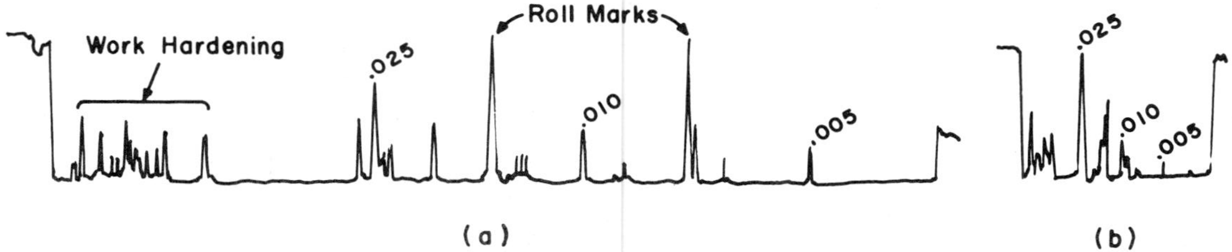

FIG. 8.11. Indications of defects obtained with collimated ultrasonic probe at speed of (a) 5.33 fpm and (b) 301 fpm. Defects and transducer on opposite sides of plate. Incidence angle = 20 degrees. Plate 3A (1055 steel). Plate thickness is 0.083 in. f = 5 MHz.

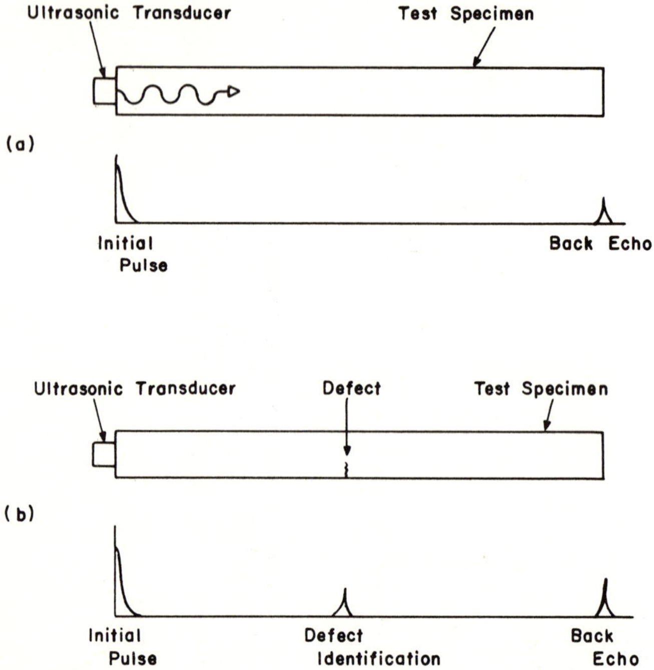

FIG. 8.12. Pulse-echo method of detecting defects in the rim of industrial ductile-iron wheels: (a) good region, (b) crack indication.

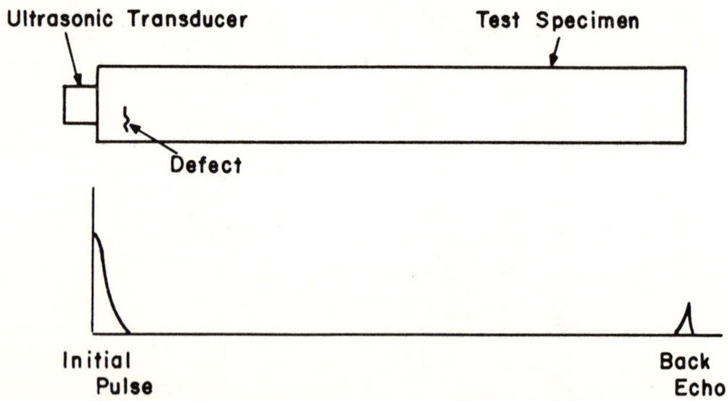

FIG. 8.13. Pulse-echo technique with defect near the transducer. Reflection from defect obscured by initial pulse. Back echo diminished as a result of "shadow" from the defect.

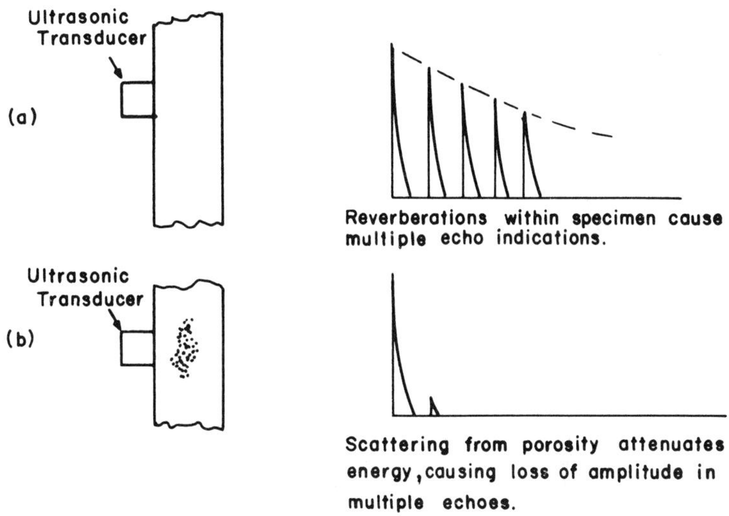

FIG. 8.14. Pulse-echo technique applied in radial direction to detect porosity in the rim of industrial ductile-iron wheels.

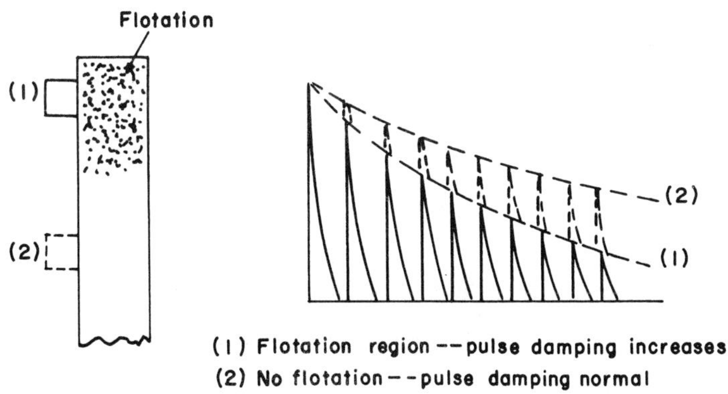

FIG. 8.15. Pulse-echo technique of detecting flotation in the rim of large industrial ductile-iron wheels: (1) flotation region--pulse damping increases; (2) no flotation--pulse damping normal.

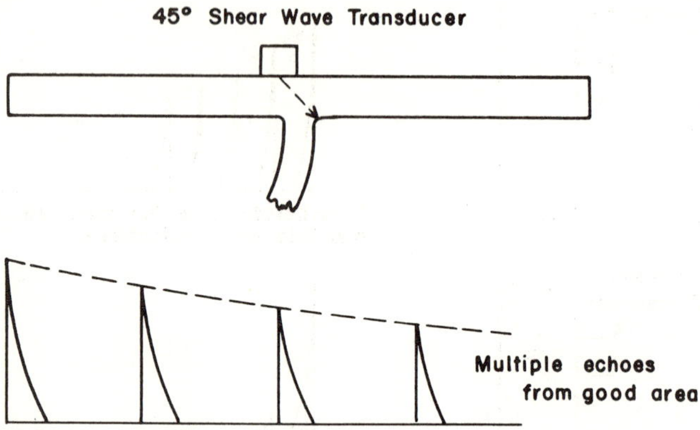

FIG. 8.16. Pulse-echo technique of locating defects in the web-rim region of large industrial ductile-iron wheels. Indication typical of no defect.

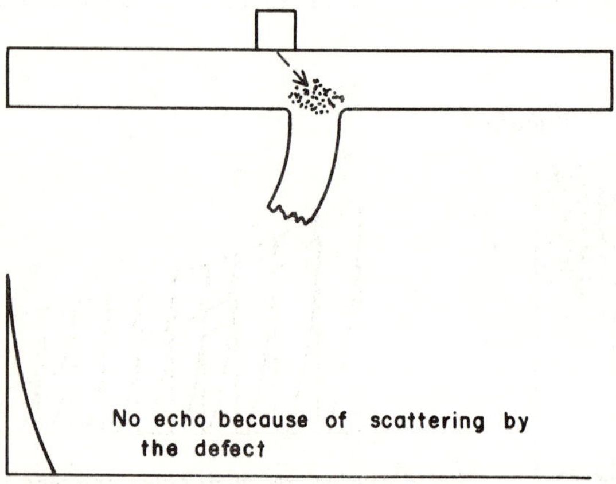

FIG. 8.17. Pulse-echo technique of locating defects in the web-rim region of large industrial ductile-iron wheels. Shrink porosity causes scattering of energy and loss of back echo.

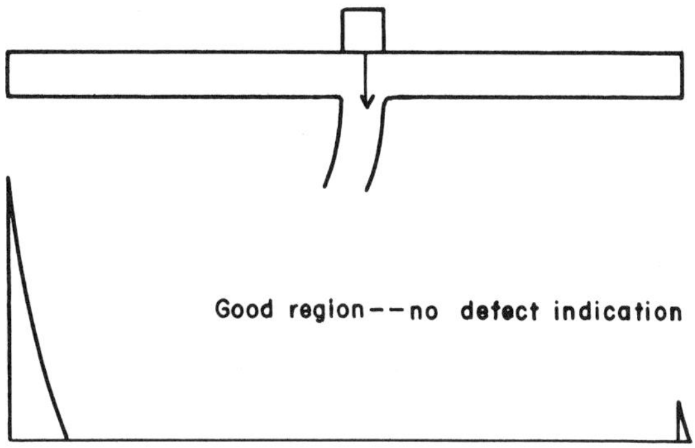

FIG. 8.18. Pulse-echo technique of locating defects in the web of large industrial ductile-iron wheels. Longitudinal pulse encounters no defect; no energy is reflected back to the transducer, which indicates a good region.

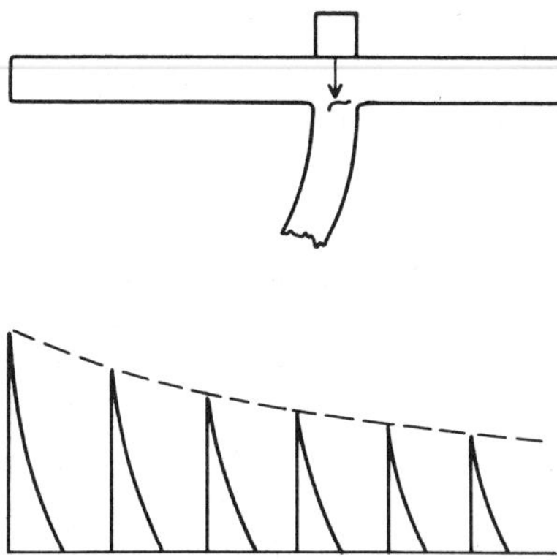

FIG. 8.19. Pulse-echo technique of locating defects in the web of large industrial ductile-iron wheels. Longitudinal pulse encounters defect; energy is reflected back to the transducer from the defect. Discontinuity below rim will give multiple echoes if discrete and properly oriented. Attenuation of multiple echoes depends upon type of defect. Spacing between echoes depends on distance to defect.

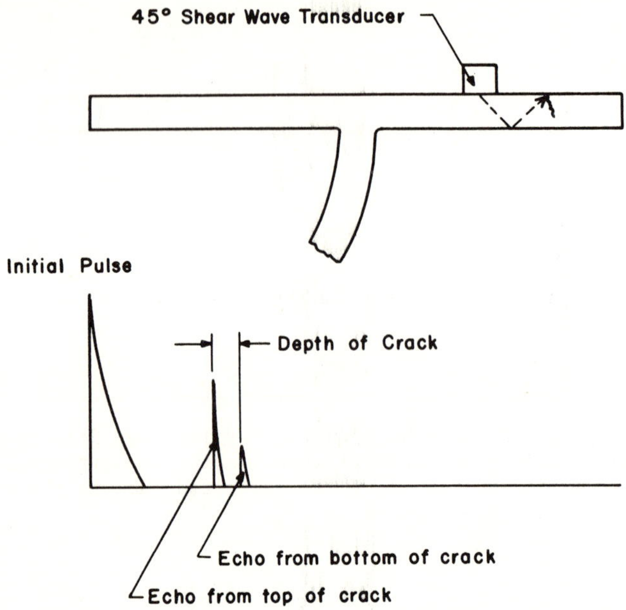

FIG. 8.20. Angle-beam shear-wave technique of locating cracks in the rim of large industrial ductile-iron wheels.

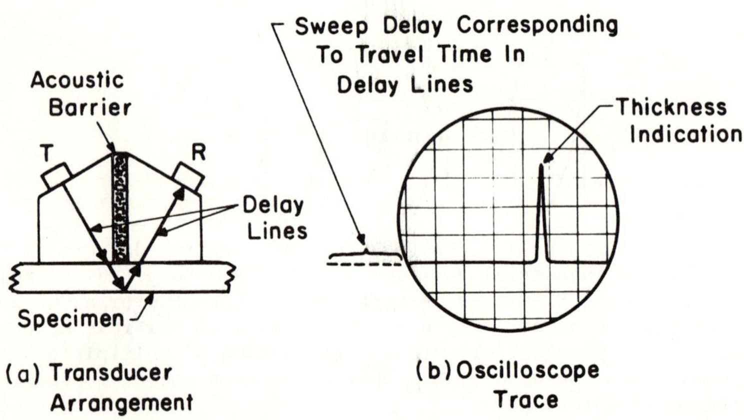

FIG. 8.21. Measurement of thickness by pitch-catch method.

REFERENCES

1. J. Krautkramer and H. Krautkramer, Ultrasonic Testing of Materials, 2nd rev. ed., Springer-Verlag, New York, Inc., 1969.

2. R. G. Aisin, Zavodsk. Lab., 33(8), 1038-1039 (1967).

3. R. N. Hafemeister, Iron Age, 176, 95-17 (1955).

4. H. L. Dunegan, Grain Size Measurements in Uranium by Use of Ultrasonics, Lawrence Radiation Laboratory, Livermore, California, July 2, 1962, 35 pp.

5. R. S. Sharpe and S. Aveyard, Appl. Mater. Research, 1(3), 170-176 (1962).

6. E. D. Nemkina, G. D. Radaev, and A. A. Zborovskii, Zavodsk. Lab., 32(1), 51-52 (1966).

7. S. Ya. Sokolov, J. Tech. Phys. U.S.S.R., 7, 160 (1941).

8. N. N. Yegorov, Sb Primeneniye Ul'traakust. i Issled. Veshchestva, (7), 169-183 (1958).

9. N. N. Yegorov, Zavodsk. Lab., 25(7), 829-833 (1959).

10. J. Koppelman, Materialprufung, 9(11), 401-405 (1967).

11. M. Fenkner, Mater. Eval., 27(1), 11-15,22 (1969).

12. L. G. Merkulov, L. A. Yakovlev, and E. K. Guseva, Defektoskopiya, (6), 33-42 (1967).

13. W. J. Bratina, J. T. McGrath, R. I. Moore, O. M. Mracek Mitchell, and R. F. Love, Proc. Fifth International Conference on Nondestructive Testing, Montreal, Canada, 1969, pp. 140-145.

14. D. Tietz, Wiss. Z. Tech. Hochsch. Otto Guericke, 10(5), 505-508 (1966).

15. Anon., Elec. Rev., 180(9), 323 (1967).

16. J. S. Oliano, ASM Tech. Report No. D8-29, 4, American Society for Metals, Metals Park, Ohio 44073, 1968, 13 pp.

17. M. W. Simmons, Tool Mfg. Engineer, 56(6), 80 (1966).

18. Anon. Ultrasonics, 4, 88-91 (1966).

19. R. H. Brockelman, Metal Progr., 90(1), 95 (1966).

20. R. H. Brockelman and K. A. Fowler, Int. J. Powder Met., 2(4), 45-53 (1966).

21. E. H. Abbe and R. D. Korytoski, Proc. 16th Meeting Metal Powder Ind. Federation, Chicago, 1960(1961), pp. 5-18.

22. R. W. Benson, Proc. Symposium on Physics and Nondestructive Testing, Southwest Research Institute, (1963), pp. 92-103.

23. G. Charlet, Rev. Franc. Mecan., (21), 75-86 (1967).

24. R. W. Benson and V. T. Realson, Prod. Eng., 30, 29, 56 (July, 1959).

25. R. W. Benson and Assoc., Inc., Monthly Progress Report No. 24 (Quarterly Report), Contract No. NAS8-20208 for the George C. Marshall Space-Flight Center of the National Aeronautics and Space Administration (June 30, 1967).

26. W. J. McGonagle and S. S. Yun, Proc. Fifth International Conference on Nondestructive Testing, Montreal, Canada, 1967, pp. 159-164.

27. F. R. Rollins, Jr., D. R. Kobett, and J. L. Jones, Technical Documentary Report No. WADD-TR-61-42, Pt. 2 (January, 1963); available from Armed Forces Technical Information Agency, Arlington Hall Station, Arlington, Virginia.

28. J. G. Rasmussen, Nondestructive Testing, pp. 103-110 (March-April, 1962).

29. N. E. Dixon, Proc. Sixth Symposium on Nondestructive Evaluation of Components and Materials in Aerospace Weapons and Nuclear Applications, San Antonio, Texas, April 17-19, 1967, pp. 16-35.

30. S. J. Klima and J. C. Freche, NASA Tech. Memo, NASA TMX-52421, October 28-November 1, 1968, November, 1968.

31. W. G. Clark, Jr., and L. J. Ceschini, Mater. Eval., 27, 180-184 (1969).

32. E. B. Norris, S. A. Viaclovsky, and A. R. Whitney, Proc. Seventh Symposium on Nondestructive Evaluation of Components and Materials in Aerospace Weapons Systems and Nuclear Applications, San Antonio, Texas, April 23-25, 1969, pp. 32-36.

33. C. K. Cole, Proc. 15th National Symposium and Exhibition, Society of Aerospace Material and Process Engineers, Materials and Processes for the 70's, Los Angeles, California, April 29-May 1, 1969, pp. 923-942.

34. S. Zirinsky, American Institute of Aeronautics and Astronautics, Fourth Annual Meeting and Technical Display, Anaheim, California, October 23-27, 1967, Paper 67-793, 14 pp.

35. B. Ostrofsky, Welding J. Research Suppl., (March, 1965).

36. D. Hagemaier, Proc. 1968 Symposium on the NDT of Welds and Materials Joining, Los Angeles, California, March 11-13, 1968.

37. B. Harris-Maddox, Ultrasonics, 1, 189-195 (1963).

38. General Dynamics Convair Division, Nondestructive Testing - Ultrasonic Testing, Classroom Training Handbook, San Diego, California, 1967.

39. A. Z. Raikhman, Zavodsk. Lab., 29(10), 1186-1188 (1963).

40. E. Scott, Ultrasonics, 4, 152-156 (1966).

41. G. Di Giacomo, J. R. Crisci, and S. Goldspiel, Mater. Eval., 28, 189-193, 204 (1970).

42. V. F. Dianov, Defektoskopiya, (5), 522-524 (1969).

43. V. A. Feoktistov and V. V. Grebennikov, Defektoskopiya, (4), 12-15 (1967).

44. A. De Sterke, Brit. Weld. J., 14(4), 183-190 (1967).

45. I. L. Grebennik, M. F. Krakovyak, V. V. Rakhmanov, I. A. Vyatskov, and I. N. Ermolov, Soviet J. Nondestructive Testing, (2), 184-187 (1967) translated from Defektoskopiya.

46. D. Hagemaier and G. J. Posakony, Mater. Eval., 26(11), 221-226 (1968).

47. D. E. Center and R. J. Roches, Mater. Eval., 27(5), 107-117 (1969).

48. Ya. F. Anikejev, Zavodsk. Lab., 29(10), 1191 (1963).

49. J. F. Lovelace and L. A. Luini, Mater. Eval., 26(10), 204-210 (1968).

50. W. C. Harmon, Mater. Eval., 24(3), 136-144 (1966).

51. T. W. McFarlan, Mater. Eval., 23(11), 553-557 (1965).

52. V. T. Bobrov, V. Ia. Kondratskii, N. A. Lebedeva, and O. R. Zaborovskii, Soviet J. Nondestructive Testing, 5, 359-362 (1968); translated from Defektoskopiya.

53. E. E. Murray, Mater. Eval., 25(10), 226-230 (1967).

54. Anon., Ultrasonics, 4, 84-87 (1966).

55. E. Mundry, Schweissen Schneiden, 19(4), 165-171 (1967).

56. B. T. Cross, K. J. Hannah, and W. M. Tooley, Proc.:1968 Symposium on the NDT of Welds and Materials Joining, Los Angeles, California, March 11-13, 1968, pp. 482-511.

57. C. E. Hohl, Proc. Fifth International Conference on Nondestructive Testing, Montreal, Canada, 1967, pp. 347-351.

58. R. W. McClung and K. V. Cook, Mater. Eval., 24(10), 573-576 (1966).

59. B. Ostrofsky and C. B. Parrish, Mater. Eval., 26(6), 106-108 (1968).

60. R. S. Sharpe and S. Aveyard, J. Iron Steel Inst., 201(10), 856 (1963).

61. J. F. Chretien and J. Huard, Bull. Inform. Sci. Tech. (Paris), (135), 35-42 (1969).

62. G. Schmitz, A. Wieczorek, and M. Levine, Technical Documentary Report No. ML-TDR-64-278, Air Force Materials Laboratory Research and Technology Division, (September, 1964).

63. R. V. Harris, Automation, 10(8), 71 (1963).

64. F. A. Silber and O. Ganglbauer, Berg-Huttenman. Monatsh., 11(9), 393-398 (1966).

65. E. M. Nusbickel and R. N. Cressman, Mater. Eval., 28(1), 1-7 (1970).

66. H. P. Hatch and K. A. Fowler, Mater. Eval., 23(10), 496-499 (1965).

67. H. J. Kopineck, G. Sommerkorn, and W. Bottcher, Stahl Eisen, 87(20), 1198-1201 (1967).

68. H. W. Grasshoff and J. Tobolski, Bander, Bleche, Rohre, 8(11), 736-743 (1967).

69. E. Isono, T. Ueno, and T. Udagawa, Nippon Kinzoku Gakkaishi, 32(10), 976-981 (1968).

70. A. E. Brown and H. L. Dunegan, Mater. Eval., 23(1), 46-47 (1965).

71. A. Matting and A. Heideman, Materialprufung, 8(5), 175 (1966).

72. S. Sasaki and K. Ono, Proc. Fifth International Conference on Nondestructive Testing, Montreal, Canada, 1969, pp. 341-344.

73. P. K. Bloch, Nondestructive Testing, 10(3), 16-18 (1952).

74. H. C. Drake, Railway Age, 128(9), 55-58 (1950).

75. G. de G. Gowan, U. S. Pat. 3,415,110 (December 10, 1968).

76. H. A. B. Wiseman and E. V. Garcia, IEEE Transactions on Sonics and Ultrasonics, SU-15(1), 57-58 (1968).

77. L. Beaujard, J. Mondet, and M. Kapluzak, Rev. Met., 62(12), 1187-1191 (1965).

78. M. Atkins and M. Druce, Iron Steel Inst. (London), 204(6), 607-608 (1966).

79. L. C. Lynnworth and E. H. Carnevale, Proc. Fifth International Conference on Nondestructive Testing, Montreal, Canada, 1969.

80. J. A. Bailey and A. Dule, Rev. Sci. Instr., 38(4), 535 (1967).

81. W. Kurz and B. Lux, Z. Metallk., 57(1), 70-73 (1966).

82. W. Kurz and B. Lux, Arch. Eisenhuettenw., 39(7), 521-530 (1968).

83. D. F. Smith and C. V. Cagle, Mater. Eval., 24, 362-370 (1966).

84. E. A. Aldridge, IEEE Transactions on Sonics and Ultrasonics, SU-14(2), 89 (1967).

85. A. E. Arave, Nucl. Appl., 6, 332-335 (1969).

86. V. A. Bobrov and N. V. Khimchenke, Defektoskopiya, (6), 50-54 (1967).

87. D. C. Worlton, U. S. Pat. 3,120,120 (February 4, 1964).

88. E. S. Proctor, Proc. Seventh Symposium on Nondestructive Evaluation of Components and Materials in Aerospace Weapons Systems and Nuclear Applications, San Antonio, Texas, April 23-25, 1969, pp. 180-187.

89. Anon., Elec. World, 39 (December 2, 1968).

Chapter 9

USE OF ULTRASONICS IN THE INSPECTION OF NONMETALS

I. INTRODUCTION

Ultrasonics is useful for inspecting many nonmetallic products. Velocity and attenuation of sound in these materials range in values from far below to much greater than those commonly observed in metals. For this reason, inspection often has to be conducted at low-kilohertz frequencies. Concrete dams, tires, or wood are examples of materials which require low-frequency examination. Certain ceramic products have excellent transmission properties and may be inspected at frequencies commonly used in inspecting metals.

Discussed in this chapter are some typical examples of the use of ultrasonics in the inspection of nonmetallic products, i.e., products made of wood, ceramics, rubber, leather, paper, and fibrous materials. Ultrasonic testing of livestock is becoming widely accepted and this is also discussed in this chapter.

II. CONCRETE

Quality and strength determine the transmission characteristics of sound through concrete. According to Deltuva (1), the strength of concrete can be determined within 10% by measuring the velocity of sound. The accuracy depends upon the aggregate used. Silicate products provide good uniformity and are easily adapted to ultrasonic testing. The velocity of ultrasonic waves in concrete increases with cure (or hardness). Cracks can be detected through their effect on energy losses.

According to Gemesi (2), measurement of the velocity of sound is the only nondestructive means for monitoring the strength of concrete during steam curing. The velocity increases with steam-curing time, passes through a maximum, and then drops to about 80% of the maximum value, where it remains constant. There is some evidence that the optimum time of steam curing corresponds to the maximum of ultrasonic velocity.

Ultrasonics has been used to examine the homogeneity of concrete in nuclear-reactor shielding prior to operation (3). The conditions of density and strength were related to ultrasonic propagation through the concrete wall.

Organic matter in sand-cement mixes can slow down the curing rate to the extent that hardening may be delayed by weeks. Jones (4) has studied the effects of soil in sand-cement mixtures by ultrasonic means. He claims a relationship between compressive strength (C) and pulse velocity (c) of the form $C = ac^3$, where a is a constant equal to 0.35 for the soil-cement tested. Savchuk and Filipchinski (5) give the following empirical formulas relating the compressive strength (C) of concrete to ultrasonic velocity (c):

$$C = c^{3.75} \text{ to } C = c^{3.85} \tag{9.1}$$

Jones (6) also has used ultrasonics to measure the thickness of concrete slabs (under favorable conditions) and to determine to some extent the severity of visible cracks in concrete.

Mikhailov and Radin (7) give the following values for determining the quality of concrete: (1) over 4500 m/sec indicates an excellent quality, (2) 4000-4500 m/sec indicates good quality, (3) 2300-4000 m/sec indicates acceptable quality, (4) lower than 2300 m/sec indicates very poor quality.

Simeonov (8) finds a similar correlation between compressive strength and velocity of sound. He reports that the velocity of sound in concrete cured under controlled laboratory conditions increased steadily with time. The velocity in specimens cured under atmospheric conditions varied irregularly, depending on the average air temperature. The velocity of sound in specimens cured under either condition corresponds well to the increase of strength measured mechanically.

These results correspond to what would be expected from a consideration of the dependence of velocity of sound on density and elastic properties. However, as Pawlowski and Raniszewski (9) warn, evaluation of the quality of concrete based upon velocity of sound must be related to the type of concrete. The velocity depends on composition, age, porosity, loading, method of reinforcing, and microcracks. Kordina (10) obtained consistent results with 27 different concretes and concluded that the ultrasonic technique of evaluating concrete is suitable for on-site quality control in the building industry. Shum (11) also recommends ultrasonic nondestructive testing for continuous production-line quality control in the manufacture of concrete products.

Testing of concrete is usually done at frequencies ranging from 25 to 100 kHz.

III. TIMBER

Decay affects the propagation of ultrasonic energy through timber (12). The velocity of sound decreases and the attenuation increases as a function of the amount of decay. The bark on opposite sides of the tree, standing or fallen, is removed and the transducers are coupled to the tree through a liquid at these points. Evaluation is based on comparison between the attenuation and velocity measured with values given in a standard table.

Recommended frequencies for inspecting wood are within the range 25 to 100 kHz.

IV. PLYWOOD

Ultrasonic through-transmission techniques have been used to detect blisters in plywood ranging in thickness from 1/4 to 3/4 in. The equipment will operate at speeds of up to 500 fpm. Automatic grading is accomplished at production-line speeds and voids are counted and marked automatically (13).

V. CERAMICS AND CERAMIC COATINGS

Ceramic materials such as porcelain, glass, and dense aluminum oxide have good ultrasonic transmission characteristics. Aluminum oxide or ceramics used in making spark plugs have a velocity of sound which is considerably higher than that in steel. Structural glass, fixed sheet glass, and polished plate glass of standard types have sound velocities that are approximately equal to that of steel. Ultrasonics is applied commercially for measuring the thickness of glass and porcelain enamels.

The velocity of ultrasonic waves in porcelain is a function of the porosity. Measurements of velocity of sound have been used to determine porosity-related mechanical strength of high-voltage ceramic porcelain insulators (14,15). Ultrasonic pulse-echo methods also have been used for detecting internal defects in these materials. Inspection of porcelains is usually done within the frequency range of 1 to 5 MHz (16,17).

Techniques similar to those for testing concrete have been used to evaluate materials used in fabricating steel-mill casting ladles (stoppers and ladle bricks). The evaluation is based upon the relationship between the velocity of sound, porosity, and compressive strength (18).

Cracks in a ceramic alter the elastic properties of the material. As a result, resonances associated with a given geometry shift in accordance with the extent of cracking. This resonance shift has been used as a basis for detecting flaws in spark-plug ceramics. A doughnut-shaped transducer formed to match the tip of the ceramic is coupled to the ceramic by oil. Resonant frequencies are located by sweeping the frequency of the driver oscillator over a range which includes at least one resonance of the specimen. Voltage peaks associated with impedance changes at resonance are displayed on an oscilloscope.

Evaluation of ceramic coatings on metallic substrates by ultrasonic methods has been tried. Ceramic coatings may vary widely in their acoustic properties. Glassy materials exhibit good transmission properties, while porous, coarse-grain structures may cause high acoustic attenuation. These differences influence the method of inspection and in some cases may rule out the use of ultrasonics.

Two types of information may be of interest in inspecting ceramic coatings. These are (1) adhesion between the coating and the substrate and (2) the quality of the coating itself.

Imperfect bonds between coatings and substrate have been detected through use of the nonlinear interaction between high- and low-frequency ultrasonic waves (19). At an unbonded region, a low-frequency wave encounters a stress discontinuity that results in a perturbation of the coating-substrate interface. The intermittent separation of the materials at this interface creates a periodic barrier to the transmission of high-frequency waves. High-frequency waves transmitted across the defective region are thus modulated by the low-frequency wave. Acoustical imaging methods have been used for detecting such discontinuities but less expensive methods based upon through transmission of ultrasonic waves could be equally effective.

Rayleigh-wave techniques have been applied to the inspection of ceramic coatings to detect defects such as cracks. This method can be used on a limited number of coatings, viz., those that are at or near their theoretical density, have fairly low attenuation, and are thick enough to permit the use of ultrasonic frequencies below 50 MHz. (The effective penetration depth of Rayleigh waves is one wavelength). Attenuation of Rayleigh waves in many coatings is so high that their use as a means of inspection is impractical.

The relative strength of fairly thick ceramic coatings sometimes has been determined by measuring the velocity of either shear or longitudinal waves in the coating or by the relative attenuation of such waves. Weak coherence of the material lowers the elastic moduli and may greatly increase the attenuation due to scattering.

VI. TIRES

Ultrasonics has been used successfully in inspecting tires

used on passenger cars, trucks, and aircraft. The technique is capable of detecting porosity and delaminations, measuring tread uniformity, and determining the state of the cure of the rubber (20). The test is based upon an ultrasonic through-transmission technique (21). Halsey (22) claims to have developed an ultrasonic technique for measuring tread thickness and tread uniformity. Because of the high attenuation due to the scattering from the fabric within the tire, the frequencies used are much lower than those normally used for nondestructive testing in metals. A technique employed by Patton and Hatfield (23) involves placing a crystal transmitting a 50-kHz beam from inside the tire to a number of receiving crystals outside the tire,which is immersed in water and slowly rotated. Faulty areas cause a drop in signal from one or more of the receivers. Defects smaller than 1/8 by 3/8 in. can be detected at a frequency of 180 kHz.

Rieckmann (24) uses a liquid-levitation technique for visualizing defect areas in tires. He places a 200-kHz ultrasonic transmitter inside the tire which is immersed in a reflection-free trough. He uses a narrow parallel beam of light, the diameter of which is small in comparison with the ultrasonic wavelength, for making the interference patterns visible. He also describes a barium-titanate hydrophone or microphone for picking up the ultrasonic wave after it passes through the tire. Rieckmann claims that the sensitivity of the method is affected if the tire has a marked tread pattern.

VII. LEATHER

The quality of leather can be ascertained by measuring the speed of sound, which varies with changes in chemical and physical structure and in fiber orientation in the leather. The velocity may be indicative of modifications of the fibrous order produced by strain, aging, and impregnating material (25). If a material produces appreciable acoustic attenuation, the velocity of sound is a function of the modulus of elasticity, the density, and

Poisson's ratio, and also of the attenuation. Velocity measurements are sufficient to determine the deviation of the quality of a sample from a control. The attenuation and frequency are of only secondary importance.

VIII. PAPER

Researchers at the Swedish Wood Center in Stockholm, Sweden, have applied ultrasonics in an interesting way to measure the thermal softening of paper products and to determine the influence of thermal autocrosslinking reactions (26). To determine the moduli of elasticity for five different papers in the temperature range -80 to +420°C , they first heated or cooled dry samples in the nip between precooled or preheated platens for 2-30 sec. This was followed by measurements of the velocity of ultrasonic waves in the material. From these measurements, and the apparent density, they were able to determine the Young's modulus of elasticity. For all samples tested, they claim that the modulus of elasticity decreased rapidly with increasing temperature in the ranges of 0 to 50°C, 100 to 240°C, and greater than 300°C. Between 240° and 300°C, the modulus of elasticity increased or remained constant. The results also indicated the second-order transitions at approximately 25 and 210°C. The latter was attributed to a glass transition in the amorphous part of the cellulous. Autocrosslinking took place at greater than 240°C in spite of short heating times, giving a thermal hardening effect in the range 240°C to 300°C. Thermal-hardening effects were separated from physical softening effects, which were less time dependent, by varying residence time in the heated nip.

IX. FIBROUS-BONDED COMPOSITES

A heterogeneous distribution of elastic properties is characteristic of certain materials such as graphite-fiber materials, filament-wound structures, and epoxy-glass laminates. Variations in the degree of cure or in the distribution of the binding agent lead to variations in the velocity of sound and in the attenuation of sound by scattering. The strength of the bond between adjacent fibers also affects the velocity and scattering. Ultrasonic measurements of thickness of reasonable accuracy may not be possible without the aid of a secondary method since such measurements are based upon velocity of sound. Resolution and defect sensitivity also are affected. However, degree of cure and quality of bonds are important and their influence on the propagation of ultrasonic waves makes ultrasonics of considerable value in the nondestructive inspection of composite materials.

Measurement of the velocity of sound generally is a good method of determining variations in the strength of composite materials. Attenuation can be correlated with delaminations, porosity, and resin content. Defects can be located by through-transmission or pulse-echo methods.

X. ANIMAL TESTING

Ultrasonic pulse-echo techniques have become an accepted means of judging livestock. Either an A-scan presentation or a B-scan presentation provides a photographic record of the fat profile. The methods used are similar to those used in medical diagnostic applications discussed in Chap. 14.

To obtain a B-scan recording, the position of the echo on the oscilloscope screen (appearing as a spot of varying intensity) is made to correspond to the location of the echo interface and moves with the position of the probe, which is liquid coupled to the skin of the animal. The spot is recorded directly onto a Polaroid film or it can be retained on a long-persistence oscilloscope screen for photographing later.

REFERENCES

1. J. Deltuva, Lietuvos TSR Aukstuju Mokyklu Mokslo Darbai, Statyba ir Arch., 1, 159-164 (1962).

2. J. Gemesi, Epitoanyag, 19(1), 31-34 (1967).

3. A. Honig and V. Zapletal, Inzenyrske Stavby, (4) 145-153 (1966).

4. R. Jones, J. Appl. Chem. (London), 6(5), 226-230 (1956).

5. A. Savchuk and A. Filipchinski, Beton i Zhelezobeton, (2) 45-52 (1958).

6. R. Jones, Ultrasonics, 1, 78-82 (1963).

7. A. V. Mikhailov and A. N. Radin, Beton i Zhelezobeton, 2, 266-268 (1956).

8. J. Simeonov, Polska Akad. Nauk, Zeszyty Probl. Nauki Polskiej, (26) 410-413 (1965) (in English).

9. Z. Pawlowski and J. Raniszewski, Polska Akad. Nauk, Zeszyty Probl. Nauki Polskiej, (26) 373-389 (1965).

10. K. Kordina, Materialprufung, 9(3), 81 (1967).

11. Y. F. Shum, Beton et Zelezobeton, 12(6), 15 (1966).

12. The Echo 23, published by Dr. J. and H. Krautkramer, Cologne, Germany, (April, 1970), pp. 319-320.

13. J. T. Collins, Instr. Soc. Am. Trans., 6(4), 303-306 (1967).

14. M. Bosek and J. Ranachowski, Polska Akad. Nauk. Zeszyty Probl. Nauki Polskiej, (26), 326-333 (1965).

15. C. Stüber, UK Atomic Energy Authority, AERE-Trans.-1077, 1967, 4 pp.

16. L. Haro and S. Hyvrylainen, Bull. Assoc. Suisse Elect. (Switzerland), 59(11), 486-490 (1968).

17. D. Weyl, Arch. Tech. Messen, (257), 121-124 (June, 1957).

18. T. Zabinska and R. Pampuch, Polska Akad. Nauk. Zeszyty Probl. Nauki Polskiej, (26), 439-448 (1965).

19. R. R. Whymark and W. E. Lawrie, Report for 1 April 1959-29 February 1960, Materials Analysis and Evaluation Techniques, Armour Research Foundation, Chicago, Illinois, May,1960, 45 pp.

20. G. H. Halsey, Rubber Age, 101(2), 70-75 (1969).

21. G. R. Kirk and L. U. Rostrelli, Nondestructive Inspection Techniques for Aircraft Tires, Final Technical Report, Southwest Research Institute, San Antonio, Texas, April, 1968.

22. G. H. Halsey, Mater. Eval., 26(7), 137-142 (1968).

23. G. R. Patton and P. Hatfield, Electron. Eng., 24, 522-525 (1952).

24. P. Rieckmann, Z. Angew. Phys., 8(8), 386-391 (1956).

25. Anon., Ultrasonic News, 1, 26 (1957).

26. E. L. Black, M. T. Htun, M. Jackson, and F. Johanson, Tappi, 50(11), (Pt. 1), 542-547 (1967) (in English).

Chapter 10

IMAGING, PROCESS CONTROL, AND MISCELLANEOUS LOW-INTENSITY APPLICATIONS

I. INTRODUCTION

Although nondestructive testing provides one of the largest markets for ultrasonic equipment, it is only one of many areas in which low-intensity ultrasonic energy is applied. Several additional applications are discussed in this chapter. Of these, imaging is the only one that might be considered to be a nondestructive testing technique. It is also of interest as a medical diagnostic tool and as a possible means of locating objects in other types of opaque media.

Other applications discussed in this chapter are process-control devices, fish locators, underwater-communication devices, delay lines, flowmeters, anemometers, pressure and temperature measuring devices, and devices for indicating the net load of a vehicle.

II. ULTRASONIC IMAGING

II.A. HISTORICAL BACKGROUND

A general definition of ultrasonic imaging is any technique of forming a visible display of the intensity and, in some systems, phase distributions in an acoustic field. This definition covers (1) the electronic-acoustic image camera, (2) B-scan systems which are occasionally used in nondestructive testing and are commonly used in medical diagnosis, (3) C-scan systems which are widely used in nondestructive testing, generally for inspecting flat materials or cylinders, (4) liquid-surface-levitation presentations, (5) liquid-crystal, photographic, and similar presentations, (6) light-refraction methods, (7) mechanical methods, and (8) acoustical holography.

The origin of ultrasonic imaging has been attributed to various scientists and the dates of invention vary widely. The similarities between acoustics optics and light optics were known to Lord Rayleigh in the nineteenth century.

In 1937, Pohlmann (1) initiated ultrasonic-optical imaging by demonstrating that small, nonspherical particles have a tendency to align themselves so that the normal to their major surfaces becomes parallel to the direction of wave propagation. Pohlmann produced a "Pohlmann" cell for imaging in which he had placed a large number of disk-shaped aluminum particles suspended in toluene. The diameter of the aluminum particles was much smaller than the wavelength in the toluene and the thickness of the particles was much smaller than the diameter. Figure 10.1 shows the details of construction of a cell based upon the Pohlmann principle. Pohlmann tried to use the cell for testing materials during World War II.

According to Suckling (2), the first published demonstration of an actual sonic image appears to be one by Barbieri in 1942 (3).

The forerunner of the schlieren technique of visualizing sonic and ultrasonic fields has been attributed to Toepler, who published papers on methods of making visible sound waves produced by a spark discharge in 1866 and 1867. The schlieren technique is based upon refraction of light as it passes through a transparent medium in which there exists a pressure gradient.

The Soviet scientist Sokolov is generally recognized as having invented the electron-ultrasonic image device which he called the ultrasonic microscope (4). There are indications that Sokolov's actual reduction of the device to practice came in 1947, although he claims that the 1947 ultrasonic microscope is a refinement of one used by him prior to World War II (5). All subsequent electron-acoustic image converters use the principles he originally proposed and thus verify the soundness of his method.

Various investigators have studied photographic techniques of one form or another, such as those involving the use of (1) sound sensitive photolayers, (2) starch-potassium iodide layers on paper, (3) thermosensitive paper, (4) phosphorescent screens, and liquid crystals. These methods, however, require intensities that are too high for most practical applications.

The liquid-levitation method was developed by Schuster according to Freitag (6). In this case, the plane of the sound image is made to coincide with the surface of a liquid so that the sound radiation pressure interacts with the force of gravity in such a manner as to cause the fluid surface to rise, in dependence upon the acoustic intensity. The sound image appears as a relief image on the fluid surface and is then converted into a light image by means of schlieren optics and made visible on a screen or ground glass plate. The sensitivity, which is quite acceptable for practical requirements, depends upon the curvature at the edges of the bulge rather than upon the total height of the bulge.

II.B. THE ELECTRONIC-ACOUSTIC IMAGE CONVERTER

Electron-acoustic image converters are modifications of the original Sokolov disclosure. Sokolov envisioned a television-camera tube in which the light-sensitive element is replaced by a pressure-sensitive element--a piezoelectric plate. Secondary electrons are given off when the plate is struck by a scanning beam of electrons. Under the influence of an ultrasonic field, electrical potentials are produced on the faces of the piezoelectric plate that are proportional to the impressed acoustic pressure and these electrical potentials modulate the secondary emission of electrons. Sokolov used quartz plates. Figure 10.2 is a simplified diagram of Sokolov's ultrasonic microscope.

Secondary emission resulting from the impingement of an electron beam on the surface of a piezoelectric plate is a function of the velocity of the primary electrons and depends upon the plate material. A secondary-emission ratio is defined as the ratio of the number of electrons leaving the plate surface to the number of electrons arriving at the surface. When the number of electrons leaving the plate equals those arriving at the plate, the secondary emission ratio is unity. If the primary-electron velocity is increased from zero, the secondary-emission ratio also increases from zero, passes through a maximum, and then decreases. For some piezoelectric materials, the maximum ratio is greater than unity and there are two velocities, one above and one below the point of maximum secondary emission, at which the ratio becomes unity. For other materials, the maximum ratio never exceeds unity.

If the secondary-emission ratio is less than unity, the excess charge accumulates on the surface of the plate, building up a potential in opposition to the scanning beam of electrons and reaching a limit where the incident beam is completely repelled.

At secondary-emission ratios greater than unity, a space charge accumulates in front of the plate surface. The space charge retards the scanning beam and leads to a loss of sensitivity toward the edges of the picture.

Starting with the basic design of Sokolov, Smyth (7) and his co-workers developed an ultrasonic-imaging camera (Fig. 10.3) which could be operated in either an amplitude-sensitive mode or a phase sensitive mode.

For the amplitude-sensitive mode a low-voltage (160-200 V) scanning electron beam is utilized. At this low voltage, the secondary-emission ratio is less than unity, which causes the electrical potential of the surface of the piezoelectric plate, which in Smyth's camera is quartz, to be brought down to that of the cathode of the electron gun in the absence of an ultrasonic field. After thus scanning in the absence of ultrasound, the ultrasound is turned on and the plate is scanned again. Electrons collect over the surface of the plate in proportion to the piezoelectric voltage present at each part of the surface. A corresponding image charge is formed on an anode located outside the tube, and the video signal is derived from the anode charges. The ultrasound is turned off and the negative charge on the piezoelectric plate is removed by flooding with gas ions. Time required to remove the charge is 0.02 sec. Smyth has also recommended the use of high-voltage scanning to remove the negative charge, followed by low-voltage scanning to bring the surface back to cathode potential, both with the ultrasound turned off. Then with the ultrasound on, the plate is scanned again at low voltage to form the image as before. This procedure avoids the need for ion recharging and thus allows the use of a high-vacuum tube.

For the phase-sensitive mode of operation proposed by Smyth (7) higher voltage beams (600 V) are utilized along with an auxiliary grid on which the secondary-emission current is collected. The inner quartz surface is stabilized at about the voltage of the auxiliary grid. The piezoelectric charge on the quartz plate tends to become neutralized and, as a result, the current to the quartz plate and to the collector grid fluctuates at the frequency of the ultrasonic waves. The output signal at each picture point is therefore proportional to the incident ultrasonic pressure.

Alternative means for reproducing the image on the piezoelectric surface have been devised to avoid some of the difficulties inherent in high-velocity scanning-beam stabilization of an insulating surface. One proposed method involves the use of a photoemissive surface deposited upon the piezoelectric surface, scanning then being accomplished by a flying spot of light (8). Mechanical scanning methods also have been used and are considerably more sensitive than scanning with an electron beam. The surface of a piezoelectric plate located in an ultrasonic field is scanned mechanically using a capacitive, noncontacting electrode or by sliding a small electrode in contact with the surface. The speed of operation of the electron-acoustic image converter more than offsets the disadvantages of lower sensitivity.

Smyth has also considered pulsed operation, which avoids cancellation effects due to reflected-wave interferences (9). However, the scanning beam and the arrival of the pulse must be synchronized and the pulse repetition rate must be low enough to avoid interferences between pulses. Only one point per pulse is obtained, and scanning the ultrasonic field to obtain one complete image may take several seconds.

The quartz plate of Smyth's ultrasound image camera is bonded by an epoxy cement to an annular plate of nonmagnetic steel. The steel plate is in turn cemented to one end of a glass tube 4.5 cm in diameter. The mating, bonded surfaces are lapped flat so that very little resin is exposed to the vacuum. An electron gun and a zirconium getter are fitted into the other end of the glass tube. An ion-trap gauze is also located in the tube near the quartz plate.

Jacobs has developed ultrasound image converters also based upon Sokolov's method (10), but with an electron multiplier built into the camera tube (see Fig. 10.4). This construction provides shielding for the low-level circuits and virtually eliminates ground loops. The effects of stray electromagnetic radiation fields from the ultrasound generator are minimized. The electron multiplier amplifies the signal currents to approximately 10^5 times greater than the threshold values before they are introduced into the associated amplifiers. In Jacobs' camera, the secondary-emission electrons produced by the scanning beam are attracted towards a positively charged electrode in the electron-multiplier structure. The electron multiplier then serves as a wide-band amplifier with good noise characteristics. Its output constitutes the video signal which is processed through a conventional closed-circuit television system. Jacobs has used both sealed ultrasonic camera tubes with quartz crystals permanently sealed to the end of the tube and tubes with interchangeable crystals which permit operation over the frequency range 1 to 15 MHz.

Jacobs and co-workers have introduced the use of color to enhance the sensitivity of the ultrasonic image converter (11). The method utilizes the NTSC system of video-signal processing, which is the standard for color-television broadcasting within the United States. The operating frequency is 3.58 MHz. The advantage of the color display is the greater sensitivity of the human eye to differential changes in color than to differential changes in brightness. The color display indicates both relative amplitude and phase of the received ultrasonic signal. Jacobs claims that at a frequency of 3.58 MHz, the minimum resolvable detail is approximately 0.6 mm without lenses and approximately 1 mm with lenses.

The minimum resolvable detail in an electron-acoustic image converter depends upon the thickness of the piezoelectric plate which, at resonance, is one-half wavelength. The thickness of the plate also is the limiting factor on the size of the image. The diameter-to-thickness ratio must be small enough for the crystal to withstand the vacuum. Decreased sensitivity and diffraction effects within the crystal at the harmonics of the crystal limit the usefulness of operation at these frequencies.

In using the electron-acoustic image converter, it is often desirable to make use of acoustic lenses. There are at least two reasons for this: (1) the sensitivity of the system may be increased by the higher concentration of energy and (2) observations may be made over a larger area by passing the image through a lens and concentrating it on the receiving piezoelectric element of the image converter. This would be particularly advantageous at higher frequencies, at which the resolution is best but at which small-diameter receiving elements are required because of the increased fragility of the piezoelectric unit, as mentioned previously. The greatest problem with acoustic lenses is image distortion due to the wavelengths involved.

Acoustic lenses used for focusing ultrasound or for imaging purposes must satisfy two conditions. The velocity of sound in the lens material must differ significantly from that of the surrounding medium and the reflection of energy at the boundary between the lens and the surrounding medium must be a minimum. This latter condition is automatically fulfilled when the acoustic impedance of the lens material exactly equals that of the surrounding medium. Liquid lenses of chloroform or carbon tetrachloride, which have acoustic impedances similar to that of water, have been used in the laboratory, but their use in industry poses somewhat of a problem. Plastic lenses also have been used. The velocity of sound in these materials exceeds that of water and also the materials present a mismatch in impedance between the lens and the water, but not to the extent of rendering them useless. Their relatively high absorption limits their use to frequencies below 15 MHz. Aluminum and similar metallic lenses have lower absorption characteristics, so they could be useful at frequencies far above 15 MHz; however, the impedance mismatch between these materials and water prevents their effective use.

Since the velocity of sound in solids is higher than that in liquids, solid concave lenses are convergent and convex lenses are divergent.

In practice the object and the image camera are located in the far field of the transmitting transducer. The field intensity is thus made fairly uniform over the object. Diffraction makes it necessary to locate the camera as closely as possible to the object. The shadow which forms the image can be carried in the sound beam for only a short distance. The discontinuity within the plane in which it appears may be considered analytically as a new source of sound with its own near field and far field. The use of lenses on the receiver side makes it possible to move the object further from the image camera. The lens also produces diffraction rings. Over 80% of the acoustic energy is contained within a central lobe and it is this central lobe that is used in producing the image. However, the distance that a shadow can be carried depends upon the relationship between the wavelength and the dimensions of the object casting the shadow and also upon the cross section of the irradiating beam.

The sensitivity of the electron-acoustic image converter can be as high as 3×10^{-9} W/cm^2, with 10^{-7} W/cm^2 being typical. The signal-to-noise ratio in the camera is a function of the transconductance of the camera, the noise resistance of the tube, the coefficient of secondary electron emission, the frequency, the electron-beam current, the collector capacitance, the stray capacitance, the input capacitance of the amplifier, the capacitance of an element of resolution of the piezoelectric plate, and the pass band required for undistorted transmission of the signal. The pass band is determined by the number of frames per second and the number of image lines which are scanned.

There exists within the receiving crystal a source of image interference which results from mode conversion. At certain frequencies, plate modes of relatively high intensity are generated and these show up as waves progressing laterally across the image. The result is that the maximum resolution obtainable is a function of frequency. The range of usable frequencies is approximately 5% below the half-wave resonance (thickness mode) of the crystal and 10% above the half-wave resonance, with operation being best at resonance (12).

Electron-acoustic image converters usually are operated using continuous waves. C-W operation is conducive to standing waves between successive reflecting surfaces. These standing waves tend to obscure the image and should be eliminated if possible. Methods sometimes used are (1) tilting the object, if it is thin and flat, to cause reflections to deflect away from the source and the receiver; (2) inserting transition layers to produce a gradual acoustic match between adjacent media (a difficult procedure that usually degrades the image); and (3) inserting an attenuator between the reflecting surfaces to attenuate the reflected signal. The third method is particularly useful between the lens and the receiving crystal, since there is little variation in the relationship between the positions and orientations of the lens and the receiving crystal.

II.C. SCHLIEREN IMAGING

Schlieren imaging has been used for many years to visualize sound fields, and it has become a useful tool for studying the propagation of ultrasonic waves. Pressure gradients in an ultrasonic wave cause density gradients. When these gradients lie in the path of a light beam, the beam is refracted. In schlieren devices, the refracted light is utilized to produce the image of the sound field. This is done by one of the following methods:

1. Interrupting the refracted part of the beam to remove it from the field and focusing the remainder of the field on an image detector, viz., ground glass screen or camera film. The image produced is dark on a light background.
2. Focusing the refracted rays on the image plane and eliminating the remainder of the beam of light from the image. The result is a light image on a dark background.

Figure 10.5 illustrates a schlieren system that has been used for direct viewing of ultrasonic waves used in nondestructive testing. The image may be viewed directly using a ground glass screen. Images can be photographed (including movies).

In Fig. 10.5, the lamp, condenser lenses, filters, first knife edge, and first collimator are mounted on a single 1-meter optical bench. The second collimator, second knife edge, camera lens, and camera back are mounted on a second 1-m optical bench. These two benches are mounted on a 10-in.-wide, 11-ft-long steel channel. The water tank is mounted completely independent of the optical system so that changes made in the tank will not misalign the optical system. The adjustment of a high-sensitivity schlieren system can be upset by alignment changes of only a few microns. Hence, the optical system must be mechanically isolated from all other parts of the apparatus.

In operation, the image of the light source is centered on the first knife edge, after which the collimating lenses, tank windows, second knife edge, camera lens, and viewing screen are centered in the resulting light beam. Maximum sensitivity and uniform field result only when optics are properly centered and knife edges are placed accurately at the focal points of the corresponding collimating lenses. When properly aligned and focused, the field can be made to go from light through gray to dark uniformly by adjusting the second knife edge further into the light beam until it intercepts it completely. In the range of gray settings, convection currents, both in the water and in the air, are clearly visible. The ultrasonic field is made most easily visible by setting the second knife edge to intercept all of the main beam, permitting only the light refracted in the ultrasonic field to pass and thus provide illumination in the viewing plane.

Great care must be taken in selecting optics which are free of defects, such as bubbles, pits, scratches, surface haze, and dust, as each of these will scatter light to such a degree as to be evident in the image. Similarly, the water in the tank must be free of particles, such as dust, corrosion products, and algae, which also can scatter light and show in the image. The windows on the tank and the collimating lenses must be accurately surfaced so that light is not deviated from the main beam except by the ultrasonic field.

The optical system also may be adjusted for bright-field operation, in which case the ultrasonic disturbance appears as dark shadows. Although not generally so sensitive as dark field, the bright-field adjustment can, in some cases, show details not evident in the dark field.

A zirconium arc lamp makes a good light source for studying continuous ultrasonic waves. For making pulsed waves visible, it is necessary to use a stroboscopic light source synchronized with the ultrasonic source. Xenon flash lamps or commercially available stroboscopes are useful for this purpose. The flash of the lamp must be synchronized with the ultrasonic pulse. This is done by using an electrical signal from the pulse generator and a circuit containing a controllable time delay to facilitate photographing pulses at any point in their travel through the optical field. Movies of pulses propagating through water have been made while varying the delay on the stroboscopic light.

Color schlieren photography has been used for many years and can be quite useful in the study of ultrasonic waves. In particular, color is useful for showing the various pressure levels in an ultrasonic field. One method makes use of a spectroscopic prism between a slit (located in the position of the first knife edge, Fig. 10.5) and the first collimating lens. This method produces colors ranging from red or from blue to green. J. H. and J. W. Waddell have developed a method of producing a complete color spectrum (13). They eliminate the spectroscopic prism and use a color-filter matrix in place of the second knife edge and a vertical slit in place of the first knife edge. They use a high-pressure mercury arc for a light source.

The color matrix used by the Waddells consists of three gelatine color filters mounted between 2- by 2-in. glass-slide covers. The three filters are the primary colors--red, green, and blue. The green filter which is the width of the projected image of the slit (approximately 1/8 in.), is located in the center, and the red and blue filters (approximately 15/16 in. wide) are located adjacent to the green filter in such a manner that no white light can pass through the matrix.

The work of the Waddells was shock-wave oriented. The system can be used for ultrasonics and where, instead of vertical orientation, horizontal positioning of the optical slit and the filter might be preferable to correspond to the orientation of the wave fronts (if the wave propagation is in the vertical direction).

The quality of the color image will depend upon the correlation between the transmission properties of filters used and the light source used. The Waddells used Wratten 25 (red), 57 (green), and 47 (blue) filters manufactured by Eastman Kodak with a BH6 mercury arc lamp.

Light optical techniques are used to make other types of ultrasonic images visible, including liquid-levitation methods and those employing nematic liquid crystals.

II.D. LIQUID-LEVITATION IMAGING

Liquid-levitation imaging refers to the production of an ultrasonic image in the surface of a liquid. The sound radiation pressure interacts with the force of gravity so as to cause the fluid surface to rise in dependence upon the acoustic intensity. The sound image appears as a relief image on the fluid surface and is then converted into a light image by means of schlieren optics and made visible on a ground glass plate or photographic film. The sensitivity, which is quite acceptable for practical requirements, depends upon the curvature at the edges of the bulge rather than upon the total height of the bulge.

Gericke and Grubinskas have developed a cell based upon the liquid-surface-levitation principle for use in the nondestructive inspection of materials (14). The cell can be coupled to an object for inspection without the need for immersing the object. Their results include good images of 8- and 16-mm-diameter holes drilled into the side of aluminum cylinders 75 mm in diameter and 50 mm in height. Attenuation in aluminum is low, which produces a good situation for this application. The technique offers several advantages. Among these are (1) potentially unlimited area of field, (2) good image resolution, (3) inexpensive equipment, and (4) sensitivity satisfactory for nondestructive testing.

In most cases, pulsed or frequency-modulated ultrasound produces the best images. Continuous-wave ultrasonic energy produces the best detail resolution, but suffers from ambiguities caused by standing waves.

The liquid-surface-levitation converter has an inherent time constant of about 0.01 t0 0.1 sec depending upon the viscosity of the conversion liquid.

II.E. ULTRASONIC IMAGING WITH LIQUID CRYSTALS

Various investigators have experimented with the use of liquid crystals for imaging ultrasonic fields. Liquid crystals are liquids having certain physical properties shown by crystalline solids but not by ordinary liquids. A technique used by various investigators (15) involves the use of a cell composed of an acoustical absorber and cholesteric liquid crystals. These crystals change color at various transition temperatures and their sensitivity to temperature depends upon the particular chemical composition. The purpose of the acoustical absorber is to increase the rate of absorption at a location near the crystals and thus to increase the sensitivity of the liquid crystal cells. Liquid crystals are subject to rapid oxidation and encapsulation retards this action.

In practice, the object to be imaged is placed in water. A beam of C-W ultrasonic energy is passed through the object and impinges on the imaging cell. The ambient temperature of the water path is kept slightly below the transition temperature. The higher amounts of energy absorbed at the high-intensity portions of the beam produces local heating to above the transition temperature and thus the image is formed.

With this cell, very high intensities (approximately 100 W/cm^2 in early experiments) are necessary to produce color change. Such an intensity is much too high for practical use in any imaging system.

A second class of liquid crystals, known as nematic crystals, can be used in which the ultrasonic energy produces a disturbance within the crystals which can be detected optically with sensitivity equivalent to that of schlieren systems or liquid-levitation systems. Nematic crystal cells provide high sensitivity, high resolution, low cost, large-area capability, and handling convenience.

II.F. PHOTOGRAPHIC METHODS OF IMAGING BY ULTRASONICS

One of the earliest approaches to imaging ultrasonic fields utilized the effects of ultrasonic energy on photographic emulsions. These early experiments have led to other similar methods involving thermal and chemical processes to produce a permanent record of the intensity distribution in an ultrasonic field.

One proponent of ultrasonic imaging by photographic methods is Arkhangelskii (16). One approach taken by this investigator involves exposing a photographic film (photodiffusion method) to light before developing the film under the influence of ultrasonic irradiation. The ultrasonic-intensity variations due to an object located in the ultrasonic field affect the rate of development accordingly. Arkhangelskii concluded that the main factor giving rise to acceleration of the development of an exposed film in an ultrasonic field is the acoustic pressure and the second most important factor is a local increase in temperature of the sensitive layer. The ultrasound promotes penetration of the liquid into the "pores" of the gel during the compression half of the cycle, but during the rarefaction half of the cycle, the liquid remains in the cell because of absorption or other forces.

Arkhangelskii found a correlation between the concentration of the developer, the exposure time, and the contrast of the image. Exposure time required increases with decreasing concentration of developer, but the corresponding contrast increases considerably with lower concentration of developer.

Arkhangelskii and Pinus investigated the influence of ultrasound on the process of diffusion of a colored electrolyte in a gelatin film (17). Exposing the entire diffusion system to ultrasound at frequencies ranging from 500 to 9 MHz at an intensity of 0.3 W/cm^2, these investigators were able to show a 15% increase in diffusion rate in a gel containing 75% water. The cause of the increased diffusion rate was traced to the acoustic pressure.

Berger and Kraska conclude that the direct action of ultrasonic waves on a photographic emulsion is influenced to a great extent by the softness of the emulsion (18). The emulsion was softened by soaking in water before the exposure, thereby making it more sensitive. Film was exposed to ultrasonic irradiation in an iodine-water solution without darkroom techniques. The iodine effected color changes in the irradiated areas, allowing the ultrasonic image to be observed as it formed.

Schilb, Bennett, and Adams have described a "bleach-out" method of producing visible images of sound fields (19). The basic materials are a dye, a bleach, and gelatine-coated glass plates. The plates are outdated photographic plates, which are fixed, hardened, and washed. They are then soaked in the dye solution for 24 hr and exposed before they have an opportunity to dry. The action of the bleach on the dye at any point occurs at a rate depending on the intensity of the ultrasound --the lighter areas corresponding to the positions of higher intensity in the field.

Eckardt and Fintelman (20) produced sound images using pigment papers which consist of a nonhardening gelatin layer with a mixture of insoluble color pigments. Such films can be used as transfer agents in the copper deprint method. They can be hardened in a 0.001% potassium permanganate solution under the influence of ultrasound, so that instead of light refinement there is sound refinement. Suitable coloring of the pigment paper can be obtained with a 5-min exposure at a frequency of 1.4 MHz and a sound intensity of 0.25 W/cm^2. Higher intensities (1.5 W/cm^2) produce detrimental heat effects.

Although it is not a photographic technique, the influence of ultrasonics on the luminescence of phosphors has been utilized to produce ultrasonic images in much the same manner as it has been to produce images photographically. An object in the path of ultrasonic rays incident on a luminescent screen casts an acoustic shadow which is transformed into a visible image by the effects of ultrasonic energy on the luminescence of the phosphors of the screen (21). The action of ultrasound is explained in part by the heating effect of the ultrasonic energy.

All photographic methods and the method based on the luminescence of phosphors under the influence of ultrasound exhibit extremely low sensitivity, that is, they require high intensities. They are slow. For these reasons, they have found no practical industrial application.

II.G. ULTRASONIC HOLOGRAPHY

Holography is a form of three-dimensional imaging conceived and named by Gabor. Gabor revealed his discovery in a paper (22) written in 1948 and published in 1949. It was first used in electron microscopy. Gabor's objective appears to have been one of overcoming problems associated with correcting for spherical aberration of electron lenses. In the new method, the problem was shifted to light optics where refracting surfaces can be figured to any shape, without limitations associated with electron optics imposed by laws of the electromagnetic field. It is necessary only to imitate the aberrations to an accuracy corresponding to the accuracy required for a given resolution.

The principle of holography, as presented by Gabor, is that if a diffraction diagram of an object is taken with coherent illumination, and a coherent background is added to the diffracted wave, the photograph will contain the full information on the modifications which the illuminating wave has suffered in traversing the object. The object can be reconstructed from this diagram by removing the object and illuminating the photograph by the coherent background alone. The wave emerging from the photograph will contain as a component a reconstruction of the original wave which appears to issue from the object. Thus, a hologram may be defined as a recording (or photograph) of two or more coherent waves. If one recorded wave is from an illuminated object and another is a reference wave, illuminating the hologram with the reference wave reconstructs twin images of the original object in three dimensions.

Gabor used a monochromatic light source placed behind the object to provide the coherent light waves. His technique was thus limited to microscopic use. The development of the laser has made available large and much more powerful coherent light sources and has been responsible for renewed interest in this technique.

Gabor explained his discovery mathematically. A coherent monochromatic wave U with a complex amplitude striking a photographic plate may be described by

$$U = Ae^{j\theta} = U_o + U_1 = A_o e^{j\theta_o} + A_1 e^{j\theta_1} \qquad (10.1)$$

where A and θ are real,

U_o is a background wave, and

U_1 is a remainder;

and the absolute value of amplitude is

$$|A| = [A_o^2 + A_1^2 + 2A_oA_1 \cos(\theta_1 - \theta_o)]^{\frac{1}{2}}.$$

During the reconstruction process, illuminating the positive hologram with the background wave, U_o, alone produces a new wave, U_s, such that

$$U_s = A_o e^{j\theta}[A_o^2 + A_1^2 + 2A_oA_1 \cos(\theta_1 - \theta_o)]^{\frac{1}{2}\gamma} \quad (10.2)$$

where γ is a power relating transmission intensity to the exposure. The best choice of γ is 2. When $\gamma = 2$

$$U_s = A_o^2 e^{j\theta_o}[A_o + (A_1^2/A_o) + A_1 e^{j(\theta_1 - \theta_o)}] + A_1 e^{-j(\theta_1 - \theta_o)} \quad (10.3)$$

Thus, if the background is uniform, the new wave U_s contains a component proportional to the original U, which is

$$A_o^2 e^{j\theta_o}[A_o + A_1 e^{j(\theta_1 - \theta_o)}]$$

It also contains a spurious component

$$A_o^2 e^{j\theta_o}[(A_1^2/A_o) + A_1 e^{-j(\theta_1 - \theta_o)}]$$

These two components constitute the twin images, the twin waves carrying equal energies.

Each part of the hologram contains the total information concerning an object. Therefore, it is possible to reconstruct the object from a section cut from the original photograph. The result is similar to viewing the object through a window.

An interesting fact is that the wave used in the reconstruction need not be the original, and this is the factor that permits the use of ultrasonic waves and the subsequent reconstruction of the image using coherent light from a laser. However, the size of the image is changed in proportion to the ratio of the wavelength of the reconstructing wave to the wavelength of the original illuminating wave.

Several methods of making acoustical holograms have been used.

One method, Fig. 10.6, makes use of liquid-levitation. The reference beam may be obtained either by reflecting a portion of the irradiating beam onto the surface or by generating a separate wave using a second transducer. The height of the bulge of the surface is critical. The height of the bulge should be small relative to the wavelength of the light. Sometimes a wetting agent is added to improve the image formation.

Other methods of recording holograms include electron-acoustic image converters and mechanical scanning.

Mechanical scanning consists of an acoustic detector (microphone or hydrophone) which produces a signal amplitude that is proportional to the intensity of the sound field at a given position. The amplified signal from the detector controls the output of a light. The acoustic detector scans the sound field and the light moves in correspondence with the position of the source, producing a record on a photographic film. This is a slow method but can be used in air as well as in liquid.

The electron-acoustic image converter is a good means of producing ultrasonic holographs, but suffers from limitations relative to frequency or the size of piezoelectric crystal that can be used in the camera.

Young and Wolfe (23) have proposed the use of deformable films on a solid substrate to produce holograms. Thermoplastic coatings were placed on suitable solid substrates and these were located at the surface of the liquid bath. The materials were either soft at the ambient temperature or softened by electrical heating. After the hologram was formed, it was preserved by solidification of the film. The image was then reconstructed by conventional coherent light optical techniques, the depth of the deformations being sufficiently shallow for suitable light reproduction. Young and Wolfe claim to obtain adequate deformations at frequencies ranging from 5 to 30 MHz and intensities between 0.1 to 1.0 W/cm^2. These intensities are too high for many practical applications, particularly medical diagnostic applications.

II.H. APPLICATIONS OF ULTRASONIC IMAGING

The nature of ultrasonic waves and their reaction to various discontinuities of acoustic impedance have made acoustic imaging of interest in two technical areas. These are nondestructive testing and medical diagnosis. The applications are restricted to those areas in which the image can be maintained in the ultrasonic beam. Diffraction effects limit the distance that a shadow can be maintained in a beam with a cross section that is large compared with the size of the obstacle. A typical distance is 8-10 diameters of the obstacle. One area of application of acoustical imaging is the inspection of flat plates such as flat fuel elements for lack of bond between core and cladding.

Acoustical holography offers an additional advantage of providing a three-dimensional display of the defect in nondestructive testing or medical diagnosis, thus enhancing interpretation.

To date, neither the electron acoustical image camera or holography has received widespread acceptance in industry or medicine.

III. PROCESS CONTROL

Ultrasonic properties of materials can be used to control various processes. In one type of process, as a material crystallizes from a liquid or syrupy state, the attenuation of ultrasonic energy increases significantly. Thus, if ultrasonic waves are caused to pass through such a medium, the decrease in signal amplitude can be used to control the energization of a relay to initiate follow-up action.

Similarly, methods based upon either the transmission properties of the materials or the receipt of echoes from inclusions have been used to indicate the solids content of liquid slurries or the presence of bubbles or other foreign particles in enclosed systems.

IV. UNDERWATER APPLICATIONS

IV.A. FISH LOCATORS

Considering the early history of ultrasonics and its relationship to underwater sound, it is not surprising to see it used commercially for locating fish and measuring water depths. Several ultrasonic fish locators are available. Their operation is based upon pulse-echo principles. These devices are widely used by commercial fishermen in certain areas of the world.

In operation, the instrument transmits a short pulse of ultrasonic energy through the water and detects echoes received from the fish and from the bottom or other obstacles. Interpretation requires a degree of skill to identify the size of fish and to discriminate between fish and other sources of echoes.

Various problems are encountered in locating fish ultrasonically. Fish near the surface will not be located. Background noise can mask out wanted signals. In rough weather, the motion of the ship or boat will cause the direction of the beam to vary severely and lose its effectiveness.

In one inexpensive unit for measuring depths of lakes or streams for use by sportsmen, a spinning arm similar to a clock is incorporated which is driven by a synchronous motor. A pulser is energized each time the arm passes the vertical position. Simultaneously, the pulser causes the transmitter to emit a pulse of ultrasound and to light a neon lamp on the arm. When an echo returns, the transducer detects the echo signal, which is then amplified and used to energize the light a second time. The clock dial is calibrated in terms of depth, and the point at which the light is energized by the echo signal corresponds to the depth at that point.

IV.B. COMMUNICATION

Ultrasonics has been used for underwater communication between ships and submarines and between surface personnel and swimmers. Underwater voice and code communication is possible to distances of approximately 7 miles in the ocean if water temperatures are favorable. Practical audio frequency ranges are approximately 150 to 3000 Hz using carrier frequencies within the range 8 to 10 kHz. Devices for use of swimmers in lakes and ponds also have been developed.

V. DELAY LINES

Ultrasonic delay lines are being used in radar and computer signal processing. A signal is stored in the form of a short ultrasonic pulse for a brief period of time and retrieved for use at a proper time for display or computer use. Two general types are used: fixed delay and variable delay.

A typical fixed type of delay line consists of a many-sided (flat) low-loss material such as quartz with a transmitting transducer attached to one side and a receiving transducer attached to another side. The pulse of longitudinal-wave ultrasonic energy reflects alternately from each of the flat sides until it finally impinges on the receiving transducer.

Techniques that provide variable acoustic delay lines include (1) the use of high frequency surface waves and (2) the deflection of laser beams. With surface waves, which travel at a much slower rate than the longitudinal waves, the time is varied by varying the position of the receiving transducer.

With the laser-beam technique, the input transducer is attached to a transparent acoustic delay bar (24). A laser beam is directed onto a rotatable mirror which directs a laser beam by way of a converging lens through the delay bar, then through a second converging lens, and finally onto a photodetector. As the sound wave encounters the laser beam, it acts as a moving diffraction grating on the beam and frequency shifts the laser beam by an amount equal to the ultrasonic frequency. The undisturbed portion of the light beam and the diffracted portion are heterodyned together to produce an output signal at the photodetector at the input frequency. The delay is accomplished by changing the rotation of the mirror. By scanning the light beam along the bar, it is possible to generate other effects. For example, if the laser beam sweeps the bar at a faster rate than the ultrasonic wave propagates, the output signal is reversed in time. If the scanning rate is such that the light-scanning rate falls between twice the velocity of sound and the velocity of sound, the output signal will be expanded. If the velocity of sound is greater than the scanning rate, the output is compressed and upshifted. When the scanning rate is exactly twice the velocity of sound, the output is a perfect time-reversed replica of the input pulse.

According to Smitz and Sittig (25), ultrasonic delay lines for digital storage, designed for bit rates near or beyond 100 MHz and capabilities around 1000 bits, compete favorably with integrated circuit storage in terms of economy, ease of fabrication, and speed.

VI. APPLICATIONS IN AIR

Several practical applications are included in this category of low-intensity ultrasonics. They include vehicle detection and speed measurement, television controls, garage-door openers, intrusion detection (burglar alarms), level detectors (liquids and solids), conveyer-type monitoring, gas-pressure measurement, guidance for blind, and air flow measurements (anemometry).

Ultrasonic sensors for use in air operate on principles involving transmission and reception of echo pulses from a target, breaking an ultrasonic beam (pulse or CW), sensing a given frequency to trigger a relay, or sensing Doppler shifts corresponding to the movement of a target.

Many patents have been issued on inventions involving the detection of vehicles by ultrasonic means. The approach of a vehicle can be detected (1) by beaming pulsed ultrasonics in a direction which intercepts the anticipated path of the vehicle and detecting echoes, (2) by causing the vehicle to break an ultrasonic beam (CW or pulsed) between a transmitting and receiving transducer, or (3) by measuring the Doppler frequency caused by the motion of the vehicle into an ultrasonic beam. The first and second methods are useful for counting vehicles. The third method can be used to determine the velocity of approach by applying the appropriate Doppler equations. A practical range for monostatic units operating in open, quiet air is 30 ft at a frequency of 26 kHz. The range increases as the frequency decreases. High-intensity broadband pulses also make possible much greater distances. High-intensity pulse-echo equipment has been used to detect an automobile in an otherwise empty parking lot at distances to 160 ft. Air turbulence (wind) places a practical upper limit on the effective distance of these detectors. Noise due to turbulence is controlled by appropriate filters and signal-cancellation techniques, but the range of velocities is limited by the degree of such filtering.

An interesting passive device which senses the approach of another vehicle from the rear has been developed by Sylvania (26). The device responds only to sounds emanating from vehicles entering a precisely controlled surveillance area by sensing engine and tire sounds. It discriminates between moving and stationary objects. It detects at a distance of 25 ft from the receiving transducer in a target area greater than a single traffic lane and at target speeds not less than 25 mph. The warning is given either audibly or by dashboard display lamp.

Television controls for switching channels from across a room are based upon generating sympathetic resonances in elements within the set in response to sound waves emitted by the control box. Pushing a button causes a striker to resonate one of the bars. The pickup unit within the television set generates a signal to operate the switching mechanism. In another method, a resonant disk is used that vibrates in flexure as it is struck (27). The resonant frequency is controlled by constraining dampers and selective location of a striking hammer. Channel selection, volume control, and color adjustment can be accomplished ultrasonically with at least 8 channels being accommodated with the frequency range from 34 to 45 kHz (28).

Similar techniques are used to open garage doors and to switch on or off lights and appliances. Some interesting interactions have occurred in households employing multiple units, such as having a coffee percolator energized when it was only intended that the garage door be closed. Sometimes low-flying jet aircraft have caused garage doors to open.

Intrusion-detection devices (burglar alarms) work either on amplitude variations of an ultrasonic signal caused by an intruder, or more commonly, they are based on sensing Doppler signals caused by the intruder. The latter devices are very effective and sensitive. Air turbulence, heat sources, and vibrations of motions of objects (walls, drapes, etc) within the room are major contributors of noise in installations employing these units. However, modern equipment contains compensating circuitry for noises due to these sources and, if properly installed, they will generate few false alarms.

Measurement of liquid level can be accomplished either from the gas phase or from the liquid phase, depending upon which appears to be the more practical approach. The measurement usually is based upon the time required for a pulse of ultrasound energy to travel from a transmitter to a free surface, reflect from the surface, and return to a receiver, which may be the same transducer used to generate the pulse. Some installations make use of the change in load impedance occurring when the liquid level falls below or rises above a fixed position marked by the location of the ultrasonic transducer. The same method can be used to discriminate between liquid level, foam, and air.

Conveyer-type monitoring includes a variety of applications. Ultrasonic switches are available that count, sort, position, or control other functions associated with a conveyer-line operation. Ultrasonic beams in the 30 to 40-kHz frequency range are broken or reflected from an object to be detected, such as a box or carton. Either the loss of signal in a broken beam or receipt of an echo by a receiving transducer is used to operate a relay, counter, or other device as the operation requires. Detection range is up to 36 ft.

Such systems have been used to detect breaks in paper webs, and to control automatic production-line operations. An example of the latter is the control of the application of cream and jam to sponge cakes (29). The cakes are split and the jam and cream are applied to the lower half. Ultrasonic sensors are used to detect the presence or absence of a cake in a pan. This is done by locating ultrasonic transducers above and below the conveyer and transmitting ultrasonic energy between them. The cake tray has a hole in the bottom through which the beam passes if no cake is present. When the receiver detects a signal indicating no cake is present in the tray, its output is amplified and operates a relay to stop the cream and jam applicator to avoid waste.

Similar systems have been used to monitor the flow of coal in mines by counting cars, indicating blocked chutes, and measuring the height of coal in bins. Measuring the height of material in bins is not restricted to use in coal mines but can be used wherever bin storage is required.

The United States Bureau of Mines has shown the feasibility of using pulse-echo techniques for evaluating coal-mine roof strata (30). Defects in the roof are indicated by echoes. One problem is the availability of compact, explosionproof equipment.

Ultrasonic detection systems for use in guiding the blind have been investigated for many years. In most cases, these operate on echo principles and provide an audible signal to the user by means of earphones. One particular method rectifies the received pulses and produces an audible signal with a pitch that corresponds to the time between successive echoes and, therefore, to the distance to the obstacle. The effectiveness of the unit is thus dependent upon the user's ability to estimate pitch.

VII. MEASURING FLUID-FLOW, PRESSURE, AND TEMPERATURE

VII.A. FLUID-FLOW

Several types of ultrasonic flowmeters have been developed for measuring both liquid and gas flow rates. All are based upon one of two principles: (1) the effect of the moving fluid on the time of travel of sound between two points or (2) the effect of the moving stream on the deflection of an ultrasonic beam that is propagating normal to the direction of flow. Ultrasonic flowmeters are capable of measuring fluid flow without altering the flow pattern, and thus they introduce little or no pressure loss.

A flowmeter developed by the British Scientific Instrument Research Association (31) is described as one that provides an electrical output consisting of pulses of constant amplitude and repetition rate (25 V and 800/sec, respectively). The width of the output pulse is proportional to the rate of flow. The pulse is averaged by an RC network, giving a direct voltage with an amplitude that varies with pulse width. The voltage from the RC circuit is applied to drive an integrator motor which is used to integrate the flow rate with respect to time.

In one method, upstream pulse travel times are compared with downstream travel times. Figure 10.7 illustrates the principle of the design. In actual practice, the transducers may be located side by side. The distance between each transmitting transducer and its corresponding receiving transducer is a constant d. If the travel time of a pulse from an upstream transmitter to a downstream receiver is T_1 and the travel time upstream is T_2, then

$$\begin{aligned} T_1 &= d/(c' + v_f \cos \theta) \\ T_2 &= d/(c' - v_f \cos \theta) \end{aligned} \qquad (10.4)$$

where c' is an equivalent velocity of sound in the fluid and includes an increment corresponding to a distance the pulse travels between the transducers and the boundary of the flowing fluid and v_f is the fluid velocity. When the fluid is stationary, $T = d/c'$.

In a variation of the meter shown in Fig. 10.7, only two identical transducers are used. Each alternates between transmitting and receiving.

Various techniques have been used to compare the upstream and downstream pulse-propagation rates. One method involves the use of a "sing-around" principle in which a pulse generator produces a short train of ultrasonic waves. The received signal is amplified and used to retrigger the pulse generator. Neglecting delay times due to the electronic system and the distance the pulse travels outside of the fluid stream, the difference between the downstream and the upstream pulse repetition rates is

$$f_1 - f_2 = (2\, v_f \cos\, \theta)/d \tag{10.5}$$

Flowmeters for measuring blood flow have been developed which are based on comparison of upstream and downstream transit times between two cylindrical transducers located around the blood vessel (32). It is claimed that this instrument can measure flow rates of 1 to 100 cm/sec. The lower limit represents a difference in transit time of 2×10^{-10} sec, or a phase difference of 0.03 degree (operating frequency between 300 and 400 kHz).

Acoustic anemometers have been developed which utilize the same principles illustrated in Fig. 10.7.

Further examination of Eq. (10.4) will show that

$$f_1 + f_2 = 2c' \approx 2c \tag{10.5a}$$

where c is the velocity of sound in the fluid. By inserting an additional transducer that senses the load impedance, providing the characteristic impedance of the liquid (ρc), it is possible to obtain a measure of the mass flow rate with the same meter. A computer determines the velocity of sound c, from the sum frequency, divides this quantity into the impedance to obtain the density, ρ, and, from the average flow rate obtained from the difference frequencies, determines the mass flow rate.

The most significant errors in meters of the types shown in Fig. 10.7 are caused by the velocity profile. The meter measures the average flow velocity along the path of the beams. For laminar flow, the meter reading is two-thirds the maximum velocity at the center of the stream. The true average is given by one-half the maximum velocity. For turbulent flow, the relationship depends upon the Reynolds number for the pipe or channel.

In a flowmeter developed by Westinghouse, an upstream traveling pulse and a downstream traveling pulse are initiated simultaneously (33). Each transmitter then becomes a receiver and detects the received pulses. The meter then measures the difference in travel time and relates this to the average flow velocity, where, for beam direction parallel to flow

$$T_1 = d/(c + v_f) \text{ and } T_2 = d/(c - v_f) \qquad (10.4a)$$

Then

$$T_2 - T_1 = 2dv_f/(c^2 - v_f^2) \approx 2dv_f/c^2 \quad v_f << c \qquad (10.6)$$

The second type of ultrasonic flowmeter is based upon the deflection of an ultrasonic beam by the fluid flow (34). A transmitting transducer located on one side of the pipe emits a continuous wave into the fluid stream. A split transducer on the opposite side of the pipe (Fig. 10.8) determines the amount of beam deflection. The outputs of the two transducers are compared by use of a sensitive differential amplifier. If there is no flow, the beam strikes midway between the two receiver sections, the two sections generate equal voltages, and the output from the differential amplifier is zero. When the fluid is flowing, the beam shifts in the direction of the flow by an amount corresponding to the flow rate, and the outputs from the two sections differ. The difference voltage corresponds to the flow rate. If the flow rate were constant across the pipe, the beam deflection, ϕ, would be given by

$$\phi = \tan^{-1} (v_f/c) \qquad (10.7)$$

Applications for ultrasonic flowmeters have included measurements of flow rates of rivers, reactor heat-exchanger fluids, gas-containing emulsions, slurries (including sewage), corrosive liquids of varying viscosities, and wind velocities.

One advantage of ultrasonic flowmeters is the fact that they introduce negligible pressure loss in a system. Also, they are inexpensive to operate and they can handle a large variation of flow rates in a wide range of pipe diameters and over a wide range of pressures.

Quantities of large particles, including gas bubbles and solids, carried in a liquid stream place a restriction on the effective use of an ultrasonic flowmeter. When the size of these particles approaches one wavelength or larger, scattering produces maximum attenuation and, therefore, loss of signal.

VII.B. PRESSURE AND TEMPERATURE

Temperature measurements by ultrasonic means are based upon factors affecting the velocity of sound in the material. If the elastic properties of a material as a function of temperature are known, a measurement of the velocity of sound in the material can be related to the temperature. In some solids, the velocity decreases at a fairly constant rate as temperature increases until it reaches a point that is slightly below the melting point. Water-cooled buffer rods are used between the transducer and the heated section.

Lynnworth has measured temperatures to above 5000°F by a technique based on a pulse-echo type of velocity measurement (35). The instrument contains a thin wire probe which is inserted into the medium (fluid or gas). An ultrasonic pulse is transmitted into the probe element, partially reflecting at a point at the beginning of the thin wire section and again at the end of the wire. The difference in time between the two echoes is a measure of the velocity of sound and, hence, of the temperature.

Lynnworth also claims that measuring the velocity of sound in gases directly enables measurement of temperature to above 30,000°F. Obviously, various problems would be associated with making such measurements.

Pressure also can be measured on the basis of the velocity of sound in gases, liquids, and some solids. Temperature compensation is necessary to eliminate errors from this source.

VIII. INDICATING THE NET LOAD OF A VEHICLE

A method of indicating the net load of a vehicle has been patented in which an ultrasonic unit is attached to the underneath side of the vehicle and a reflector is mounted on the axle (36). A meter, calibrated in weight, records the time it takes for an ultrasonic pulse to echo from the reflector to the receiver. The time of travel of the pulse decreases with the compression of the spring of the vehicle. Two elements primarily affect the accuracy of such a system. The first is associated with the ultrasonic system and the associated electronics and this can be small. The second is associated with the condition of the spring. Weakening of the spring will affect the calibration of the unit.

A second unit can be used on a trailer. A meter located inside the cab of a truck indicates to the driver the load on the trailer.

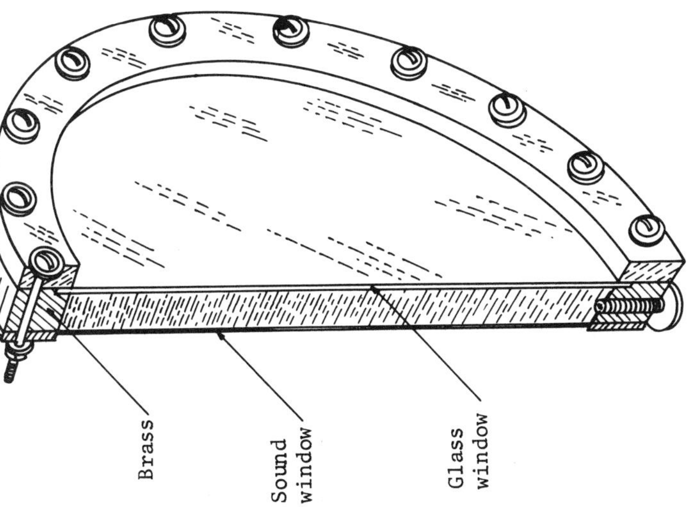

FIG. 10.1. Pohlmann-cell construction

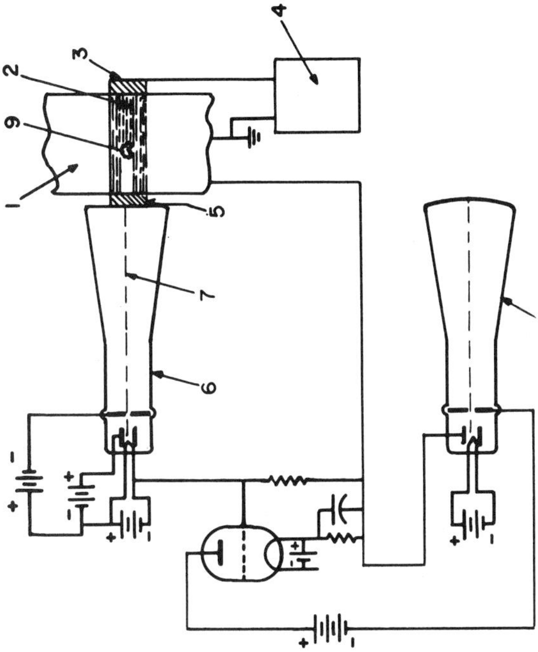

Fig. 10.2. Electron ultrasound camera by Sokolov: (1) object under examination (2) ultrasound beam, (3) source transducer, (4) source oscillator, (5) receiving transducer, (6) television iconoscope "as used by Zworykin," (7) scanning electron beam, (8) picture-reproducing television tube scanned in sympathy with the camera 6, (9) defect.

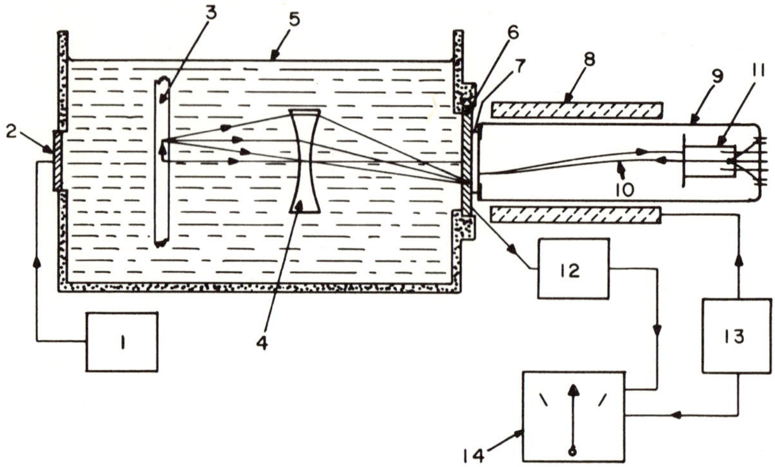

FIG. 10.3. Ultrasound image camera of Smyth. (1) 4 MHz oscillator, (2) ultrasound transmitter, (3) material under inspection, (4) acoustic lens, (5) water, (6) signal plate, (7) quartz plate, (8) deflection and focus coils, (9) ultrasound camera tube, (10) ion-trap mesh, (11) electron beam, (12) electron gun, (13) video-frequency amplifier, (14) deflection generator.

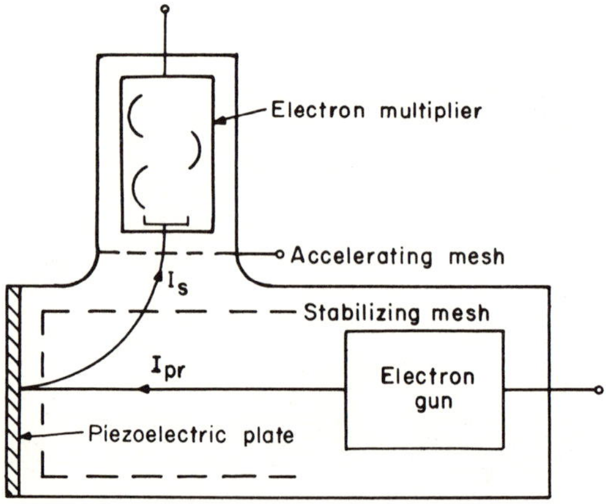

FIG. 10.4. A schematic diagram of Jacobs' ultrasonic camera tube.

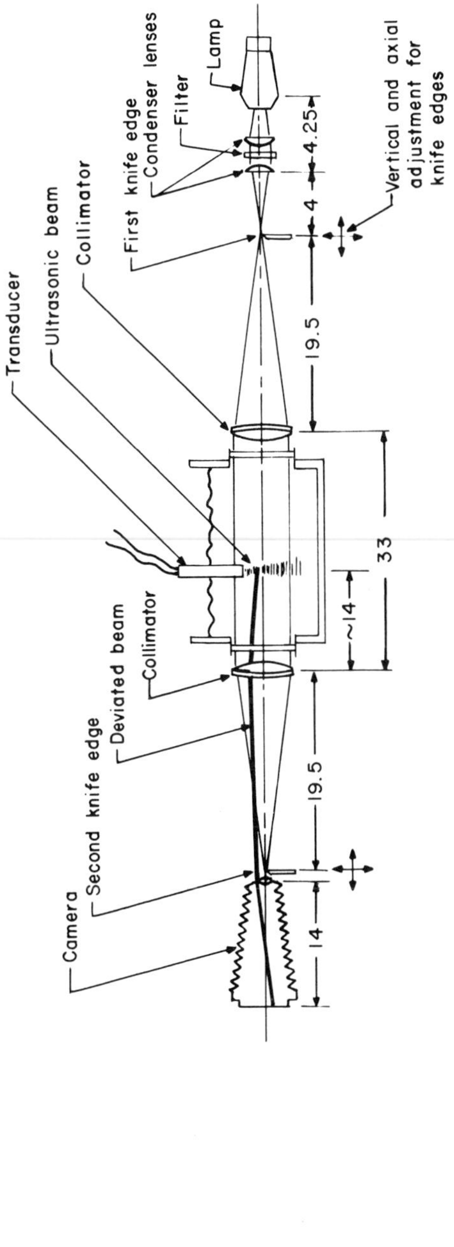

FIG. 10.5. Sketch of schlieren apparatus for viewing ultrasonic fields. Dimensions are in inches.

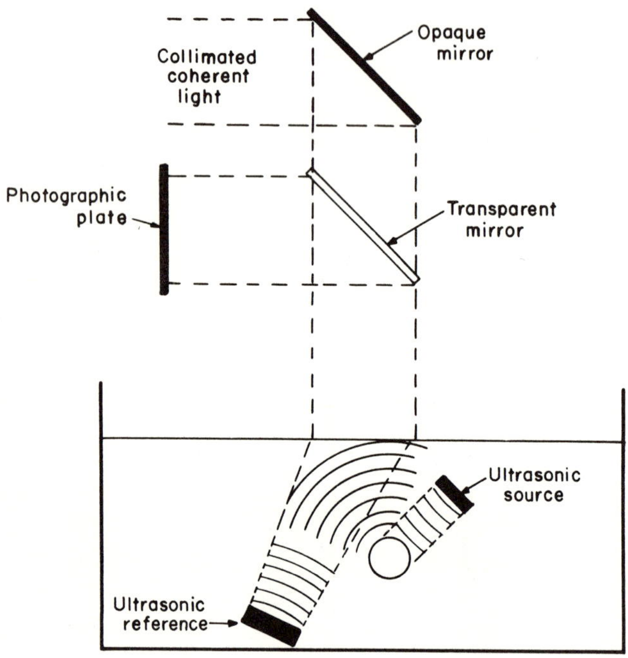

FIG. 10.6. Liquid-levitation ultrasonic holography.

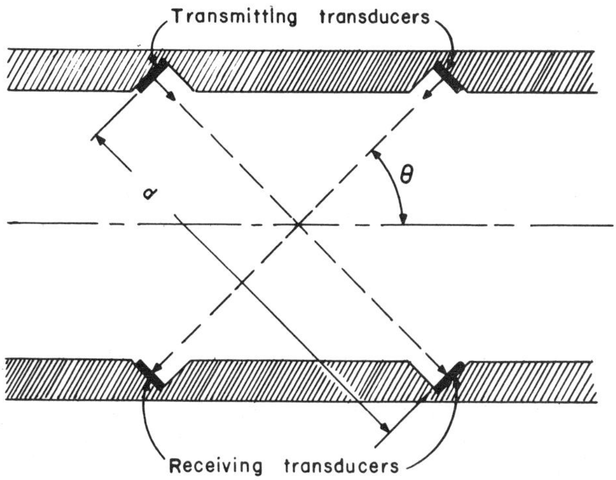

FIG. 10.7. Flowmeter based on comparing upstream and downstream pulse propagation rates.

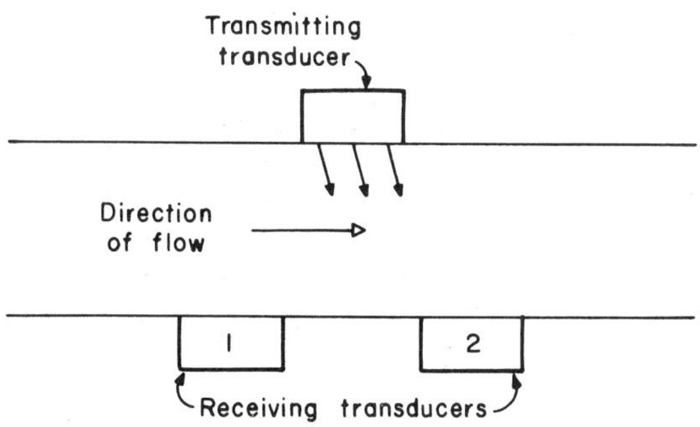

FIG. 10.8. Beam-deflection ultrasonic flowmeter.

REFERENCES

1. R. Pohlmann, Z. Phys., 107(7-8), 497 (1937).

2. E. E. Suckling, J. Acoust. Soc. Am., 29(1), 146-148 (1957).

3. O. Barbieri, Alta Freq., 11, 383 (1942).

4. S. Ya. Sokolov, U. S. Pat. 2,164,185 (July, 1937).

5. S. Ya. Sokolov, Zavodsk. Lab. 14(11), 1328-1335 (1949).

6. W. Freitag, Jena Jahrbuch, Scientific Publications, VEB Carl Zeiss, Jena, Part I, 1958, pp. 228-274.

7. C. N. Smyth, F. Y. Poynton, and J. F. Sayers, Proc. IEE, 110(1), 16-28 (1963).

8. R. A. Fetland, U. S. Patent 2,919,574 (January 5, 1960).

9. C. N. Smyth, Proc. of a Symposium on Ultrasonic Imaging, Mine Advisory Committee, National Advisory Committee, National Academy of Sciences, National Research Council, April, 1965, pp. 39-58.

10. J. E. Jacobs, Proc. Symposium on Physics and Nondestructive Testing, (October 2-4, 1962), Southwest Research Institute, San Antonio, Texas, pp. 59-74.

11. J. E. Jacobs, K. Reimann, and L. Buss, Mater. Eval., 26(8), 155-158 (1968).

12. J. E. Jacobs, Proc. Symposium on Ultrasonic Imaging, Mine Advisory Committee, National Academy of Sciences, National Research Council, (April, 1965), pp. 59-75.

13. J. H. and J. W. Waddell, Res. Develop., 30, 32 (1970).

14. O. R. Gericke and R. C. Grubinskas, J. Acoust. Soc. Am., 45(4), 872-880 (1969).

15. W. H. Sproat and S. E. Cohen, Mater. Eval., 28(4), 73-76 (1970).

16. M. Ye. Arkhangelskii, Soviet Phys.-Acoust., 9(3), 301-302 (1964).

17. M. Ye. Arkhangelskii and G. N. Pinus, Soviet Phys.-Acoust., 6(3), 276-281 (1960).

18. H. Berger and I. R. Kraska, J. Acoust. Soc. Am., 34(4), 518-519 (1962).

19. T. Schilb, C. Bennett, and C. E. Adams, J. Acoust. Soc. Am., 29(1), 145-146 (1957).

20. A. Eckardt and D. Fintelman, Naturwissenschaften, 42(20), 555-556 (1955).

21. B. B. Kudryavtsev, A. N. Medbedev, and A. P. Ponomarev, Sb. Primeneniya Ultraakust i Issled. Veshchestva, (9), 139-145 (1959).

22. D. Gabor, Proc. Roy. Soc. (London), A197, 454-487 (1949).

23. J. D. Young and J. E. Wolfe, Appl. Phys. Lett., 11(9), 294-296 (1967).

24. M. J. Riezenman, Electron. Design, 12, 28 (1969).

25. F. M. Smitz and E. K. Sittig, Ultrasonics, 7(3), 167-170 (1969).

26. Anon., Electromech. Design, 14(11), 4-5 (1970).

27. L. J. Sienkiewicz and H. T. Goldstein, U.S. Patent 3,372,669 (March 12, 1968).

28. G. Rossteutscher, Radio Mentor (Germany), 35(5), 329-333 (1969).

29. Anon., Ultrasonics, 5(4), 198 (1967).

30. C. E. Mongan, Jr., and T. C. Miller, Bureau of Mines Report of investigations 5617, 1960.

31. H. C. Bertova, Electron. Eng., 32(389), 442-443 (1960).

32. M. G. Haugen, W. R. Farrall, J. F. Herrick, and E. J. Baldes, Proc. National Electronics Conference, 11 465-475 (1955).

33. Anon., J. Acoust. Soc. Am., 41(2), 535-536 (1967).

34. H. E. Dalke and W. Welkowitz, Instrum. Soc. Am. J., 7(10), 60-63 (1960).

35. L. C. Lynnworth, Mater. Eval., 27(3), 60-66 (1969).

36. E. C. Nolan, U. S. Patent 3,428, 138 (February 18, 1969).

Chapter 11

APPLICATIONS OF HIGH-INTENSITY ULTRASONICS--BASIC MECHANISMS AND EFFECTS

I. INTRODUCTION

High-intensity applications of ultrasonics are those which produce changes in or effects on media, or the contents of these media, through which the waves propagate. Various mechanisms may be activated by the ultrasonic energy to promote the effects but the mechanisms involved are not always known or understood. Most of the effects probably can be related to the following:

1. Heat: As ultrasound progresses through a medium, energy is lost to that medium in the form of heat. Losses vary according to the nature of the medium. At certain interfaces, absorption may be high because of factors such as shear (friction) across the interface.

2. Stirring: Intense ultrasound will produce violent agitation in a liquid medium of low viscosity and disperse material by resulting currents of liquid or accelerations imparted to the particles.

3. Cavitation: Many of the effects associated with ultrasonics occur in the presence of cavitation. Emulsification of otherwise immiscible liquids occurs under this condition.

4. Chemical effects: Chemical activity, especially oxidation reactions, may be accelerated, sometimes many-fold, under the influence of ultrasonically produced cavitation. The effects have been variously attributed to heat and to mechanical rupture of chemical bonds. In some cases, the effect is a result of mechanical mixing or of dispersion of saturated layers that ordinarily form at an interface between the participants in the reaction.

5. Mechanical effects: Stresses developed in an ultrasonic field can cause ruptures to occur in materials. They may also cause relative motion between surfaces which produce selective absorption at these surfaces, as in ultrasonic

bonding of materials. Stresses developed in cavitation-bubble walls can cause severe erosion of surfaces.

6. Electrolytic effects: There are indications that when two metals separated on the electrolytic scale by even a small amount are exposed to intense ultrasonic irradiation in water, an accelerated galvanic action may be induced which causes electrolytic corrosion.
7. Diffusion: Ultrasonic energy promotes diffusion through cell walls, into gels, and through porous membranes.
8. Vacuum effects: During the low-pressure phase of each cycle, boiling in liquids may be induced and fluids may be drawn into tiny pores.
9. Cleansing: Sometimes the observed effects may result from acoustically eroding a protective coating from a surface so that reactions between two materials may occur that would not be possible otherwise.

Although the mechanisms may not be understood and unexpected effects are often produced, all effects produced have logical explanations. Basic mechanisms involved in high-intensity applications and effects produced are discussed in this chapter. Cavitation, which plays an important part in so many reactions, is discussed more thoroughly in Chap. 2.

II. MECHANICAL EFFECTS

II.A. CAVITATION

Some of the principles of cavitation are discussed in Chap. 2. Many of the effects produced by ultrasonics in liquids are associated with the production of cavitation. Ultrasonic cavitation is the formation of bubbles or cavities in liquids during the low-pressure portion of a wave cycle. The bubbles may be filled with gas coming out of solution in the liquid under reduced pressure or with vapors of the liquid itself. Extremely high stresses may be associated with the formation and collapse of cavitation bubbles. At the same time, free chemical radicals are produced, and even very tough metals are eroded.

II.B. DISPERSIONS, HOMOGENIZATION, AND EMULSIFICATION

The production of various types of solid/liquid and liquid/liquid dispersions by ultrasonic means has many applications both in the laboratory and in industry. The basic mechanism appears to be cavitation. The energy released as the bubbles collapse imparts high stresses and accelerations to particles in close proximity to the bubbles. The particles are thus propelled into the fluid medium, and at extremely high intensities, the particles may be fractured or shattered.

The choice of equipment is based upon two fundamental considerations: the difficulty of the task and the quantity to be processed.

The earliest experiments using ultrasonics to emulsify immiscible liquids were performed with transducers of X-cut quartz crystals. Energy was coupled from the radiating surface of the transducer through transformer oil to the specimen. Quartz crystals are capable of producing high intensities at frequencies ranging from 100 kHz to approximately 8 MHz. Large crystals are expensive. The high operating voltages associated with high ultrasonic intensities often lead to arc-over between electrodes and concurrent damage to the crystal.

The development of ceramic piezoelectric materials has made possible high-intensity ultrasonic energy at frequencies above 100 kHz at lower voltages. These materials have the advantage of being formable into shapes for focusing to obtain intensities which will produce dispersions and emulsions. Bowl shapes have been used for treating small quantities at high intensities, and cylindrical shapes for continuous flow systems. However, these types of transducers have found limited use as dispersers and emulsifiers compared with systems employing stepped mechanical amplifiers and "liquid whistles".

Very high intensities may be produced in small volumes for analytical purposes, using transducers equipped with stepped mechanical amplifiers. Systems using the stepped horn are sold extensively for biological investigations. Dispersions which require several minutes for suitable preparation in ball mills and high-speed blenders may be prepared within a few seconds by ultrasonic means. Applications range from dispersion of bull semen for evaluation to disruption and homogenization of cells from various organs of the body.

The so-called "liquid whistle" is adapted to large-volume production of emulsions and dispersions (1). The whistle is based on a principle proposed and developed by R. Pohlmann. A high-pressure, flat jet of liquid impinges on the edge of a thin blade. The blade is supported at the nodal points and located in a resonant cavity. The unstable condition created by the jet on the edge of the blade causes the blade to vibrate in resonance. The intensity of the vibrations is sufficient to cause localized cavitation and thus produce dispersions or emulsions at a rapid rate. These systems are adaptable to production installations, a typical system handling from 300 to 350 gph. Systems have been designed for handling abrasive materials. Liquid whistles should not be expected to break down large solid particles other than agglomerates.

The liquid whistle is claimed to have a long operating life. However, failure of the blade by fatigue can be a problem. This is especially true when treating viscous materials, if the system becomes "starved" for fluid supply.

A typical commercial use of the liquid whistle is in the preparation of resin-bonded solid-film lubricants (2). These lubricants consist of molybdenum disulfide, graphite, and epoxy resins. Production rates with a 3-hp motor powering the liquid-whistle system are claimed to be 2½ times those with a 12-hp colloid mill.

Dispersions and abrasive emulsions have been prepared with "liquid whistles" during the manufacture of waxes and polishes.

Whistles for dispersing abrasive materials are designed with abrasion-resistant parts. One whistle reportedly handles 300- to 350 gph of waxes and polishes. The whistle also has been used to produce DDT emulsions (3) and margarine emulsions (4-6). A sulfite liquor has been found to be a good emulsifying agent for use in the preparation of stable DDT emulsions.

The permanence of an emulsion prepared by ultrasonic means depends upon the relative characteristics of the liquids. Some oil-water emulsions may remain stable for weeks without the addition of an emulsifier. Mercury in water without an emulsifier usually separates within a few minutes, but if it is carefully handled, it may last several hours.

Marboe and Weyl (7) have shown that the dispersion of mercury in a liquid by ultrasonic means requires the presence of strong forces between the metal and the medium. For example, mercury cannot be dispersed in paraffin oil at 25°C under the influence of ultrasonic energy without strengthening the forces between the two liquids. When these forces are strengthened by adding stearic acid to the paraffin oil, the mercury disperses into the paraffin oil, resulting in a stable emulsion. In the latter case the incompatible liquids are linked together by a molecular film of mercury stearate.

Similar results are obtained with metallophilic groups. It is difficult to produce an emulsion between mercury and deaerated (by boiling) distilled water. However, mercury/distilled water emulsions are readily produced after the water has been aerated. The air oxidizes the mercury under the influence of ultrasonic energy, thus producing highly polarizable cations of Hg^{+} and Hg^{2+}. The necessary binding forces between the mercury and water are provided by the resulting ionic character of the mercury surface. The Hg^{2+} ion in deaerated distilled water can be produced by adding a small quantity of $HgCl_2$ solution to the water; then, when ultrasonic energy is applied to the combination, a mercury/water emulsion results. Emulsification of mercury in water can be hindered or prevented by adding KCl solution to the water. The KCl changes the Hg^{2+} ion into a complex anion of the type $(HgCl_4)^{2-}$.

Marboe and Weyl (7) claim that polar molecules exert image forces upon mercury so that it can be dispersed in polar liquids such as nitrobenzene but it cannot be dispersed in nonpolar liquids such as carbon tetrachloride.

II.C. AGGLOMERATION AND FLOCCULATION

Intense sonic and ultrasonic energy can cause agglomeration and precipitation of some particles in fluid media. Thus, several proposals have been made to use ultrasonics to precipitate smoke from industrial chimneys and fog from airports. In general, high-intensity sirens operating at frequencies between 1 and 30 kHz have been considered, and these induce the desired agglomeration and precipitation. The mechanisms of smoke agglomeration are discussed more extensively by Bergmann (8) and by Hueter and Bolt (9).

High-intensity ultrasound applied to liquid suspensions of fibrous materials, such as paper pulp, causes the particles to become entangled and to agglomerate.

II.D. PRECIPITATES AND SOLS

Kukoz and others (10) have investigated the effects of temperature, ultrasonic energy, and rate of mechanical mixing on the fractional composition and particle size of nickel hydroxide precipitated from aqueous solutions of $NiSO_4$ and NaOH. Particle size was influenced more by mixing rate than by temperature. The finest particles were produced by a combination of ultrasonic irradiation and intense mixing while saturating the medium with gas. The ultrasonic dispersion was greatest at the moment of precipitation.

Some Indian scientists have investigated the effects of ultrasonic energy on the stability of cobalt ferrocyanide (11) and nickel ferrocyanide sols (12). In both cases, ultrasonic energy increased the stability of the sols, with or without oxygen. The increased stability was attributed to the attrition of the particles due to cavitation in the case of nickel ferrocyanide and to an increase in $(OH)^-$ ions in the case of cobalt ferrocyanide. Both sols were negatively charged.

II.E. ENHANCEMENT OF HEAT TRANSFER

Ultrasonic energy can be beneficial in increasing heat transfer but the effects are not limited to ultrasonic frequencies. Lemlich (13) has noted an improvement in the coefficient of heat transfer from wires vibrated in transverse modes at frequencies between 39 and 122 Hz. Lemlich used electrically heated wires 0.0253, 0.0396, and 0.0810 in. in diameter and obtained heat-transfer coefficients with vibration that were as much as four times the coefficients obtained without vibration. He noted an increase in coefficient with both amplitude and frequency, but the effects were independent of the direction of vibration. He proposed a "stretched film" theory to explain his observations. Lemlich arrived at the following empirical equation which is applicable to higher than 10% increases in heat transfer in air and other diatomic gases:

$$\frac{h}{h'} = 0.75 + 0.0031 \frac{\overline{R}_e^{\,2.05} (\beta \Delta t)^{0.33}}{(G_r)^{0.41}} \qquad (11.1)$$

where h is the heat-transfer coefficient with vibration [Btu/(hr)(ft)(F)],

h' is the heat-transfer coefficient without vibration [Btu/(hr)(ft)(F)],

$\overline{R}_e$ is Reynolds number computed from the diameter of the wire and the average vibrational velocity (dimensionless),

β is the thermal coefficient of volumetric expansion (F^{-1}),

Δt is the difference between the surface temperature of the wire and the ambient temperature (F), and

G_r is the Grashof number (dimensionless).

Lemlich makes an approximate extension to include other fluids, such as air, by the following equation:

$$N_u = \left[0.75 + 0.0022 \frac{\bar{R}_e^{-2.05}(\beta\Delta t)^{0.33}}{P_r^{1.54} G_r^{0.41}}\right]\left[0.63 + 0.35(G_r \times P_r)^{0.17}\right]^2 \tag{11.2}$$

where

N_u is Nusselt's number ($\approx h/h'$)(dimensionless), and

P_r is the Prandtl number (dimensionless).

Fand (14) concludes that the increase in heat-transfer rate from a heated horizontal cylinder subjected to transverse horizontal vibrations is caused by thermoacoustic streaming. Both Fand and Westervelt (15-17) state that there is a critical intensity below which vibrations applied to a cylinder in a fluid produce no effect on heat transfer. Fand gives this "critical intensity" as being approximately 0.3 ft/sec for both vertical and horizontal transverse vibrations of a heated horizontal cylinder in air. The "intensity" is the product ξf, where ξ is the sinusoidal displacement amplitude of vibration and f is the frequency of vibration. According to Fand, the ratio of displacement amplitude to cylinder diameter ξ/D is an important parameter in all fluid dynamical problems involving transversely vibrating cylinders. This ratio is included in Reynolds number [Eqs. (11.1) and (11.2)] and therefore it is taken into account in all equations involving Reynolds number.

Westervelt (16-17) proposes that modifications in heat transfer by sound are caused by modifications of the convective flow in the inner-streaming boundary layer. The critical intensity is determined by the ratio of particle-displacement amplitude to the acoustic-boundary-layer thickness ξ/δ_{ac} and it is reached when $\xi/\delta_{ac} = 1$. Increase in heat transfer occurs when $\xi/\delta_{ac} > 1$. In terms of critical sound-pressure level, $N_{sp\ell}$, in air

$$N_{sp\ell} = (136 + 10 \log f) \text{ db} \qquad (\xi/\delta_{ac} > 1) \tag{11.3}$$

where f is frequency in kHz. At a higher critical pressure level, the inner-streaming boundary layer collapses and the action near the surface becomes vigorous and chaotic. Westervelt believes that this activity tends to greatly increase the mixing of hot fluid near the cylinder with cooler fluid away from the cylinder and thus causes an additional and significant increase in heat transfer.

Each of the different hypotheses advanced by the previously mentioned investigators indicate that whether the heated surface is vibrated or whether it is stationary and located in an intense acoustic field, heat transfer should improve by virtue of the acoustic vibration when the intensity of vibration exceeds a critical value. In studying heat transfer in forced convection, steam condensation, and nucleate boiling, the United States Office of Saline water (18) has found that, of the two methods, vibrating the heated surface in water is the more effective. The difference in the results obtained may be due to the difficulty of controlling the experimental variables, for, if the same relative conditions are obtained in and about the inner-streaming boundary layer, the end result by either method should be the same. The Office of Saline Water attributes the improved heat transfer to combined axial and oscillating cross flows.

Several investigators have studied the effects of ultrasonic vibrations on heat transfer. Any difference between the heat-transfer mechanisms at ultrasonic frequencies and those at low frequencies, if such a difference exists, might be related to the wavelengths involved. At ultrasonic frequencies, the "stretched film" theory of Lemlich may be less applicable than the modification of the convective flow in the inner-streaming boundary layer proposed by Westervelt. The latter theory should apply regardless of frequency.

Bergles and Newell (19) investigated the influence of ultrasonic energy on heat transfer to water flowing in annuli consisting of two concentric stainless steel tubes. The inner tube was heated electrically. Ultrasonic energy at 70 and 80 kHz was directed through the water toward the heated tube from cylindrical transducers mounted on the outer tube. Improvements of 40% in heat transfer were obtained and this improvement was attributed to cavitation near the surface of the heated tube.

As the temperature of the liquid increases, the intensity of ultrasonic energy at which vaporous cavitation occurs decreases. Since the liquid is hottest near the source of heat, cavitation is quickest to form in this region. However, when the temperature reaches the boiling point, the additional stimulus from ultrasonic energy becomes insignificant. Bubbles formed during the boiling reflect and scatter the ultrasound and thus prevent its acting at the heated surface. It would be impractical to overcome the effects of boiling by increasing the intensity of the ultrasound because cavitation would occur nearer the source, and thus cause even worse scattering. An alternative is to vibrate the heated surface and thus avoid having to transmit energy through a medium in which the attenuation is so affected by temperature. The results obtained by Bergles and Newell confirm this. They show that ultrasonic agitation was most effective in the improvement in heat transfer at low flow rates and at temperatures below which surface boiling becomes well established. Low flow rates are more conducive to formation of cavitation because cavitation is time dependent.

In studies on effects of ultrasonics in a nucleate boiling system (20) Schmidt, Torok, and Robinson obtained results that can be explained in a similar manner. They applied ultrasonic energy at 26 kHz to a flash boiling system and concluded that the major contribution of the ultrasonic energy in increasing heat-transfer rate was to increase the turbulence in the system by cavitation. When the bubble density in the fluid from other sources is high, the effectiveness of cavitation is overshadowed and actually degraded by scatter and attenuation of the ultrasonic energy. Therefore, Schmidt, Torok, and Robinson found that at large values of parameters which control bubble density--properties of the fluid, degree of superheat, fluid velocity, and system pressure--the effect of ultrasonics on heat transfer was not detectable. However, at low values of superheat (approximately 2°F) and low flow rates (0.6 to 2.3 lb/min), heat transfer increased 50 to 60%.

McCormick and Walsh (21) claim that ultrasonic energy gives a threefold improvement in heat transfer and suggest its use in cooling high-power tubes.

Cheng (22) claimed to obtain an average increase of 80% in the "coefficient of thermal conductivity" between hot and cold water. The increase between water and lubricating oil was 20%. He believes these values are far below the theoretically possible maximum increase.

II.F. DIFFUSION THROUGH MEMBRANES

Diffusion of ions and liquids through membranes and cell walls can be accelerated ultrasonically. Woodford and Morrison (23) attribute the effect to disruption of the boundary layer at the phase interface. They have found that ultrasonic irradiation increased the diffusion rate of Sennoside A through a cellulose membrane into water irrespective of temperature conditions under which diffusion took place. Gilius (24) also found a correspondence between the rate of diffusion of sugar and water through sugar-beet membranes under the influence of ultrasound. The increased transfer rate continues after irradiation is stopped. Water passes more easily through plasmolized membranes, while sugar passes more easily through fresh membranes. These observations are in agreement with those of other investigators (25, 26).

Fairbanks and Chen (27) have obtained substantial increases in the amount of flow of crude oil through porous sandstone and of water through porous stainless steel filters under the influence of ultrasound at 20 kHz.

The diffusion of hydrogen, generated in a cell during electrolysis of 0.1 N sulfuric acid solution, through an Armco iron membrane 0.17-cm thick to a compartment filled with glycerol was enhanced under the influence of ultrasound at 24.5 MHz and 3 W/cm^2

when irradiation was applied from the glycerol compartment (28). However, diffusion was hindered when irradiation was from the opposite side of the membrane.

The rate of diffusion of salt from 5%-saline solution through a cellophane membrane into distilled water was increased by as much as 100% when ultrasound was applied in the direction of diffusion (29). Diffusion was slower when irradiation was in a direction opposite that of diffusion. The increased rate was proven to be not a result of damage to the membrane or of temperature rise in the medium. Increased diffusion was greatest when irradiation was in the direction of the force of gravity. Frenzel, Hinsberg, and Schultes (30) obtained similar results with ultrasound promoting diffusion of potassium oxalate through a parchment membrane. The increase in diffusion over that without ultrasound was 69% with irradiation in the same direction as diffusion; when the two directions were in opposition, the increase was only 35%.

III. CHEMICAL EFFECTS

The chemical activity promoted by ultrasonic energy may be the result of many different mechanisms. These may include intimate mixing of reactive substances, production of localized heat, production of free ions associated with the production of cavitation, and stress-associated molecular breakdown. Among other effects, ultrasound promotes oxidation reactions, promotes polymerization or depolymerization depending upon the nature of the molecule being treated, affects the potential of polarized electrodes, increases the rate of crystallization, and facilitates extraction of oils.

The mechanisms involved in the chemical effects of ultrasonics have been widely studied and many interesting applications of these effects have been made. The applications, in general, have been on a laboratory scale. However, chemical effects may be involved in many industrial applications of ultrasonics. Ultrasonics may increase the rate of a reaction occurring under normal circumstances or it may induce reactions that would not occur under normal circumstances.

III.A. DEPOLYMERIZATION

Depolymerization of certain types of high-molecular-weight polymers occurs under the influence of ultrasonic energy. The initial structure of the molecule, rather than its size, determines whether it will be broken by this means. Degradation of the molecules continues to a limiting value and cannot be carried further. The mechanism appears to be cavitation. Cavitation is difficult to achieve in viscous materials corresponding to high-molecular-weight polymers. For this reason, dilute solutions of the polymers have been used for studying ultrasonic depolymerization.

Newton and Kissel (31) fragmented Tobacco Mosaic virus at 7 MHz and observed the results by electron microscopy. At an intensity implied to be below cavitation level, polymeric forms of the virus were fractured into predominantly monomeric lengths and this was accompanied by partial fragmentation of a small percentage of the monomers, the amount of fragmentation being a function of time. Breakage of the monomer occurred most frequently at a distance of 175 to 205 mμ from one end, which appears to indicate a fundamental weakness at this location in the monomer. At high intensities and correspondingly high cavitation activity, the monomers were extremely fragmented. The extent of fragmentation was relatively independent of time. A large number of the fragments were spherical in appearance. The virus had lost 95% of its infectivity during intense treatment.

Brett and Jellinek (32) and Weissler (33) demonstrated the role of cavitation by comparing ultrasonic depolymerization of polystyrene under conditions which were conducive to cavitation with those which inhibited cavitation. Brett and Jellinek ultrasonically irradiated 1% solutions of polystyrene in benzene at constant intensity and temperature and at a frequency of 500 kHz. Cavitation activity was controlled by pressurizing the samples either with nitrogen or with mercury. Increasing pressure decreases the tendency of a liquid to cavitate under the influence of ultrasonics. When pressure was applied through nitrogen, depolymerization occurred even at 450 psi at a rate only slightly lower than that obtained at atmospheric pressure. When the pressure was applied through a mercury column, the rate of depolymerization decreased as the pressure increased in accordance with the effect of pressure on cavitation. At 450 psi, very little depolymerization was evident. The upper limiting pressure, of course, depends upon the intensity of the ultrasonic field.

Weissler (33) ultrasonically irradiated a 1% solution of polystyrene in toluene at 400 kHz. When the solution was treated without prior degassing, cavitation activity was apparent and depolymerization occurred as expected. When the solution was degassed by boiling prior to treatment, there was no visible cavitation during irradiation and depolymerization did not occur.

Weissler obtained similar results with hydroxyethyl cellulose dissolved in ordinary water and in degassed water. He further attempted to prove that degradation is not due to oxidants produced during ultrasonic irradiation of solutions containing nitrogen or oxygen. Depolymerization occurs at approximately the same rate, even when purified helium is the only gas present.

If cavitation alone is responsible for depolymerization under the influence of ultrasonic energy, varying frequency would affect depolymerization only as it affects cavitation. Langton and Vaughan (34) compare results of depolymerizing polystyrene at 1 MHz with results obtained at lower frequencies by other investigators. They were similar. Langton and Vaughan proposed an equation to describe the rate of degradation as follows:

$$dP_t/dt = -k(P_t - P_\infty) \tag{11.4}$$

where P_t is the degree of polymerization after t seconds of irradiation,

P_∞ is final limiting degree of polymerization, and

k is a constant.

Ultrasonic energy is equally effective at 20 and at 250 kHz in depolymerizing dextrin.

Results of depolymerization studies conducted by Head and Lauter (35) illustrate the importance of molecular structure to the susceptibility of the molecule to rupture under the influence of ultrasonic energy. They treated 1- and 2%-aqueous solutions of Acacia (gum arabic), 0.5%-aqueous solutions of Viscarin, 0.25%-aqueous solutions of Agar, 0.5%-aqueous solutions of locust bean gum, 0.5%-aqueous solutions of tragacanth, 1%-solutions of karaya gum, and an 0.5%-aqueous solution of 4000-c.p.s. Methocel (Dow methyl cellulose) at 275 kHz. Viscarin, Agar, locust bean gum, and 4000-c.p.s. Methocel are linear polymers with molecular weights ranging from 140,600 to 310,000. These materials depolymerized at comparable rates. Acacia (gum Arabic) has a molecular weight of 250,000; however, it is not a linear polymer and did not depolymerize. Karaya, having a much larger and more complex molecule (molecular weight 9,500,000), apparently did not depolymerize. On the other hand, tragacanth, with a molecular weight of only 5000, which has a complex structure appears to polymerize (characterized by an increase in viscosity) to some maximum and then to depolymerize to less than half its initial molecular weight.

According to Hanglein (36), when poly(methyl methacrylate in benzene solution is degraded ultrasonically in the presence of iodine, long-chain free radicals are produced that react rapidly with the iodine in the absence of oxygen. Spectroscopic studies with I^{131} show that for each ruptured bond, two atoms of iodine are used and these are incorporated in the fragments. The rate of depolymerization is somewhat greater in iodine-containing solutions than it is in iodine-free solutions. In the absence of iodine, 20% of the polymer radicals either recombine or combine with radicals from other macromolecules. In the presence of oxygen, the reaction is inhibited.

Kossler and Novobilsky (37) studied the ultrasonic degradation of polychloroprene both fresh and aged in air. They concluded that crosslinking occurs by the formation of labile peroxidic bonds that prevail, in the first stages of aging, over crosslinks of another kind. Infrared spectra showed an increased content of C=O groups in the first fraction. The number of bonds broken by ultrasound per unit time decreases with increasing concentration. The addition of gases retards the depolymerization according to the sequence air, nitrogen, oxygen, and carbon dioxide.

Degradation of yeast ribonucleic acid (RNA) has been observed at a frequency of 0.8 MHz under cavitational conditions when a 0.005% solution of RNA was treated (38). No degradation occurred at 2.4 and 4.0 MHz. The degradation of RNA was attributed to the formation of hydrogen peroxide and nitrous acid in the presence of cavitation, which occurs at 0.8 MHz but did not occur at 2.4 and 4.0 MHz. Higher concentrations of RNA (0.025% and 0.2%) retarded degradation. The pH, which drops to 3.3 in distilled water at 0.005% RNA solutions, does not drop below 6.5 in 0.2% solutions.

Decomposition of the benzene ring has been observed when phenol and benzene in aqueous solutions have been treated at 0.8 MHz (39). The decomposition has been attributed to two different reactions: (1) the benzene ring is oxidized, which leads to successive formation of phenol, monosubstituted benzenes, quinones, and unsaturated acids (muconic, maleic); and (2) direct splitting off of parts of benzene molecules occurs with formation of formaldehyde; under proper conditions the latter condenses with phenol to form resins. In treating phenol solutions in concentrations of 48 mg/liter ultrasonically, decomposition of the ring occurs only as a result of oxidation, according to the reference. The process occurs in acid, neutral, or weakly alkaline media. In a strongly alkaline medium, the process is severely inhibited. The rate of decomposition of phenol depends on its initial concentration in the solution; the amount of oxidized phenol increases with increase in concentration to a certain point above which an increase in concentration has no effect on reaction rate. With increase in sound intensity and temperature, the reaction rate increases. Prolonged ultrasonic treatment of a phenol solution in a strongly acid medium (in contact with air) results in formation of small quantities of mononitrophenols.

The inhibition of bromine release from bromoform under the influence of ultrasonics has been observed when water, chloroform, diethylene glycol, butyl alcohol, glycerol, ethyl alcohol, ethylene glycol, diacetone alcohol, methyl alcohol, Dioxan, xylene, acetone, and benzene were added to aqueous solutions of the bromoform (40). The inhibition has been attributed to certain physical properties such as surface tension, viscosity, vapor pressure, and internal pressure which were affected by these nonelectrolytes. All of these factors affect the intensity of cavitation.

Shaw and Rodriguez (41) have shown that samples of polysiloxane solutions with different initial distributions approach a uniform distribution after 120 min of ultrasonic degradation. Dilution of the polymer to 1.0 to 0.01 g/dl more than triples the degradation per gram of polymer but it requires 10 times as much energy per gram to degrade from an average molecular weight of 10^6 to 10^5. However, doubling the power input (36 to 72 W) nearly doubles the degradation rate per gram of polymer.

Degradation of agar-agar solutions in water and methyl sulfoxide at concentrations ranging from 0.1 to 2 g/l showed that degradation is more extensive at the higher concentrations (42). A reversible viscosity decrease was observed at concentrations greater than 2 g/l and this was attributed to the degradation of the gel. At concentrations of less than 0.3 g/l, only slight degradation of agar-agar occurred. The viscosity of agar-agar decreased very rapidly during the first stages of ultrasonic treatment, and later converged toward a limiting value of viscosity. The gas in contact with the solution influenced the rate of degradation. Air produced the most extensive degradation; in nitrogen degradation was slower and not so extensive; for sulfur dioxide, the quantity (t_u - t_v) was linear and an increase in the sulfur content of agar-agar was noted. Here, t_u is the ultrasonic treatment time and t_v is the time of flow through an Ostwald viscosimeter. Argon was the most suitable gas for study of degradation. Temperature (16, 20, and 36°C) had no effect on variation in viscosity in the concentrations of 0.1 to 0.3 g/l.

After 100 min of ultrasonic radiation, the pH decreased from 6.98 to 3.96 at concentrations of 0.8 g/l. These authors propose that degradation proceeds according to a four-step mechanism, as follows: (1) cleavage of the 1,3 and 1,4 glucosidic bonds in agar-agar, yielding macroradicals, (2) recombination of these macroradicals, (3) reaction of water with the macroradicals OH + agar-CO→agar-COOH, and (4) transfer between agar-agar and radicals of water.

Witekowa (43,44) attempts to explain chemical action of ultrasonics in various quantities. In explaining the kinetics of the reduction of acidic potassium permanganate solutions in an ultrasonic field, Witekowa claims that acidic MnO_4^- solutions are reduced to Mn^{++} in several minutes, whereas in the absence of the field they remain unchanged for days. The reaction occurs according to the equation $-d[KMnO_4]/dt = k_o$, where k_o is a constant. The value of k_o increases with pH and temperature, and at 10 to 30°C the average energy of activation is found to be 10.6 kcal/mole. Witekowa explains the reaction as being dependent on the concentration of excited OH^- ions. The latter depends on the initial OH^- concentration (i.e., on pH), frequency, and power of the ultrasonic field. At a suitable frequency the OH^- radical may be partly liberated from the surrounding water molecules and become more active.

In explaining the kinetics of oxidation of aqueous potassium ferrocyanide, $K_4[Fe(CN)_6]$, Witekowa (44) claims that these solutions are oxidized rapidly in an ultrasonic field to potassium ferricyanide, $K_3[Fe(CN)_6]$. The rate constant increases with temperature, hydrogen-ion concentration, and the square root of ionic strength. The activation energy at 10 to 30°C is 6.88 kcal/mole. The reaction $[Fe(CN)_6]^{4-} + (H^+) \rightarrow [Fe(CN)_6]^{3-} + \frac{1}{2}H_2$ is supposed to be rate determining. The mechanism is complicated by the possible formation of nitrous acid and nitric acid in an ultrasonic field.

Zaskal'ko (45) studied the degradation of 7%-polyisobutylene solution in oil at 100°C. He showed that there is more degradation of long polymer chains than of the short chains. After 5 hr exposure to ultrasound, 81.5% of the polyisobutylene of molecular weight 85,000 was degraded, but there was no loss of polyisobutylene of molecular weight 2000.

III.B. POLYMERIZATION

Not only can ultrasonics be used to depolymerize materials but it can also produce the opposite effect, that is, polymerization. Hastings and Bolten (46) have shown that applying ultrasonics during the polymerization of styrene produced a sharper distribution of the molecular weight than that obtained by degradation of the polymer in toluene and less polymer of low molecular weight was present. An increase in molecular weight and sharpening of distribution compared with photoinitiated polymer was achieved under some conditions.

Graft copolymerization of cellulose of vinyl monomers in the presence of salts of Group VIII metals, cerium or Cu^{++}, was accelerated by ultrasonic treatment, as observed by Okamura and others (47). Treating a mixture of 438.1 mg of viscose sponge, 8 cc of methyl methacrylate, and 2 cc of 0.1 mole cerium nitrate treated ultrasonically at 420 kHz and 200 W at 40°C for 2 hr produced 1525 mg of graft copolymer extracted with benzene. A similar run without ultrasonics yielded 4309.5 mg of copolymer from 454.3 mg of viscose.

The intrinsic viscosity of polymers obtained by the polymerization of methyl methacrylate with Grignard reagents and benzoyl peroxide dropped markedly upon ultrasonic irradiation at 500 kHz (48). Osawa and Igarashi believe that in the transition state in polymerization the growing polymer end and the catalyst form bonds which are stronger than the C-C covalent bonds in the main chain and are not broken by ultrasonic energy. The stereoregularity of the polymers is not affected. Polymers polymerized with Grignard reagent and benzoyl peroxide are predominantly isotactic and syndiotactic in structure, respectively.

Romanov and Lazar (49) studied the modification of a copolymer of atactic polypropylene in an ultrasonic field. They prepared a 15% solution of atactic polypropylene in styrene and exposed the mixture to ultrasonic radiation at 12 W/cm^2 in a nitrogen atmosphere for 11 hr to produce a highly viscous solution. They prepared another copolymer chemically by initiating polymerization of styrene with ozonized atactic polypropylene. After the two copolymers had been isolated and precipitated out, they discovered that the chemically prepared copolymer has more of the character of atactic polypropylene, while the copolymer prepared by ultrasonic irradiation has more of the character of polystyrene, although from the elemental analysis, the opposite would be expected. They concluded that the chemically prepared sample is a graft copolymer, and the ultrasonically prepared sample is a block copolymer.

Shiota and co-workers (50) show that ultrasonics increases the rate of polymerization and the molecular weight in the production of polyamide, an effect they attribute to an increase in terephthalyl chloride, p-$C_6H_4(COCl)_2$, and ethylenediamine, $H_2NCH_2CH_2NH_2$, in the presence of cavitation. Interfacial polycondensation occurs at lower ultrasonic frequencies such as 200 to 2000 kHz. They observed that the optimum concentration of the reagents appears to be independent of the ultrasonic irradiation.

III.C. CATALYSIS

Ultrasonic energy acts as a catalyst to some chemical reactions. When a chemical reaction occurs in the presence of both a catalyst and ultrasonic irradiation, the activity usually increases but some reactions may be inhibited by the effect of the ultrasonics on the catalyst.

The application of ultrasound in hydrolysis of metallic peroxyacetate, using hydrochloric acid as catalyst, produces an increase in the reaction rate with increasing intensity (51). Ultrasound has no effect on activation energy and produces no extra hydrogen ions.

The rate of oxidation of cyclohexanone in air in the presence of magnesium stearate and benzyl alcohol decreases with ultrasonic irradiation (52). Ultrasonic irradiation does not influence the rate of oxidation in the presence of adipic acid alone, or if no acid is added. With adipic acid-benzyl alcohol mixtures, the effect is immediate. Apparently the active catalyst is a benzyl alcohol-magnesium stearate complex which is decomposed by ultrasound favoring a less active, but more stable adipic acid-magnesium stearate catalyst.

Menzhulin and Ponomareva (53) compared the reduction of selenium in sulfuric acid solutions with the reduction of copper, copper-zinc couple, zinc, iron, aluminum, and magnesium with and without ultrasound. Without ultrasound, reduction of copper and copper zinc is relatively rapid. The other metals were rather inactive, with no reduction in aluminum being detected after 3 hr. Ultrasonics accelerated the rate of reduction in all cases, with the reduction of copper and copper-zinc being the most pronounced. Ultrasonics also caused a finer crystallization of the deposits. The authors attribute the effects of ultrasound on the reduction of selenium to the destruction of the deposited film and to the acceleration of anodic and cathodic processes by ultrasound.

Parthasarathy was able to show that ultrasonics promoted esterification of ethyl alcohol-acetol without the use of a catalyst (54).

Prakash and Prakash (55) review the decolorization of aqueous solutions of Rhodamine B, bromocresol green, Congo red, Aniline Blue Chrysoidine, and Brilliant Green by ultrasound at 1 MHz. Decolorization is a unimolecular process and there appears to be no single explanation; oxidation, reduction, and decomposition reactions are all involved.

III.D. PRECIPITATION

Ultrasound has been found to cause precipitates of dissolved or suspended materials in liquids. Examples are calcium carbonate (in the form of aragonite crystals) (56) and certain metallic compounds such as nickelous oxide hydrate (57). Under the influence of ultrasound, nickelous oxide hydrate precipitates at room temperature. When this process occurs in an alkaline battery, the electrical characteristics of the battery remain unimpaired.

High-purity nickelous oxide hydrate is produced at low temperatures without the use of excess base in the reactor.

III.E. METALS

Certain chemical and chemically related effects of ultrasound are beneficial in certain metal-processing applications. Ultrasound accelerates nitriding in a molten bath with ammonia bubbling through it. The depth of the nitrided layer and its microhardness are increased under such treatment (58). The ultrasound treatment (1) promotes mixing of the liquid, (2) removes reaction products from the surface of the metal, (3) cleans the surface of the metal and activates it, permitting surface-active substances to be adsorbed, and (4) promotes diffusion of nitrogen into the metal as a result of cavitation.

When steel is located in dilute sulfuric acid, the application of intense ultrasound accelerates the dissolution of the steel. The accelerated activity probably is due to the removal of hydrogen bubbles which screen the surface from the acid (59).

III.F. CRYSTALLIZATION

Ultrasonic energy initiates crystallization processes and produces effects relative to size and character.

Abramov (60) confirms the effectiveness of ultrasound in generating crystallization centers in transparent organic substances such as betol, naphthalene, azobenzene, salol, and thymol. Ultrasonics produced crystallization centers at lower supercooling temperatures and in shorter times than are required with the usual processes; however, the increase in the threshold of metastability differed with different materials. Depending on the nature of the material and the conditions for crystallization, dispersion of growing crystals and increased growth rate were effected by ultrasound with either dispersion or increased growth rate being the predominant process. Cavitation was a factor in both processes.

Ultrasonics applied to solidifying metals also results in uniform grain refinement, as has been shown by several investigators.

Small crystals of hydrocortisone have been obtained by ultrasonic treatment of saturated solutions of isopropyl alcohol, ethyl alcohol, and 50% by volume blend of isopropyl alcohol and ethyl alcohol (61). Particle size was independent of treatment time and distribution was similar at various coolant temperatures. Crystals obtained from saturated solutions in ethyl alcohol were smaller than those obtained from isopropyl alcohol or the mixture of isopropyl alcohol and ethyl alcohol. The degree of saturation affected the particle size and distribution. It also affected the yield. Smaller crystals and greater yields generally were obtained from higher levels of saturation. The extent of crystallization increased with decreasing temperature. Weak cavitation produced larger crystals, and low pressures increased yields.

Kapustin (62) suggests that ultrasonics induces crystallization by changing the density of crystal nuclei in the melt. He found that in crystallizing o-chloronitrobenzene by means of ultrasonic energy, no crystallization was induced at temperatures of 20 to 28°C, but crystallization centers appeared either immediately or within a few seconds of the application of ultrasound near the limit of the metastable region (18 to 20°C). High-intensity ultrasound seldom produced crystallization.

III.G. SONOLUMINESCENCE

The weak emission of light that sometimes accompanies cavitation in an ultrasonic field is called sonoluminescence. Several theories as to the origin of sonoluminescence have been proposed, but the validity of none of them has been proven satisfactorily. The two most common theories are (1) that electrostatic discharges occur as a result of the formation of electrical charges on the walls of the cavitation bubbles and (2) that temperatures sufficiently high to cause the vapors of the liquid to becom incandescent occur as a result of microshocks associated with the collapse of cavitation bubbles. There appears to be some justification for both theories. Grandchamp and Prudhomme (63) have shown a drastic reduction of sonoluminescence and oxidation reactions in the presence of hydrogen and also in the presence of ether. They attribute the reduction of these reactions in the presence of hydrogen to an inhibition of the formation of electrical charges on the walls of cavitation bubbles. The reduction in the presence of ether is attributed to the fact that ether accepts free radicals and also increases the size of the cavitation bubbles. Both materials lower the intensity of cavitation bubbles, which results in a correspondingly lower number of free radicals being produced. The proven appearance of free radicals in the presence of cavitation and experiences with induced electrolytic erosion in the presence of high-intensity cavitation seems to make the electrostatic-discharge theory a plausible one. The effect is similar to the phenomenon of lightning in nature.

Hickling (64) claims that sonoluminescence is largely thermal in origin, with the spectra showing a direct relation between luminous intensity and the temperature generated inside the collapsing bubbles. He explains the strong dependence of the luminous intensity on the nature of the gas dissolved in terms of thermal conduction. On this basis, if the bubbles are small, loss of heat from the bubble into the liquid can significantly reduce the temperatures attained during collapse, with a corresponding reduction in luminous intensity. Small cavitation bubbles are produced under two conditions: (1) the intensity is low, in which case, luminescence is minimal or nonexistent, or (2) no gas is present in the liquid, the liquid has a low vapor pressure, and the surface tension of the liquid is high so that it is difficult to produce cavitation. On the other hand, dissolved gases and liquids with high vapor pressures lower the cavitation threshold and weaken the intensity of the cavitation. In the latter case, luminescence will not occur. This appears to contradict the thermal-conductivity explanation of luminous intensity.

Parke and Duncan (65) observed a broad band in the visible spectrum while ultrasonically irradiating aerated solutions of bromine in water and aerated CS_2 in water. Intensity of sonoluminescence increased with time of irradiation up to equilibrium values, which varied with ultrasonic intensity. Oxidation increased linearly with time in solutions of oxygen in water, oxygen and potassium iodide in water, and air, potassium iodide, and carbon tetrachloride in water. The oxidation rate in these three materials increased linearly with both ultrasonic intensity and intensity of sonoluminescence. Oxidation also increased linearly with time in solutions of argon in water and nitrogen in water. Neither the addition of salts, except potassium iodide, nor of sulfuric acid affected luminescence or oxidation. However, lowering the pH of solutions of O_2-KI-H_2O resulted in a threefold increase in the rate of iodine liberation without changing luminescence.

IV. METALLURGICAL EFFECTS

Certain beneficial effects have been demonstrated by applying ultrasonic and low-frequency vibratory energy to molten and solidifying metals. Among these effects are degassing, reduction of grain size, and uniform mixing of different metals.

Degassing occurs by the same mechanisms that cause degassing in other liquids. During the low-pressure phase of each cycle, gas is released from solution. The effect is only partially reversible, so that only part of the gas returns to solution during the high-pressure phase of the cycle within the time available. As the process continues, small bubbles coalesce and rise to the surface. Combined vacuum and ultrasonic degassing results in considerable improvement in the rate and effectiveness of either method used separately.

Reduction of grain size is obtained under certain conditions. Ultrasonics and low-frequency vibratory energy effects the formation of crystallization centers in supercooled liquids. If the distribution of energy throughout the melt is such that the entire melt is subjected to intensities above a certain threshold level, uniform and refined grain sizes will result. Refinement is independent of intensity above the threshold level. In molten metals refinement appears to be most effective in alloys with low alloy content and which are predominantly solid solutions in structure. Grain refinement is best when the melt is cooled slowly with the irradiation continuing until solidification is nearly complete.

Ultrasonics tends to emulsify or mix liquids and disperse particulate matter uniformly throughout the affected volume. This phenomenon has been demonstrated in molten metals by applying ultrasonics to disperse oxides, graphite, or other metals through a melt. Reportedly, a better distribution of lead in aluminum is obtained by applying vibratory energy at 500 Hz than is obtained by the chemical lead-chloride process (66).

Although many beneficial metallurgical effects have been shown, the commercial application of ultrasonics in foundry processes has not been realized. The hindrances to this application are inherent in the methods and materials involved in generating the energy in the melt and in containing the molten metals. Various means have been used to generate energy in moderately large batches. These include (1) vibrating the mold using eccentrics, aerodynamic, or other mechanical means of producing vibrations, and (2) generating vibrations within the melt by electromagnetic induction. On a small scale, ultrasonic vibrations may be introduced through ceramic rods. Low-melting-temperature alloys can be treated in Pyrex beakers placed in a pan to which is coupled a suitable transducer. Coupling to the bottom of the beaker is accomplished through a molten layer of metal.

The greatest problem with most materials used to contain molten metals or to couple energy to the metal is erosion and contamination of the melt. Kudel'kin and others (67) found that in treating various steels and alloys, the percentage of nonmetallic inclusions was higher in those materials treated ultrasonically than it was in the controls. Otherwise, ultrasonically treated materials showed superior properties in uniformity of deformation and plasticity indexes, while other mechanical properties were similar to those of the controls.

REFERENCES

1. P. Alexander, Paint Manufacture, 21, 157-161, 175 (1951).
2. Anonymous, Soap and Chemical Specialties, 39(10), 175 (1963).
3. H. C. Wang, Yao Hsueh Hsueh Pao, 12(5), 346-348 (1965).
4. S. Witkowski, Tluszcze Jadalne, 10(2-3), 56-60 (1966).
5. S. Witkowski, Tluszcze Jadalne, 10(4), 129-135 (1966).
6. S. Witkowski and M. Otowski, Tluszcze Jadalne, 10(5), 192-203 (1966).
7. E. C. Marboe and W. A. Weyl, J. Appl. Phys., 21, 937-938 (1950).
8. L. Bergmann, Der Ultraschall, Hirzel Verlag, Stuttgart (1954).
9. T. F. Hueter and R. H. Bolt, Sonics, Wiley, New York, 1955.
10. F. I. Kukoz, E. M. Feigina, and N. Ya. Avdeev, Issled. Obl. Khim. Istochnikov. Toka, pp. 35-45 (1966).
11. S. Prakash, O. Prakash, and S. Prakash, Bull. Chem. Soc. Japan, 38(9), 1426-1430 (1965).
12. S. Prakash, O. Prakash, and S. Prakash, J. Indian Chem. Soc., 44(6), 508-512 (1967).
13. R. Lemlich, Proc. Second ICA Congress--Sound and Man, Abstract JD6 (1956).
14. R. M. Fand, J. Acoust. Soc. Am., 34(12), 1887-1894 (1962).
15. W. P. Raney, J. C. Corelli, and P. J. Westervelt, J. Acoust. Soc. Am., 26(6), 1006-1014 (1954).
16. P. J. Westervelt, J. Acoust. Soc. Am., 32(3), 337-338 (1960).
17. P. J. Westervelt, J. Acoust. Soc. Am., 35(4), 618 (1963).
18. Anon., U. S. Office of Saline Water--Research and Development Progress Report 65, (August, 1962), 141 pp.
19. A. E. Bergles and P. H. Newell, Jr., Internat. J. Heat and Mass Transfer, 8(10), 1273-1280 (1965).
20. F. W. Schmidt, D. F. Torok, and G. E. Robinson, J. Heat Transfer, 89(4), 289-294 (1967).

21. J. E. McCormick and T. W. Walsh, Research Technology Briefs, 3(3), 7-14 (1965).

22. H. Cheng, English Transl. of Wu Li Hsueh Pao (Chinese People's Republic), 20(4), 368-373 (1964). AD-639161.

23. R. Woodford and J. C. Morrison, J. Pharm. Pharmacol., 21(9), 602-606 (1969).

24. I. Gilius, Liet. TSR Aukst. Mokykly Mokslo Darb., Chem. Chem. Technol., 9, 199-205 (1969).

25. G. Tamas and T. Tarnoczy, Magyar Fiz. Folyoirat, 3, 543-552 (1955).

26. T. Tarnoczy, Magyar Fiz. Folyoirat, 4, 67-74 (1956).

27. H. V. Fairbanks and W. I. Chen, Ultrasonics, 7(3), 195-196 (1969).

28. V. V. Kuznetsov and N. I. Subbotina, Elektrokhimiya, 1(9), 1096-1098 (1965).

29. F. Baumgarte, Arztliche Forschung., 3(21), 525-530 (1949).

30. H. Frenzel, K. Hinsberg, and H. Schultes, Z. Ges. Exp. Med., 96, 811-816 (1935).

31. N. Newton and J. W. Kissel, Arch. Biochem. and Biophysics, 47(2), 424-437 (1953).

32. H. W. W. Brett and H. H. G. Jellinek, J. Polymer Sci., 21, 535-545 (1956).

33. A. Weissler, J. Appl. Phys., 21, 171-173 (1950).

34. N. H. Langton and P. Vaughan, Brit. J. Appl. Phys., 8, 289-292 (1957).

35. W. F. Head, Jr., and W. M. Lauter, J. Am. Pharm. Assoc., 46, 617-621 (1957).

36. A. Hanglein, Z. Naturforsch., 10b, 616-622 (1955).

37. I. Kossler and V. Novobilsky, Collection Czech. Chem. Commun., 28, 578-584 (1963).

38. M. S. Levinson and V. P. Nefedov, Izv. Sibirsk. Otd. Akad. Nauk SSSR, Ser. Biol. Med. Nauk, (1), 145-147 (1964).

39. Yu. Yu. Lur'e, P. F. Kanzas, and A. A. Mokina, Vestn. Tekhn. i Ekon. Inform., Nauchn.—Issled. Inst. Tekhn—Ekon. Issled. Gos. Kom. Sov. Min. SSSR po Khim., (6-7), 54-56 (1961).

40. J. D. Pandey, S. Prakash, and S. Prakash, J. Phys. Chem. (Leipzig), 228(3-4), 272-276 (1965).

41. M. T. Shaw and F. Rodriguez, J. Appl. Polymer Sci., 11, 991-999 (1967).

42. Cr. Simionescu and V. Rusan, Celul. Hirtie (Bucharest), 15(10), 369-373 (1966).

43. S. Witekowa, Roczniki Chem., 36, 693-702 (1962).

44. S. Witekowa, Roczniki Chem., 36, 703-711 (1962).

45. P. P. Zaskal'ko, G. I. Kichkin, and V. L. Lashkhi, Khim. Tekhnol. Topl. Masel, 14(6), 46-49 (1969).

46. G. W. Hastings and R. F. E. Bolten, Soc. Chem. Ind. (London), Monograph No. 20, 274-281 (1966).

47. S. Okamura, K. Hayashi, T. Sasaki, and T. Shioda, Japanese Pat. 16,155 (July 26, 1965).

48. Z. Osawa and N. Igarashi, J. Appl. Polymer Sci., 9(9), 3171-3176 (1965).

49. A. Romanov and M. Lazar, Plaste Kautschuk, 10(8), 470-471 (1963).

50. T. Shiota, Y. Goto, S. Tazuke, K. Hayashi, and S. Okamura, Kobunshi Kagaku, 22(239), 186-192 (1965).

51. J. W. Chen and W. M. Kalback, Ind. Eng. Chem. Fundam., 6(2), 175-178 (1967).

52. K. A. Chervinskii and V. N. Mal'tsev, Ukr. Khim. Zh., 32(1), 69-71 (1966).

53. Yu. N. Menzhulin and E. I. Ponomareva, Tr. Inst. Met. i Obogashch., Akad. Nauk Kaz. SSSR, 5, 24-28 (1962).

54. S. Parthasarathy, M. Pancholy, and T. K. Saksena, Current Sci. (India), 31, 500-502 (1962).

55. S. Prakash and S. Prakash, J. Sci. Ind. Res. (India), 24(12), 629-630 (1965).

56. R. Cypies, Bull. Centre Belge Etude et Document, Eaux (Liege), (35), 52-55 (1957).

57. N. P. Gaivoronskaya, L. A. Kukoz, F. I. Kukoz, M. F. Skalozubov, and S. F. Sorokina, Akust. Magn. Obrab. Veshchestv, Novocherkassk., 51-57 (1966).

58. G. Z. Zemskov, E. N. Dombrovskaya, V. T. Yarkina, L. K. Gushchin, and A. K. Parfenov, Metalloved. i Term. Obrabotka Metal, (1), 52-55 (1964).

59. A. P. Kapustin and M. A. Fomina, Dokl. Akad. Nauk SSSR, 83(6), 847-849 (1952).

60. O. V. Abramov, Rost Nesoversh. Metal. Krist., Dokl. Vses. Soveshch., Kiev, 326-337 (1965, published 1966).

61. R. M. Cohn and D. M. Skauen, J. Pharm. Sci., 53(9), 1040-1045 (1964).

62. A. P. Kapustin, Kafedra Chshchei Fiz., 88, 53-56 (1954).

63. Cl. Grandchamp and R. O. Prudhomme, Acta Polytech. (Prague), 4(1), 21-30 (1967).

64. R. Hickling, J. Acoust. Soc. Am., 35(7), 967-974 (1963).

65. A. V. M. Parke and T. Duncan, J. Chem. Soc. (London), Pt. 4, 4442-4450 (1956).

66. E. A. Hiedemann, J. Acoust. Soc. Am., 26(5), 831-842 (1954).

67. V. P. Kudel'kin, M. M. Klyuev, S. I. Filipov, E. N. Milenin, and I. I. Teumin, Izv. Vyssh. Ucheb.Zaved., Chern. Met., 12(11), 64-70 (1969).

Chapter 12

APPLICATIONS OF HIGH-INTENSITY ULTRASONICS BASED ON MECHANICAL EFFECTS

I. INTRODUCTION

In the strictest sense, all high-intensity applications of ultrasonic energy are based upon mechanical effects. They all result from particle motions. However, as a result of the particle (or mechanical) motion, certain secondary effects are produced. Some of these effects are strictly mechanical, such as the dispersion of particles in a liquid. Some are combinations of mechanical effects working in cooperation with certain secondary effects such as the mechanical eroding of an oxide coating from a metal surface, which exposes the surface to chemical attack which the ultrasonic energy also accelerates.

The applications discussed in this chapter are those in which the mechanical effects are the primary factors in the process, although others may also be active. In Chapter 13, those applications are discussed in which chemical factors are of primary importance in the processes, although it is recognized that mechanical effects are always present.

There are many high-intensity applications of ultrasonic energy. They include cleaning, soldering, plastic bonding, spot welding of metals, liquid atomization, materials forming, foam control, emulsification of liquids, compaction of powdered metals and plastics, acceleration of chemical activity, accelerated diffusion, alleviation of frictional effects, drying, depolymerization, fragmentation and dispersion of particles, control of grain size in solidifying metals, and dust and smoke precipitation. Of these, cleaning probably is best known and offers the most general market for high-intensity ultrasonics. However, there is a good market for soldering, plastic bonding, machining, and atomization equipment.

II. CLEANING

The market for ultrasonic cleaners has benefited from the development of ceramic transducer materials and power transistors.

The equipment required is of relatively simple design and ranges in size from pints to several gallons. It is used by military, industrial, medical, and domestic personnel.

II.A. PRINCIPLES OF ULTRASONIC CLEANING

The effectiveness of ultrasonic energy in cleaning materials may be attributed to phenomena which accompany cavitation. These include (1) development of stresses between the cleaning fluid and the contaminated surface, (2) agitation and dispersion of contaminant throughout the cleaning fluid, (3) increase of the attractive forces between contaminant and cleaning fluid, (4) promotion of chemical reactions at the contaminated surfaces in some cases, and (5) effective penetration of pores and crevices.

Cavitation forming at a liquid/solid interface imposes such severe stresses on the solid surface that in time the surface itself may be severely eroded. When a surface containing a contaminant is exposed to cavitation, these stresses operate to disperse the contaminant. The intensity of the stress when cavitation occurs is a function of the vapor pressure of the liquid, the gas content of the liquids, and the adhesive force between the liquid and the surface.

Particles of contaminant removed from the surface are immediately subjected to the violent activity of the cavitation bubbles. These particles are thus propelled with high initial accelerations and dispersed throughout the solvent.

The agitation also provides a scrubbing action which promotes the removal of contaminant. Such contaminant may be loose, solid particles or it may be materials which will dissolve or emulsify in the cleaning fluid.

Ions are produced in and near the walls of cavitation bubbles. When ions from the cleaning fluid show a preferential attraction for ions from the contaminant, the net attractive forces are greater than those that would exist between the liquid and the contaminant in the absence of ultrasonic agitation. As a result, the contaminant is more readily removed by the cleaning fluid.

In some cases, ultrasonic cleaning may be attributed to the promotion of chemical reaction. When the contamination is a part of the material to be cleaned, cavitation erosion alone is insufficient to perform the cleaning operation at a practical rate. This situation is typical of heat-treated specimens. Heat-treat scale is often removed in a hot, acid-pickling operation. It may be removed at a much faster rate in a lukewarm, ultrasonically agitated pickling bath.

When a contaminated surface is located in an ultrasonic cleaning bath so that it is perpendicular to the direction of the ultrasonic stress, contaminants in pores, cracks, and crevices extending from this surface are withdrawn and replaced with solvent. Other orientations which provide suitable components of stress which are directed into the pores are also effective. Formation of cavitation bubbles over the pores provides a strong "vacuum cleaner" action.

II.B. FACTORS WHICH AFFECT THE CLEANING OPERATION

The effectiveness of an ultrasonic cleaner depends upon the cleaning liquid and the ability to provide sufficient energy at the contaminated surface to promote the desired cleansing. This section is devoted to those factors which affect the distribution of power in the cleaning tank. The choice of solvent will be considered in a later section.

The amount of power required to remove contaminants from a surface varies with the contaminant and the solvent used. Since cavitation is associated with the cleaning rates for which ultrasonics is noted, the intensity at the surface must exceed the threshold of cavitation. The threshold varies with the solvent but intensities within the range 0.5 to 6 W/cm^2 will include the threshold of cavitation of solvents most generally used. Commercial equipment is usually designed to draw approximately 200 W of power from the line for each gallon of tank capacity. This is true regardless of whether piezoelectric or magnetostricitve materials are used in the transducer.

The conversion efficiency of the electronic generator and of the transducer determines the amount of power available to the cleaning solution. Several factors must be considered in order to make best use of the available power.

The ideal ultrasonic cleaner would maintain a high-intensity field over all surfaces to be cleaned under all load conditions. This ideal, best fulfilled in a diffuse field, is generally impractical because of the nature of acoustic waves, especially reflection of energy from boundaries between media of different acoustic impedances. Certain methods are used in attempting to approach the diffuse field conditions, for example: (1) placing the radiating surfaces at an angle to the surface of the tank and to any other large reflecting surface in order to avoid establishing standing waves, and (2) introducing several frequencies simultaneously into the cleaning tank. Generally, tank-mounted transducers in cleaners are suspended from the bottom of the tank and operate at discrete frequencies. Their effectiveness is influenced by the presence of standing waves resulting from surface reflections. Such cleaners operate best when their solvent depths are multiples of half-wavelengths.

Operation of ultrasonic cleaners at discrete frequencies has some advantages. High-intensity cavitation forms in local regions rather than throughout the cleaning tank. This concentration of cavitation permits operation at relatively low power input per unit volume of solvent. The disadvantage is that parts to be cleaned must be located in the high stress region of the solvent. Intensities in other regions may not be sufficient for satisfactory cleaning. Materials closest to the radiating surface will receive the greatest benefit from the cleaning activity. These surfaces may shield materials located behind them with respect to the radiating surface. Diffuse fields produced by placing the radiating surfaces at angles to large surfaces distribute the high stress regions through more of the tank. The effect is to increase the active volume, and the power required to obtain cleaning action is correspondingly higher.

The multifrequency system represents an attempt at approximating uniformity of field. With a multiplicity of wavelengths, high stress regions are brought close together. Destructive interference by reflections from the surface of the solvent or from large surfaces placed in the tank is less of a problem than it is with single-frequency operation. If one wavelength is cancelled by destructive interference, activity continues at the remaining frequencies. The intensity of the energy at each frequency must be above the cavitation threshold. The cavitation threshold increases as frequency increases. The total energy consumed is necessarily higher than that for single-frequency units, but the additional power of the multifrequency units is generally better distributed.

II.C. TYPES OF ULTRASONIC CLEANERS

The basic equipment required for ultrasonic cleaning includes the transducer, tank, and electronic generator. Applications may very from the simple use of the basic components to large automated installations complete with conveyers, additional rinses, heaters for the solvents, dryers, mechanical prewashers, and solvent-recovery units. Figure 12.1 is a view into a typical commercial-type ultrasonic cleaning tank.

Both magnetostrictive and piezoelectric transducers have been used widely in the past. Transducers may be obtained either in immersion form or attached directly to the cleaning tank.

Immersion transducers are separate, completely sealed units designed so that they can be placed in any suitable tank in any desired position. A number of immersion transducers can be arranged to focus energy for increased intensity.

Most ultrasonic cleaners are constructed with the transducers attached to the tank. In early piezoelectric models, the transducer was a piezoelectric slab epoxy bonded to the bottom of the tank. This construction has certain drawbacks. The epoxy bond may deteriorate with age, with an accompanying loss of acoustic coupling to the cleaning bath. The total power-handling capability of the epoxy bonded, freely suspended slab is limited by the strength of the slab and by that of the bond.

Improvements in coupling, power-handling capabilities, and efficiency are obtained with piezoelectric transducers by relatively simple design considerations. A discrete frequency system can be designed using a sandwich construction. In this design, the transducer plus the bottom of the tank constitutes a half-wave vibrator. The piezoelectric slab (or pair of slabs) is located at the node sandwiched between two metal plates, preferably with the node on the center plane of the slab. The metal plate nearest the bottom of the tank is welded or silver brazed to the tank. The welding or brazing must be free of unbonded areas or voids, which hinder the coupling of energy to the solvent in the tank. The metal backup plate is bolted to the mounting plate which has been welded to the tank and torqued to produce a bias compressive stress across the piezoelectric element which corresponds to that for maximum operating efficiency recommended by the manufacturer of the element. If the elements of the transducer, including the bottom of the tank, have been properly matched acoustically, the transducer will operate effectively when the cleaning solution is an integral number of half-wavelengths deep.

Magnetostrictive transducers can be silver brazed or welded to the bottom of the cleaning tank. As with the piezoelectric systems, the weld should be free of voids and the transducer should operate as a half-wave system.

Magnetostrictive transducers often appear as a laminated stack suspended from the bottom of the tank. The stack is surrounded by a coil and a separate iron yoke completes the magnetic system.. Although this design has the advantage of low cost construction, it is less efficient than other magnetostrictive designs. The air gap inherent in this design produces the major mmf drop in the magnetic circuit.

A second and more efficient magnetostrictive transducer consists of a laminated stack with a window punched from the center. The coil is wrapped about the two legs on either side of the window. Best operation occurs when the coil is centered over the velocity nodal plane.

Magnetostrictive transducers require a bias magnetic field. Otherwise, they vibrate at twice the frequency of the driving voltage, but at low efficiency. The amount of bias affects the efficiency.

The bias flux may be supplied by permanent magnet or by passing direct current through the coils. The advantages of permanent-magnet biasing are: (1) no separate power supply is necessary and (2) no additional provision for isolating direct current from the output transformer of the power oscillator and alternating current from the power supply is necessary. The magnet must be capable of enduring high-frequency mmf fluctuations to which it might be subjected. Care must be exercized to prevent overbiasing.

Advantages of biasing with direct current in the coils are: (1) the bias field can be adjusted for optimum performance of the tranducer and (2) the bias magnetic path is identical with that of the alternating field. The alternating current must be electrically isolated from the direct-current source, usually by means of a high inductance in the d-c branch, and the direct current is isolated from the output transformer of the high-frequency source by means of blocking condensers in the a-c branch.

II.D. ELECTRONIC GENERATORS FOR ULTRASONIC CLEANERS

Recent developments in semiconductors have benefited the ultrasonic industry. Generators are more compact and more rugged, and can be more reliable than their electron-tube predecessors.

The electronic generator basically consists of an oscillator stage, an amplifier stage, and an output stage.

Good impedance matching between all components of the system is important for maximum power transfer. Matching must include a consideration of the acoustic impedance provided by the cleaning bath as well as consideration of each element of the transducer and of the electronic generator. The acoustic impedance of the load reflected into the electronic system as electrical impedance depends upon the type and density of the cleaning fluid, the temperature of the bath, the type of cleaning load, the depth of the fluid, and the cross-sectional area of the radiating surface. All of these factors are variable except the cross-sectional area of the radiating surface of the transducer. Their influence on the reactive component of the impedance reflected into the transducer has its effect on the resonance frequency of the transducer. The oscillator is tuned to the resonant frequency of the transducer/tank combination, either manually or by feedback control of the oscillator frequency. Tuning becomes less critical with decreasing mechanical Q. A low-Q installation will operate effectively over a range of frequencies with little loss in efficiency.

II.E. CHOICE OF ULTRASONIC CLEANING FLUIDS

Two obvious considerations in the choice of cleaning fluids for use in ultrasonic cleaners are (1) chemical compatibility of the cleanser with the materials to be cleaned and the materials of the container and (2) the effectiveness of the cleanser in removing the soil. There are several important factors that determine compatibility and effectiveness in the presence of intense ultrasonic fields. Stresses in the walls of cavitation bubbles are concentrated and can be extremely severe. These stresses are controlled in part by the vapor pressure of the liquid. As vapor pressure increases and surface tension decreases, the maximum stress associated with cavitation decreases. Higher stresses are associated with smaller bubble sizes. The intensity of the stress waves associated with the collapse of cavitating bubbles is a function of the ratio of maximum bubble size to minimum bubble size. Bubbles in liquids with high vapor pressure or in liquids containing a large amount of dissolved gases tend to grow under the influence of the ultrasonic field, and the ratio of expansion to contraction under the alternating stresses diminishes rapidly.

Three types of cavitation are sometimes recognized or are claimed to be possible. One type results from entrapped or dissolved gases. Under the influence of ultrasonics, these gases coalesce into larger bubbles and rise to the surface of the liquid. The second type is associated with vapors of the liquid being cavitated, in which the bubbles contain vapors of the liquid, The third type is associated with the cohesive strength of the liquid and might be described as a tearing apart of the liquid followed by violent collapse of bubbles thus formed.

Dispersed materials of solid particles or gas bubbles, small point-protrusions, and rough surfaces form nuclei for the formation of cavitation bubbles. Therefore, the stresses that might be associated with cavitation in a given solvent are a major consideration in the choice of solvent. Its importance is related to the possible effects produced on the contaminated surface. For instance, high-intensity cavitation will erode plated and many coated surfaces. The effect can be minimized or eliminated by the choice of high-vapor-pressure solvents. Sometimes the desired effect is accomplished by blending solvents to obtain the desired properties of solubility and nondestructive stress levels. For example, high cavitation intensities are possible in water. Dissolving a small quantity of high-vapor-pressure material, such as ethyl or methyl alcohol, in the water can decrease the intensity of the cavitational stresses considerably. A blend of the two materials makes it possible to take advantage of the solvent properties of the water without the destructive intensities being applied to the contaminated surface. One might consider the alcohol as being representative of a mechanical fuse.

The chemical classification of the solvents and contaminants is a major consideration in the choice of solvent, i.e., whether they are polar or nonpolar. In general, polar solvents are chosen to remove polar contaminants and nonpolar solvents to remove nonpolar contaminants. According to the dictionary, a polar chemical is one that is capable of being ionized as NaCl, HCl, or NaOH. The polar group also includes such materials as alcohol, acetone, acids, aldehydes, water, nitriles, and amines, which are chemically active. A distinguishing characteristic of polar materials is a high dielectric constant. Nonpolar substances are characterized by low dielectric constant and include materials such as hydrocarbons and their derivatives--benzene, carbon disulfide, gasoline, kerosene, and chloroform, which are comparatively inert. For a given polarity of solvent best penetration of the contaminants is obtained at lower molecular weights.

The possible effect of solvents, particularly acids, on the material to be cleaned may be a factor. Acids may cause hydrogen embrittlement and the effect is accelerated under ultrasonic excitation. In addition, the cavitation may produce a chemical reaction between the solvent and a base metal. The author has twice had the experience of having used acidic solvents recommended for cleaning certain stainless steel items which were supposedly unaffected by these chemicals only to have a rapid chemical action initiate that dissolved the specimen. Commercially available cleansers of the strong-mineral-acids group contain wetting agents and inhibitors. The purpose of the inhibitor is to minimize hydrogen embrittlement and base-metal attack. The inhibitors usually are organic materials that may break down at high temperatures.

Other effects caused by solvents include dissolving of base material, swelling of materials such as elastomeric coatings, and cracking and distorting of certain plastics. These are the necessary considerations for choosing a solvent for material compatibility.

Other considerations in the choice of solvents for ultrasonic cleaning include ability to contain the fumes (for example, the fuming of an acid, such as hydrochloric acid, is accelerated under the influence of ultrasonically induced cavitation), toxicity, ability to dissolve the contaminant, and its ultrasonic transmission properties. Materials that cavitate at a very low intensity (low cavitation threshold) inhibit the production of high-intensity cavitation and limit the transmission of ultrasonics through the liquid to low intensities.

Cleansers used most frequently for ultrasonic cleaning include aqueous materials (including household detergents), acidic chemicals, alkaline chemicals, and hydrocarbons. Commonly used solvents and recommended applications are listed in Table 12.1. Most manufacturers of ultrasonic cleaners have prepared literature on recommended solvents which can be obtained upon request.

II.F. PROCEDURES FOR ULTRASONIC CLEANING

The operation of bottom-mounted transducers is affected by the depth of solvent in the tank and by the loading in the tank. When resonance in the tank coincides with the resonance of the transducer, maximum agitation occurs (i.e., when the depth of the solvent is a multiple of half-wavelengths at the driven frequency). When the depth is an odd multiple of quarter-wavelengths, the reflected wave is 180° out of phase with the transmitted wave and maximum destructive interference occurs. When the tank is loaded in such a manner that the standing-wave pattern is destroyed, the depth of loading becomes less significant and the acoustic impedance of the load becomes primarily a function of the characteristic impedance of the fluid in the tank. The effective cleaning volume is, then, a function of the total energy that is converted to acoustic energy in the fluid.

TABLE 12.1

Commonly Used Solvents for Ultrasonic Cleaning

Solvent[a]	Base Materials	Contaminants	Remarks
HCl with inhibitor	Ferrous alloys	Oxides, heat treat scale	Plastic containers preferred
H_2SO_4 with inhibitor	Ferrous alloys	Oxides, heat treat scale	Plastic containers preferred
HNO_3 with inhibitor	Stainless steel, some aluminum	Oxides, heat treat scale	316 stainless steel or plastic are acceptable container materials
Phosphoric acid with inhibitor	Iron, steel, brass, zinc, aluminum	Light scale and oxides, some drawing compounds, oil, grease	316 stainless steel is an acceptable container
Water	Any material not damaged by water	Loose soils, water-soluble soils	Stainless steel containers
Detergent and water	All metals and many other materials	Oils, greases, loose soils, etc.	Stainless steel containers
Soap and water	All metals and many other materials	Oils, greases, loose soils, etc.	Stainless steel containers
Trichloroethylene	All metals	Oil, grease, buffing compounds, etc.	Stainless steel containers

Tetrachloroethylene	All metals	Oil, grease, buffing compounds, etc.	Stainless steel containers
Acetone	All metals	Oil, some plastic-base cements (polar)	Stainless steel containers
Benzene	All metals	Oil, grease	Stainless steel containers
Xylene	All metals	Oil, grease (slightly polar)	Stainless steel containers
Freons[b]			
Trichlorotrifluoroethene (Freon TF)	Most materials	Circuit boards, etc., where halogenated hydrocarbons may not be safe; fluxes, loose soils, oils, other organic compounds	Stainless steel, aluminum, or plastic containers can be used; easily distilled and recycled
Tetrachlorodifluorethane	Most materials	Circuit boards, etc., where halogenated hydrocarbons may not be safe; fluxes, loose soils, oils, other organic compounds	Stainless steel, aluminum, or plastic containers can be used; can be used to 150F but expensive
1,1,1-trichloroethane	Electrical and electronic assemblies	Flux, dust, etc.	Stainless steel containers

[a]Wetting agents are recommended for most solvents used in ultrasonic cleaners.

[b]Du Pont trade-mark.

Sometimes it is desirable to clean specimens in a basket or in a container filled with solvent that differs from that in the tank. In either case, care should be taken to obtain good transmission into the container. Wire mesh baskets may be composed of wire of such a scattering cross section that transmission of energy to the parts to be cleaned is inhibited. Considerable energy may be lost in going through glass beakers with thick bottoms. Sometimes it is advantageous to use plastic bags, such as polyethylene, to contain the fluid and part, being careful not to puncture the bag. A container consisting of a stainless steel cylinder with a thin bottom of metal or plastic can be made so that transmission losses through the bottom are negligible. The buildup of gas bubbles on the bottom of a container will form a barrier to the transmission of ultrasonic energy and should be prevented.

Soft glass containers and thick thermoplastic materials should not be used in the presence of strong cavitation. The glass will fracture under these conditions. The plastic will be damaged and perhaps punctured by intense cavitation.

The arrangement of parts in a cleaning bath should be such that all areas to be cleaned are accessible to the cavitation. If parts are acoustically shielded by other heavier parts, they should be rearranged or removed for cleaning later. Large parts may have to be reoriented occasionally to obtain full coverage.

The choice of frequency depends upon the size of parts that the unit is designed to clean. Low frequencies are less subject to scattering and losses due to loading than are high frequencies, while high frequencies produce a greater number of stress maxima in a given volume.

When properly used, ultrasonic energy provides an effective means of accelerating cleaning rates and of cleaning materials that are otherwise difficult to clean. It is especially valuable in removing debris from pits and crevices.

Two novel cleaners are shown in Figs. 12.2 and 12.3. Shown in Fig. 12.2 is a strip-cleaner for cleaning components such as assemblies of soldered detonator wires. The strip is fed through a flooded cavity where it is exposed to ultrasound and cleaned. Solvent is fed into the cavity through a manifold and discharged through slits into a reservoir for reclamation and recirculation. This cleaner utilizes a magnetostrictive transducer.

Shown in Fig. 12.3 is a continuous dry cleaner used for cleaning belts, such as sanding or grinding belts, which move over the output surface of the transducer assembly. It is claimed that this device increases belt life by as much as three to four times.

II.G. METHODS OF EVALUATING ULTRASONIC CLEANERS

The rapid growth of the market for ultrasonic cleaners has brought the recognition of a need for suitable techniques of effectively evaluating these cleaners. The Ultrasonic Manufacturers Association has attempted to develop a standard for this purpose. Ultrasonic cleaners are usually rated according to watts per gallon, but the power rating actually refers to the line power rather than to the power density in the cleaning tank. The actual measurement of power density and total power within a cavitating ultrasonic bath is difficult. Listed in Table 12.2 are a number of techniques that have been used for measuring an ultrasonic field or for evaluating an ultrasonic system. Some of the techniques are applicable to intensities below the cavitation threshold only. Few of the others can fulfill the requirements of a general standard.

Effective cleaning and processing in liquids at medium and high intensities is usually associated with the occurrence of cavitation. Cavitation bubbles so disrupt the normal propagation of high-intensity waves that a true measurement of the power density in a bath can be obtained by few means. If one can assume that all of the energy is absorbed within the liquid, calorimetric methods can sometimes be used to determine the power radiated by the

TABLE 12.2

Methods of Measuring Ultrasonic Activity in Treatment Systems

Probe methods	Sonochemical	Calorimetric
Acoustic pressure sensitive	Chlorine release	Substitution
Velocity sensitive	Bromine release	Thermal isolation
Acoustic absorption	Sonoluminescence	Constant flow
Displacement sensitive	Chemical Oxidation	Acoustic dilatometer
Feel	H_2O_2 formation	
Bouncing beads	Depolymerization	Other
Calorimetric expansion		Visual
Noise sensitive	Surface properties of cleaned sample	Complex transducer electrical impedance
Rayleigh disk	Water break	Reaction of radiation pressure
Radiometer	Coefficient of friction	Fountain or surface distortion
Hot-wire microphone	Surface electrical conductivity	Volume flow rate
Thermocouple in lossy matrix	Adhesion of evaporated or sprayed on metal film to nonmetallic surface	Heat transfer
Acoustic fluxmeter	Thermally induced oxide film (400°C) (metals)	Complex mechanical impedance
Signal to noise	Plating effectiveness (metal)	Light scattering
Total noise	Standard soil	Volume change
Subharmonic measurement	Radioactive tracer	Change in conductance
Semiconductor thermo-resistors	Reflectometer	Attenuation by scattering
Photoelastic	Light transmission	Printing
Emulsion probe	Chemical tracer	Biological cell breakage
Electrolytic probe	Fluorescent dye	
Erosion	Chemical dye	
Foil (heavy)	Graphite-coated ceramic	
Soil removal		
Strips or rods		
Foil (light)		
Dispersion		

transducer. However, calorimetric measurements are limited to systems of certain designs only.

All reactions or ultrasonic effects in some way relate to the frequency and intensity of the impressed field. The most popular methods of evaluating ultrasonic cleaners are based on one or more of these effects and may be classified according to whether they operate by (1) sonochemical activity, (2) erosion effects, and (3) probe reactions.

Cavitation tends to produce free chemical radicals at and near the bubble walls. The quantity of these radicals in a given chemical mixture depends upon the intensity of the stresses involved and the total wall areas of all radical-producing bubbles. This phenomenon was investigated at one time by the UMA as a possible basis for a standard (1). In the UMA proposed standard, 200 ml of a filtered 0.1 M potassium iodide solution saturated with carbon tetrachloride is placed in a cell--a 4 x 6 in. polyethylene bag with a 0.004-in. wall thickness.

The solution in the tank to be evaluated consists of tap water which has been aged for 12 hr at room temperature or ultrasonically agitated for 30 min in order to stabilize the gas content of the liquid. A wetting agent is added to the water in concentrations that range from 0.005 to 1%. The evaluation is performed at temperatures between 78 and 80°F.

The cell is placed in the tank so that the liquid level in the cell is even with that in the tank. The cell may be suspended from a scanning mechanism for giving an integrated evaluation of the activity in the tank, or it may be held stationary to indicate the activity at any desired location. (The UMA proposed standard gave the design of a traversing mechanism for scanning a tank.) The final reading is reduced to a units-per-minute basis (mg Cl_2/liter-min).

The UMA technique has been called a chlorine-release method because, when the test solution is subjected to intense ultrasonic radiation, free radicals of chlorine are released from the carbon tetrachloride at or near the surfaces of cavitation bubbles, as described previously. The chlorine replaces the iodine released from KI to form KCl and I_2. The iodine thus released is soluble in the KI solution and can be measured either by titration or by spectrophotometric means. The measurement must be performed very carefully. If the titration method of analysis is used, the titration solution must be standardized each day that it is used. (The UMA recommended titration solution is a 0.0028 N sodium thiosulfate solution prepared with distilled water. A starch indicator solution can be made by dissolving, by boiling, 0.1 g of potato starch per 100 ml of water.) Titration or spectrophotometric measurements should be made within a few minutes after exposure.

The linearity with time of the chlorine release in a constant-intensity field has been proven empirically (2). Similarly, the rate of chlorine release appears to be proportional to cavitation "intensity," i.e., the actual power density within the cell.

To overcome the objection that little, if any, ultrasonic cleaning is done at 80°F, Jacke (3) has obtained results similar to those obtainable with the chlorine release by using CBr_4 instead of CCl_4 at 150°F. The bromine release is slower than the chlorine release, but this approach does hold promise as an evaluation technique at common cleaning temperatures.

A popular and simple method of evaluating the activity in ultrasonic cleaning tanks consists of immersing a strip of aluminum foil mounted on a wire frame in a vertical position in the tank. The evaluation is based either upon the amount of pitting and erosion that occurs within a certain length of time or upon the time required to erode holes in the foil. The foil-erosion method provides a good indication of the distribution of the cavitation within the tank. However, there are problems in its use that prevent its becoming a suitable standard. Foils of even the same brand may exhibit different degrees of hardness and toughness to cavitation erosion. Particles of the foil resulting from cavitation erosion contaminate the solvent and must be removed. Even the method of mounting the foil on the wires and the wire sizes and types themselves may affect the formation of cavitation pits in the foil.

Several types of probe techniques have been developed. These may include pressure probes (most of which are useless in a cavitating field), particle-velocity probes, emulsion probes, weight-loss probes (an erosion technique), noise-measurement probes, thermoelectric or heat-sensitive probes, and electrolytic probes.

Hydrophones can be calibrated to measure actual acoustic pressure at intensities below the cavitation level. At intensities above the cavitation threshold, pressure-sensitive probes can be used in measuring noise associated with cavitation. Such probes are easy to use. The noise measurements correlate fairly well with the chlorine-release technique to an intensity that corresponds to the formation of a continuous cavitation bubble between the bottom of the tank and the cleaning fluid. At this point, the noise-measurement values continue to increase but the chlorine release drops. Gas films forming on the surface of the probe and spurious noises from other sources in the unit and not related to cavitation adversely affect the meter indications.

Relative to the tank, the intensity of the ultrasonic field and its distribution at the operating conditions are perhaps the most important factors controlling the effectiveness of ultrasonic cleaning. The intensity within the tank is controlled primarily by the radiation from the transducer. The most effective transfer of energy occurs when maximum displacement occurs at the transducer/solvent interface, which is usually the bottom of the tank. The intensity at this surface is a function of the displacement times the frequency, and it is primarily responsible for the cavitation activity within the liquid load.

Velocity-sensitive probes, or displacement-sensitive probes, are of little value in measuring acoustic fields at intensities above the cavitation threshold. However, they can be used to determine the displacement of the radiating surface and thus to evaluate the performance of the transducer. Probes of this type may operate on magnetic principles, capacitance principles, or piezoelectric principles. The methods based on magnetic principles involve variations in the length of gap in a magnetic circuit caused by the ultrasonic vibrations. Similarly, the methods based on capacitance principles involve the change in capacitance between plates of a condenser (one of which may be the radiating surface) due to vibrations induced in one of the plates. Piezoelectric displacement probes may be bimorphic and the charges developed on the faces of the piezoelectric elements are functions of the displacement of the elements.

A probe for indicating the amplitude of vibration of the radiating surface has been used for evaluating transducers (4). It is a resonant accelerometer with a bandwidth of approximately ±1 kHz, which must be brought into contact with the surface to be measured. The amplitudes at various points on the bottom of the tank are measured and recorded. Any later measurements must be made at the same points. Comparative evaluations between cleaners are limited to those of exactly the same frequency. The reaction of the probe on some types of tank designs and the problem of maintaining a satisfactory acoustic coupling to the surface are objectionable.

Emulsion probes consist of two immiscible liquids. The cavitation activity is evaluated by the rate at which the liquids are emulsified. The probe may be a small plastic bag containing distilled water and a piece of absorbent material saturated with oil. Or the oil-saturated material may be immersed directly into the cleaning fluid. Although this method does not provide an absolute measurement of "cavitation activity," it does provide a rapid and visual indication of the potential effectiveness of the cleaner.

Weight-loss probes are in a category similar to the foil-erosion technique. A probe of material is located in an ultrasonic field and the cavitation activity is determined on the basis of the amount of material lost by erosion during a specified period of time. Various materials have been used, with lead being the most common.

Thermoelectric probes consist of thermocouples embedded in materials that are good absorbers of ultrasonic energy. The acoustic intensity is determined by the heating rate in the absorber.

Regardless of the type of probe used, the effect that inserting it into the field has upon the field itself will have a corresponding effect upon the accuracy of the measurement. The ideal probe would be small compared with the wavelength of ultrasound and the dimensions of the tank. Large probes which are acoustically mismatched to the surrounding liquid may cause undesirable reflections, which in turn may cause either destructive or constructive interferences and consequent errors.

None of the previously mentioned techniques is entirely acceptable to all manufacturers and consumers of ultrasonic cleaners. A method that is gaining acceptance, however, is based upon the removal of graphite stains from aluminum oxide blocks. The face of the block is coated with a dark layer of pencil-type graphite and located within the cleaning tank. The tank is filled with water and operated for a predetermined length of time. The block is then removed, allowed to dry, and then compared with a set of blocks rated from 1 to 10 according to the amount of light reflected from their surfaces (i.e., according to the amount of graphite removed).

III. MACHINING, FORMING, AND JOINING

III.A. MACHINING

Ultrasonic machining is one of the earliest of the commercial high-intensity applications of ultrasonic energy. The earliest method, and one that continues to find useful acceptance in industry, was based upon techniques described by Farrer (5) and Balamuth (6). In this method, a slurry consisting of abrasive particles in a low-viscosity liquid is washed over the end of a tool shaped to conform to the desired geometry of the impression to be made. The axial motion of the tool tip and the resulting cavitation impart high accelerations to the abrasive particles and thus erode the work away. The method is especially useful for machining brittle ceramic materials, but it also has been used for machining some powder-metal parts. In the latter application, evidences of a highly accelerated electrolytic erosion have been observed on occasion. The effect may be an adverse one in that the tool rather than the work is "eaten" away.

Shown in Fig. 12.4 is a typical early model ultrasonic-machining assembly consisting of a half-wave magnetostrictive transducer silver brazed to a half-wave Monel coupling horn. Fig. 12.5 is a photograph of a 2 kW magnetostrictive transducer attached to a rectangular, wedge-shaped horn of stainless steel. A second half-wave horn is attached, by means of a fine-thread stud, to the coupling horn; this second horn is used as a toolholder. Toolholders usually are Monel or stainless steel and are designed to resonate at the transducer frequency with the tool attached. The tool is usually made a fraction of an inch longer than the length required to match the transducer and removed when it wears to a fraction of an inch shorter than the resonant length. Equations for designing horns are given in Chap. 4.

Typical materials for making horns include Monel (66 Ni, 28 Cu, 5 Fe, plus Mn, Si, C), Titanium 318A (6 Al, 4 V, 90 Ti), Admiro 5 (mainly copper and zinc, with small amounts of Ni, Mn, and Al), 304 Stainless steel, and other stainless materials consisting of 1.3 C, 12.5 W, 1 Mo, 4 V, 5 Co, 4.3 Cr, and remainder Fe.

For best operation the combination of toolholder and attached tool should resonate at the resonant frequency of the transducer. To accomplish this, the length of the horn should be such that the acoustic impedance at the junction is equal in magnitude in both the tool and the toolholder at the resonant frequency. The diameter of the selected horn preferably should be only slightly larger than that of the tool. The length of the tool usually is less than 1 in.

at 20 kHz, although holes 0.030 by 1½ in. deep have been drilled into glass at low driving power. The reason for the limitation in length is that the stress at the junction increases as the tool length increases, and when this stress exceeds the strength of the bond, the tool falls off. Joining at a point of maximum stress in a horn should be avoided. In tapered horns, maximum stress occurs near the small end, which is the end to which the tool is attached. In stepped horns (with two quarter-wavelength sections) or cylinders, the point of maximum stress is located at the center of the length of the horn. Brazed joints should be free of defects and preferably filleted. A rule of thumb for the selection of lateral dimensions of the tool is 0.2 of a wavelength of the material of the toolholder or smaller. Drill rod is used most often as tool material.

Some deep holes are drilled using the slurry method with half-wave tools attached to the half-wave toolholder. Such drilling is done only occasionally because of the difficulty of pumping slurry to the tool tip.

Best machining is accomplished when the driving frequency is at the resonant frequency of the transducer and tool assembly. Tuning was done manually in older models. Some attempts at feed-back from the transducer to control the oscillator frequency have been made with magnetostrictive transducers, but this has not been entirely satisfactory under load conditions. Later models use ceramic elements in a sandwich-type construction in place of the magnetostrictive transducer and these are more readily adapted to feedback control of the driver frequency.

Abrasive slurry usually is applied by means of a recirculating pump; however, it is sometimes advisable to use an eye dropper or other means of application. Satisfactory slurries consist of abrasives of boron carbide, aluminum oxide, garnet, silicon carbide, and diamond dust in such liquids as water, kerosene, or other low-viscosity liquid that is compatible with the work and that presents no health hazard. Abrasive-grit sizes are usually 220 mesh or smaller, the smaller sizes producing better surface finishes and the larger sizes producing faster machining.

A slurry of one part by volume of water to one part by volume of abrasive is best for shallow holes. Slightly thinner slurries are better for deep holes because they penetrate the hole better. In thicker slurries, the mobility of the particles is restricted and the machining is hindered. Thin slurries are less effective because of the smaller number of cutting edges available. The abrasive should be harder than the work to be machined.

Conventional ultrasonic machining units operate with weights, or other means of controlling the pressure at the tip of the tool, applied to force the tool into the hole. The speed of drilling is affected by the applied force. Too much force can be as detrimental to the machining process as too little force. It may prevent free flow of abrasive and change the reflected impedance, causing the unit to operate at less than optimum frequency. In machining difficult materials it is often advisable to relieve the pressure periodically to permit free flow of abrasive. Best pressures usually are determined empirically.

Tool wear is a function of the heat treatment of the tool, the abrasive slurry used, and the relationship between hardness and toughness of the tool and the work. Martensitic-type tools show good wear resistance as do hard alloy tools.

Small hand-held ultrasonic drills have been constructed for machining small holes. A skilled person can drill holes 0.001 in. in diameter to depths of 1/16 in. A drop of slurry is placed on the surface of the material to be drilled and the operator feeds the tool into the hole as it is being drilled. This operation is extremely difficult in opaque materials. The fact that the eye can follow the progress of the drill into the work makes it much easier to drill very small holes in transparent material.

A laboratory model of a hand-held drill with bits and attachments for drilling deep holes is shown in Fig. 12.6. The tools and toolholders use threaded attachments. The transducer consists of lead-zirconate-titanate plates in a sandwich construction designed to resonate at 28 kHz.

Dry ultrasonic machining has been accomplished without slurry by using chisel-type tool tips. Since the tool can become imbedded in the work, it is necessary to continually change its position to present new shearing surfaces. This operation requires fairly soft work material and especially designed tool tips that are hard, tough, and sharp. Energy concentration at the cutting edges can quickly heat and melt the tip to a dull bead.

Materials most often machined by conventional ultrasonic machining techniques using abrasive slurries are glass, ceramics, stones, occasionally brittle metals, and sintered materials.

Combining ultrasonics with electrolytic or spark-erosion machining of conducting materials results in faster work removal than is accomplished using electrolytic or spark-erosion machining only. However, the combination is seldom used for economic reasons.

Ultrasonics also has been used with chemical etching to machine materials to controlled shapes. The application of ultrasound increases the etching rate by a large factor which depends upon the material to be machined and the chemical being used.

Other methods include (1) revolving a workpiece and ultrasonically energizing a conventional drill bit to drill materials otherwise difficult to machine, (2) energizing a lathe tool to obtain improved surfaces and reduce chatter in materials that have a tendency to chatter normally, and (3) energizing reamers to increase feed rates and improve surface finish. There are reports of improved grinding effects under the influence of ultrasonic energy.

Ultrasonic rotary drills with conventional diamond tips are commercially available. These units are capable of drilling holes in ceramic materials as well as in metals. A commercially available rotary-type ultrasonic drill is shown in Fig. 12.7. Several typical specimens machined with the rotary drill are shown in

Fig. 12.8. The bit is water cooled but no slurry is required. This equipment is primarily of value for machining ceramic-type materials but shows promise for improved machining of certain metals as well. The effectiveness of the imposed ultrasonic forces in improving the machining effectiveness is a function of the ratio of maximum vibratory tip velocity to the peripheral velocity of the tool relative to the work surface. The effectiveness is diminished as this ratio decreases below unity. Increasing the ratio above unity should be beneficial if the cutting edges of the tool remain in good condition. At very high ratios, heating is a factor and careful cooling is required to prevent damage to the tool.

III.B. METAL FORMING

A plate of soft-annealed copper as thick as 1/8 in. can be reduced to a thin foil by forcing it between a flat tip of an ultrasonically vibrating tool and an anvil when the spacing between the anvil and the tool is fixed. The force necessary to maintain a continuous feed rate can be applied easily by hand. This example illustrates the reduction in drawing force that can be accomplished with the application of ultrasonic energy to drawing dies and other forming operations.

The drawing force necessary for some forming processes can be reduced by more than 50% by using ultrasound. Langenecker claims that drawing forces can be reduced by 80% at drawing speeds of 100 ft/min (7). Two factors believed to improve formability of metals with the application of ultrasonic energy are (1) the softening effect on the crystals (volume effect) and (2) the reduction of frictional forces between the workpiece and the tool. The potential advantages are the increase in metal formability and the beneficial influence of ultrasonic energy in forming difficult-to-form materials such as case-hardened steel and tungsten. Other benefits that have been claimed, especially for drawing operations, are increased area reduction per pass, better size control, improved inside-diameter surface finish (tubing), greater tool life, and minimized chatter.

Balamuth (8) proposes a thermal-equivalent theory to explain the volume effect relative to drop in force required to draw materials. The theory is based upon the internal thermal energy of a solid being present in the form of incoherent atomic vibrations, the combination corresponds to an effectively increased kinetic energy content per atom. This argument suggests that ultrasonic energy introduced into a solid should have the same effect on the behavior of the metal as elevating the temperature of the material by an equivalent amount. Balamuth then claims to show experimental evidence of a drop in drawing force under ultrasonic excitation that exceeds the dynamic force applied. Reduction in apparent static yield stress occurs under the influence of high-intensity ultrasonic energy for strain rates up to 100/sec.

Prolonged exposure of material to intense ultrasound before, during, and after forming can produce undesirable and irreversible effects.

Langenecker (9) also explains plastic deformation with the aid of ultrasonics from the total-energy viewpoint. Acoustic strain energy interacts with materials in several time-dependent ways. Any resonance absorbs energy in the form of higher dynamic strains. Absorption due to internal friction or absorption at lattice defects or grain boundaries raises the total thermal energy of the material. Some acoustic energy is converted to residual strain energy through plastic deformation.

Lehfeldt (10) states that the lowering of the yield point under the influence of ultrasound is a spurious effect rather than a specific change in the material.

Pozen and others (11) investigated the metallurgical effects of high-intensity ultrasound on a specimen of carbon steel. The microstructure of sections cut from the maximum stress zone revealed grain boundaries between pearlite and ferrite less sharp than those in untreated material.

Results of superimposed ultrasonic vibrations on compressive deformation have been investigated by Izumi and his co-workers (12). The flow stress of compression can be reduced ultrasonically, but the effectiveness is dependent upon the acoustic impedance, Young's modulus, melting point, work hardenability, and stacking-fault energy. Temperature rise in the material with increased intensity also is material dependent, as one would expect. Ultrasonics influences the hardness distribution of the material in accordance with the distribution of the stresses in the field.

Kralik and Weiss (13) found that work hardening and tensile strength of wires of face-centered cubic metals increased, especially when these materials (copper, gold, silver, and nickel) were treated at 20 kHz at various temperatures. However, work-hardening response to ultrasonic treatment decreased for solid solutions of face-centered cubic metals. They suggest that the work-hardening effect might be related to some as yet unexplained mechanism involving dislocation loops and vacancy agglomerations in the materials.

Langenecker (14) claims that yield strengths of metals such as zinc and aluminum are greatly increased by a few minutes' application of ultrasonic energy at an acoustic pressure exceeding 2.5×10^7 dyn/cm^2. The ultrasound oscillates and distributes the crystal-lattice dislocations similarly to work hardening, but without change in shape. He claims yield strength in shear is increased 50% by radiation at 4.5×10^7 dyn/cm^2 and as much as 150% at 5×10^7 dyn/cm^2. Fracture along grain boundaries occurs at stresses exceeding 5.5×10^7 dyn/cm^2.

Multiplication of dislocations in nickel, copper, and aluminum (as well as lithium fluoride and sodium chloride crystals) was found by Pines and Omel'yanenko (15) to begin at a certain threshold amplitude of vibrations, and this threshold decreases with increasing temperature. Above the threshold amplitude, the dislocation density increases with the ultrasonic intensity. The dislocation density also increases linearly with time, initailly, but afterward tends towards saturation. Saturation level increases both with increasing temperature and with increasing amplitude of vibration.

Chachin and Skripnichenko (16) found the use of ultrasonics during the tempering of cutting blades increased their wear resistance over that with the normal treatment. Specimens of an alloy steel containing 0.87% carbon, 4% chromium, 6.3% tungsten, 2.5% molybdenum, and 1.7% vanadium were subjected to intense ultrasonics in a salt bath at 560°C for periods from 15 to 60 min. Comparisons of hardness, impact and bending strength, red hardness, and wear resistance of these specimens with those of specimens similarly heat treated but without ultrasonics showed that (1) hardness increased with a 15-minute application of ultrasonics over that of usual tempering; (2) impact strength and bending strength after a 60-minute application of ultrasonics correspond to values obtained after a 120-minute normal tempering period; and (3) ultrasonics applied during tempering also increases the red hardness and wear resistance of this alloy. These effects probably are associated both with internal mechanisms and with mechanisms associated with heat transfer.

Schmid (17) states that the application of ultrasonic energy during a static test causes softening of a metal, but if the insonation is superimposed prior to the static deformation, the sample is usually hardened by an amount which depends upon the intensity of the ultrasound, temperature, and amount of prestrain. Increasing prestrain is accompanied by decreasing hardening effect of ultrasonics until, at a certain level of prestrain, softening occurs. Schmid believes decrease in hardening may be due to recovery and recrystallization. Recovery occurs for polycrystalline materials if the insonation of the samples is stopped before fracture occurs. Increases in hardening are less when insonation continues to fracture due to recrystallization. According to Schmid, ultrasonics induces the same types of dislocation structure as does alternating stress at low frequencies.

Ultrasonics has been investigated as a possible means of

enhancing many types of forming operations, including wire and bar drawing, tubing drawing, sheet-metal rolling, densification of powder-metal parts, and grinding.

Experiments with 18.7 kHz ultrasonic vibrations applied to a drawing die for mild-steel wire with the oscillations in the axial direction conducted by Winsper and Sansome (18) indicated that the only reduction in force was attributable to the superposition of ultrasonic force on the drawing force rather than to heat and reduction of friction. No reduction in drawing force was observed for drawing speeds equal to or greater than the peak velocity of the oscillations. Specimens drawn under the influence of ultrasound revealed no changes in surface finish or mechanical properties.

Pohlman and Lehfeldt (19) claim that internal friction can be reduced by ultrasound and that considerable reduction of the external friction between the die and the workpiece can be achieved in wire- and tube-drawing operations. Their experiments were with copper, spring steel, and other steel alloys. Their work also showed that the surface conditions and the microhardness of copper wires were similar to those of wires drawn without ultrasound.

Kralik (20) reports that if wires are drawn through orifices located at the node of a rod vibrating at 20 kHz, the vibration direction being perpendicular to the direction of drawing, the rotational symmetry of the drawing texture is destroyed.

Severdenko and Clubovich (21) found that vibration amplitudes of 0.012 to 0.02 mm at 23 kHz applied to a wire-drawing die for copper wire caused a 50% reduction in drawing force, but it also increased the elongation with a corresponding reduction in quality, estimated to be approximately 15%. The diameter of standard copper wire was reduced from 1.57 mm to 1.25 mm.

In the previously mentioned wire-drawing work, the transducers were attached to the die and the die vibrated. Lorant (22) has taken a different approach to draw wires of high-purity copper, aluminum, and nickel to a diameter of 0.0007 in. under the influence of ultrasound. The wire-drawing machine, using 11 dies, was immersed in a small ultrasonically agitated bath of liquid. Reduction was 30% per die at 300 m/min. He claims the following benefits: (1) reduction in wire breakages, (2) increases in the life span of drawing dies, and (3) closer dimensional control.

Maropis and Clement (23) report substantial reductions in bar-drawing forces for rods of 6061-0 aluminum alloy, AISI 4340 alloy steel, and 6Al-4V-Ti alloy, which they attribute to the reduction of coefficient of friction between the die and the material. Power input to the transducers was as high as 7 kw at a frequency of 15 kHz, vibrating the die in the axial direction. Stock of 11/16-in. initial diameter was reduced in area 15% per pass at 60 ft/min. At 4.5-kw input to the transducers, drawing force was reduced 15 to 20% for the aluminum alloys and 4.5 to 6.5% for the steel alloys. Surface finish was reportedly improved and microhardness and microstructure were not affected. Increasing the ultrasonic intensity was claimed to result in corresponding increases in the magnitude of the effectiveness.

In tube drawing, a reduction by 36% in the time required to draw 316 stainless steel tubing has been claimed through the application of ultrasonic energy 24). Thin-walled tubing was drawn to 0.1825-in. OD and a wall thickness of 0.01 in. at tolerances of 0.001 in. using a fixed-plug mandrel. Longer mandrel life and improved surface finish were also claimed. Additional claims (1) of drawing 1/2-in. OD 25% silver-alloy tubing with a wall thickness of 0.001 in. held to 12 μ in. inside surface have been made for ultrasonic plug drawing (25), (2) of 321 stainless steel bellows tubing being held to a tolerance of 0.0003 in. on a 0.003-in. wall, and (3) of a Fe-Ni-Co tube with a 0.015-in. wall drawn to a minimum inside corner radius of 0.015 in. Reduction of drawing forces is claimed to be as high as 10%.

Increased area reduction with ultrasonics reduces the number of drawing steps. Alloys in addition to those previously mentioned that have been formed with the aid of ultrasonics include 7075 aluminum alloy, 6061-T6 alloy tubing, 1100 aluminum (impact extruded), 3003-H14 alloy covers, 304 stainless steel, phosphor bronze, titanium and beryllium copper, Inconel 718, Kovar, brass, and nickel-plated steel. Forming operations also include flaring required to form stainless steel-tubing connections and production of brass cartridge cases (26).

Applying ultrasonic energy to conventional lathe tools and to grinding operations is also reported to be beneficial (27). The poor surface quality, often observed when cutting light-metal alloys at low speeds, can be eliminated by ultrasonically agitating the tool. The effect drops sharply with increasing cutting speed, the ratio of maximum vibratory velocity to cutting speed again being important. Quality improvement is dependent upon vibratory motion of the tool in the direction of main cutting forces. Nothing was gained by vibrating the tool in the direction of traverse.

According to Balamuth (28), grinding with ultrasonic assistance nearly always produces results superior to those obtained with regular grinding. Fatigue properties are not impaired, burning is almost totally absent, and with light cuts, vibrating the wheel does not materially affect the surface finish. Vibration-assisted grinding with heavy cuts produces a generally superior surface finish. Chatter is reduced, resulting in a better surface finish. It is further claimed that ultrasonically assisted grinding takes about a third as much power on the spindle as does an identical but unaided grind.

Ultrasonically assisted rolling has been more difficult to implement. However, Cunningham and Lanyi (29) report that a 5% load reduction was attained by attaching a magnetostrictive transducer to the workpiece. A similar load reduction was attained when the transducer was designed integral with the roll, but the system was plagued with short roll life and electromechanical difficulties.

III.C. DEBURRING

Ultrasonics has been used to remove burrs left by forming or machining. The technique can be used on precision parts where fine deburring is required if tolerances are not less than 0.0001 in. The method is based upon the fact that ultrasonic energy accelerates particular reaction rates, especially in the presence of cavitation. When cavitation erodes a metallic surface, the pits formed may be fairly rough at the opening. If the erosion is due to accelerated chemical attack, the pits may be smooth at the entrance.

Small burrs not only form nuclei to encourage the formation of cavitation, but they also present maximum surface area per unit of volume to the attacking fluid. For this reason, ultrasonics applied to an acid will accelerate the dissolution of small metallic burrs from such items as gear wheels. Ultrasonic deburring has been used on such materials as stainless steels, mild steels, and copper and nickel alloys. Its use requires careful formulation of the deburring bath and good control of temperature and time to avoid excessive erosion of the part.

The items to be deburred are placed in a cleaning-type tank prepared with an acid-resistant lining. The tank contains an acidic slurry formulated specifically for the type of metal to be deburred. The unit is then turned on for a predetermined time, with the temperature of the bath set at a predetermined level. The deburring cycle is followed by ultrasonic cleaning in an alkaline bath to neutralize the acid, and this is followed by a clean-water rinse.

Ultrasonic deburring does not remove heavy burrs without excessively eroding the part. It should be used only on those parts where the economics of the operation justify its use and not on inexpensive, nonprecision parts.

III.D. COMPACTION OF POWDER METALS AND SIMILAR MATERIALS

The use of ultrasonics in compacting metal powders makes

possible higher densities with greater dimensional stability. The feasible application of ultrasonics to powder-materials compaction is limited to small parts with thicknesses of only a fraction of a wavelength ($\approx 0.03\lambda$) in the material of the compact.

Two methods of applying the ultrasonics are possible. For compacting loose or porous metals and plastics (thermoplastic), in which the displacement between particles is relatively free, the compact is best located at the end of a half-wave tool--a position corresponding to minimum acoustic impedance. This method is acceptable for low-melting-point materials; however, protection against oxidation usually is necessary. The pressure required is usually very low. Even 304 stainless steel specimens 1/2 in. by 2-1/4 by 1/4-in. pressed in low-thermal-conductivity dies reach a red-hot temperature within a few seconds at 1-kw input to the transducer. Under similar conditions, silver will melt within a few seconds.

Transverse strengths of various iron compacts prepared in this manner without presintering or reduction of oxides and using only dead-load clamping pressure showed improvements ranging from 7.5 to 15 times the green fiber-stress values for the materials. The corresponding increases in density were between 8 and 10%. Higher strengths were associated with higher intensities. Massive oxides were present near the pores and outside surfaces. These oxides may have been partially produced during ultrasonic excitation but there is some evidence that these particles result from agglomerations of oxides already present in the compact. The frequency of the applied ultrasonic energy was 20 kHz, the specimen dimensions were ¼ by ½ by ¼ in., and the power to the transducer was 500 to 1000 W.

On the other hand, if the specimens have been precompacted to approximately 75% theoretical density or greater, particle motion within the compact is restricted, corresponding to a high acoustic impedance. The specimen is then located at a high-impedance point in the vibratory system. In a uniform bar, this corresponds to the velocity node and stress antinode. Shown in Fig. 12.9 is a basic design of an ultrasonic system for producing powder-metallurgy parts. The arrows indicate the points at which the external force may be applied. This particular design was for use in a 75-ton laboratory press. With the pressure applied between nodal points, little energy is lost to the press. The energy loss within the specimen causes considerable heating and again a protective atmosphere may be necessary to inhibit oxidation. Compactions of various alloys to 100% theoretical density have been produced with the system, with power input to the transducer ranging from 1200 to 1650 W.

Compacts of iron specimens which were hydrogen annealed before green pressing to remove the inherent oxides and also presintered in a hydrogen atmosphere showed improved characteristics after ultrasonic compaction in the system shown in Fig. 12.9. Transverse strengths of these specimens were 1.5 to 2.5 times greater than those of samples prepared from the as-received powders. Increases in strength were obtained with both increased intensity and increased clamping pressure. In one instance, doubling the pressure resulted in a 66% increase in strength. Metallographic examination indicated that only a nominal amount of oxides was present.

Lehfeldt and Pohlman (30) indicate some improvement in density for a given punch load at low densities, and some reduction of temperature for equivalent density in hot pressing of metal powders with superimposed ultrasonic vibrations.

Pokryshev and Marchenko (31) found an initial intensive shrinkage in the first 5 to 10 min of sintering hot-extruded iron powder in the presence of ultrasonics, after which time the shrinkage rate decreased. This was attributed to an acceleration by the ultrasound of the transition from nonstationary to stationary flow in the powder. They, too, found that ultrasonic irradiation effectively reduces the temperature of hot extrusion by up to 20%.

Sonic vibrations in the range 5 to 15 kHz at displacements of 3 to 6 μ were found by Pines and his co-workers (32) to activate sintering and interparticle diffusion in various copper powders when the amplitude of vibration exceeded a critical threshold for each type. Powder compacts of 25 to 55% porosity were used for these studies. Sintering of compacts of 50%-Cu-50%-Ni powder showed that sonic vibrations stimulate the process of heterodiffusion.

Hoffmann (33,34) has applied ultrasonics to precision casting as used in dentistry and jewelry making. Ultrasonic energy was applied to promote flow and improve properties of castings from molds prepared by lost wax process. Ultrasonics promotes flow by reducing the apparent surface tensions and by driving the molten metal into the various regions of the mold to completely fill the mold. In addition, it eliminates dissolved gases and generally improves the quality of the product.

III.E. SOLDERING

One of the first commercial applications of high-intensity ultrasonics was in soldering difficult to solder materials, namely aluminum. In the earliest model, a soldering tip was incorporated in the working end of a transducer. The transducer was a magnetostrictive Permendur stack. The assembly was housed in the form of a conventional soldering gun and operated in much the same manner.

In later models the transducer and soldering tip were separate, and outside sources of heat such as a hot-plate were relied upon to melt the solder.

Soldering pots and soldering baths of various sizes can be obtained which contain a built-in heater and an attached transducer (see Fig. 12.10).

When soldering aluminum, the joints are immersed in a puddle of solder so that surfaces to be bonded are completely protected from the atmosphere. The soldering tip is then brought into contact with the workpiece. The ultrasonic vibrations erode the protective oxide coating, bringing the solder into direct contact with the base metal, to which it becomes attached.

Not only does the ultrasonic energy disperse the oxide coatings but it also promotes flow into crevices and cavities by reducing the apparent surface tension of the solder and by the imposed stresses that promote flow. For this reason it is also applied to other soldering applications in production lines to assure better coverage.

III.F. WELDING

When two materials are placed in intimate contact and pressure is applied normal to the interface, ultrasonic vibrations producing shearing stresses at the interface cause heating as a result of local absorption. This phenomenon has been used to produce spot and seam welds in metals such as aluminum, brass, 304 and 321 stainless steel, copper, zirconium, titanium, niobium, commercial iron, Chromel, nickel, molybdenum-0.5% titanium, tantalum, gold, and platinum. Applications include bonding both similar and dissimilar metals, in areas ranging from very small wires in the fabrication of microelements to the seam welding of metal plates to 0.2 in. in thickness (25 kw). All materials which can be pressure welded probably can be ultrasonically welded.

The primary mechanism of bonding appears to be associated with formation of a thin molten film in the interface (35). However, there is also some indication of a solid-state bonding. A galling effect has been observed in experiments with similar materials in which the ultrasonic intensity was believed to be too low to cause melting. According to Ginzburg (36), the welding process occurs by diffusion in the welding zone at low amplitudes and high applied pressure.

Optimum conditions for formation of welded joints by ultrasonics depend on the amplitude of vibration of the welding tip and applied static pressure. Joints can be made either at high vibration amplitudes and low applied static pressure or at low amplitudes and high applied pressure. Welding at high amplitudes and low pressures is similar to friction welding. The choice of method depends primarily upon the materials to be joined. Very ductile materials will yield under ultrasonic strain without interfacial sliding even at low pressures. Hard brittle materials will tend to fatigue under strains and pressures required for welding.

Typical ultrasonic welder arrangements are shown in Fig. 12.11. Figure 12.11a is a diagram of a seam welder with a sliding or rotating tip. Either spherical or wheel types of tips are used. Figure 12.11b is a diagram of the typical spot welder in which the longitudinal motion of the coupling horn is converted to flexural motion in the welding tip. In both cases, shearing stresses are generated in the specimen in a direction parallel to the transducer axis. In a third type of arrangement, a torsional-type welding tip is used to apply shear in a plane normal to the axis of the welding tip. Torsional types of ultrasonic welders have been used to produce hermetic seals in propellant containers (37). The cover material was 0.005-in.-thick 3003-H19 aluminum alloy. Welding time was 0.5 sec with an input power of 5400 W and 1500-lb clamping force to produce hermetic seals. Other appplications of torsional-types of bonders are described in a patent issued to Jones and DePrisco (38).

Removing the oxide film prior to welding aluminum does not affect the strength of welds in this material. The oxide layers are broken up and often forced to the outer perimeter of the weld by the ultrasonic energy.

Cracking often occurs at the edges of spot welds made in various combinations of heat-resisting alloys. These cracks probably are due to fatigue failure caused by high cyclic stresses occurring in these areas. Fatigue cracking is minimized by holding weld times to a minimum. Another problem encountered occasionally is the welding of the tip to the material to be welded.

Some additional typical applications of ultrasonic welding that have been reported include the bonding of 0.008-in. anodized aluminum to 0.015-in.-thick brass within 0.35 to 0.45 sec (39). Applied pressure and power were not reported. However, this reference does claim that the microstructure of the weld showed a good quality of contact which was free of fragments of the anodized layer. Shin and Gencsoy (40) claim good results in bonding low-carbon steel to polypropylene, acrylics, and glass by using aluminum foil interleafs. The presence of oxygen, aluminum, silicon, or magnesium as a main constituent in the nonmetallic materials enhanced the strengths of the welds. These materials easily form atomic or ion bonds with aluminum-foil interleaf.

Varga (41) recommends aluminum sheet thicknesses of 0.01 to 0.8 mm for spot welding and 0.009 to 0.6 mm for seam welding ultrasonically.

Pfaelzer and Frisch (42) compared results of welding copper in air and in a vacuum at 10^{-10} Torr. Welds produced under vacuum were much stronger than those produced in air. According to these authors, strength of welds depends on the interfacial areas which have not slipped relative to each other.

In a related area, several studies have been performed with respect to control of grain size and quality of welds by applying ultrasonics and low-frequency vibrations to the welding bath during arc welding of aluminum and other alloys. Russo (43) applied low-frequency vibrations to 18-mm-thick plates of low-alloy Cr-Ni-Mo steel being welded manually using an eccentric device at frequencies ranging from 30 to 50 Hz and amplitudes ranging from 0.01 to 0.1 mm. The grain size of the welds were considerably reduced. Treatment at 20 kHz during crystallization of an Al-Mg-alloy ingot 18 mm in diameter and 10 mm high produced appreciable reduction in grain size of the initial structural grain and a more regular distribution of the second phase along the grain boundaries. Similar results have been obtained by applying ultrasonic energy into the weld pool, using tungsten, molybdenum, or ceramic rods to transfer the energy from the transducer into the pool. A problem occasionally encountered was the loss of fragments of the transmission line into the weld metal.

Although it is not primarily a welding technique, a method of heading aluminum rivets developed at The Ohio State University is worthy of mention here (44). The sonic riveter consists of a full-wave, high-Q, piezoelectric transducer operating at 10 kHz. A half-wave catenoidal horn is attached to the transducer by means of a threaded stud. The actual riveting tool is loosely coupled to the transducer assembly so that it receives only intermittent impacts from the horn. The impacting rate is from 1000 to 3000 i impacts per sec. By this method, energy taken from the driving unit during impact reduces the amplitude of vibrations. Between impacts, the amplitude returns to maximum. The maximum efficiency of such a system would be obtained when the repetition rate of impacting corresponded to the exact time required for the vibrations of the horn tip to regain their maximum amplitude.

Ultrasonic equipment for plastics bonding is smaller and requires considerably less power than its counterpart for metals bonding. Losses at the interface are considerably higher than transmission losses through the plastic material and, for this reason, the heat for bonding is generated directly at the bond area. Thus the problems of heating the interface by thermal conduction or of heating the mating surfaces to above the bonding temperature prior to clamping are eliminated by the ultrasonic method. Bonding occurs when the temperature of the materials reaches the tacky level.

Joints are designed to make the interface lossy to promote rapid bonding. This is usually done by designing energy concentrators into at least one of the mating surfaces and by designing the parts so that shear stresses are produced between mating surfaces. Energy concentrators may consist of a small ridge running along the center of one of the surfaces. Shear stresses are promoted by forming the parts so that mating surfaces contact at an angle, at least over part of the surfaces to be bonded. The energy is introduced in a longitudinal-wave mode, since application of a shear mode (as in welding metals) would lead to surface distortion and transmission of energy would be poor.

There are certain advantages to ultrasonic bonding of thermoplastics over adhesive bonding or thermal methods in addition to those mentioned previously. For example, since there is no addition of chemicals as in the case of adhesive bonding, materials that must meet Food and Drug Administration regulations may be joined ultrasonically. Ultrasonics also has an advantage over thermal methods for joining thick materials.

Ultrasonic bonding of thermoplastics is best accomplished when the relative motion of the parts is restricted to avoid chatter between them. The parts should be located in a jig in such a manner that proper alignment can be maintained and that the motion of the part opposite the tool is well damped to avoid excessive vibration and loss of control. A typical industrial installation for assembling plastic products is shown in Fig. 12.12.

The terms near field and far field are often used to describe the distance from the welding tip to the bonding surfaces. Near-field bonding refers to the condition in which the mating surfaces are within a short distance (1/4-in. at 20 kHz) of the welding tip such as is the case in lap welding thin sheets. Far-field bonding refers to the condition in which the bondline is at a distance from the welding tip (greater than 1/4 in. at 20 kHz), as might be the case in bonding two hemispheres.

In addition to bonding plastic parts, the same equipment has been used for inserting metal parts in plastic (staking). A hole is drilled or molded into the plastic part and the metal is driven into the hole ultrasonically. During insertion (which usually requires less than 1 sec) material encapsulates the metal piece and fills flutes, undercuts, threads, etc.

Kolb (45) lists typical plastics and their weldability by ultrasonic means. This reference also includes recommended joint designs for best bonding. A list of materials and their welding properties can be obtained from manufacturers of ultrasonic bonding equipment. Materials that have excellent ultrasonic-bonding properties in both the near field and the far field and for staking include polystyrene, methyl methacrylate, polycarbonate, and SAN. Materials with excellent near-field bonding properties and staking properties and good far-field properties, include ABS and acrylics. Nylon, polysulfone, acetal, and phenoxy display excellent staking qualities, good near-field bonding qualities and good-to-fair far-field characteristics. Nylon fibers can be bonded, but in oriented fibers, ultrasonic bonding produces a disruption of the molecular orientation, thus weakening the material at the joint.

Olefins (polypropylene and polyethylene), especially filled materials, have fair bonding properties and excellent staking properties. The unfilled materials exhibit poor bonding qualities.

Other materials that have been ultrasonically bonded successfully include cellulose acetate butyrate, cellulose propionate, cellulose acetate, acrylonitrile-butadiene-styrene copolymers, polyurethane foam, polyvinyl chloride, cellulose propionate, cellulose nitrate, cellulose acetobutyrate, polyacetal resin, poly (phenylene oxide), and polyester (Mylar).

The bonding qualities of vinyls range from fair to poor.

Typical applications include splicing magnetic tape and splicing films. Ammonium nitrate fertilizers have been sealed in polyethylene-lined bags by ultrasonic means (46).

Oxidation reactions can occur during ultrasonic bonding of plastics (47). Carbonyl groups from peroxides which form during local thermoxidation have been detected in polystyrene and Lucite bonds. Graft and block polymers formed in the seam during ultrasonic bonding, according to these observations.

In related areas, ultrasonics has shown promise as a means of reactivating adhesives. It is especially effective in promoting wetting of surfaces, including pores, and in some cases, forming bonds through contaminants. Reactivation of adhesives by ultrasonics is limited to small parts and areas.

IV. LIQUID ATOMIZATION AND DROPLET FORMATION

Atomization of liquids and control of droplet formation by means of ultrasonics have many practical applications. There are two general types or classes of droplet formation that are of interest: (1) formation of mists or fogs with some control of droplet size distribution and (2) formation of droplets in a regular and orderly pattern by controlling the breakup of jets. Formation of mists or fogs has been applied to atomizing medical inhalants, atomizing fuels for efficient combustion, atomizing industrial paints for use with electrostatic painting, dispersing cleansers for effective use in cleaning large tanks, and production of metallic and other types of powders. Controlling the breakup of jets to form droplets of uniform size and in an orderly pattern has found use in certain unique cases in which it is desirable to have the droplets impinge on a surface in controlled patterns. For instance, the droplets can be passed through an electric field and deflected according to a designed pattern, just as a stream of electrons is deflected when passing through the electrostatic or electromagnetic field of a cathode-ray tube.

Aerosols have been produced in several ways. Fogs can be produced in a bowl operating within the range 0.4 to 2 MHz when the focal region is at or near the surface of the liquid. The ultrasound also produces a fountain which is indicative of the intensity. Liquids passed over an ultrasonically vibrating surface are sprayed into droplets according to the frequency and amplitude of vibration. Liquids injected into an active zone of a high-intensity stem-jet whistle are efficiently and rapidly dispersed into droplets. Controlled droplet formation from jets is accomplished by passing the liquid through a small nozzle built into a transducer vibrating in a longitudinal node at a frequency corresponding to the frequency for maximum instability.

The production of aerosols by ultrasonic means has been attributed to at least two mechanisms: (1) production of capillary waves on the surface of the liquid and (2) cavitation. Droplets are hurled from the crests of the capillary waves and from the walls of cavitation bubbles that break open at the liquid surface. To a degree, the particle sizes are functions of the excitation frequency. If the conditions are correct for optimum atomization, the frequency dependence applies over a wide range of frequencies. However, if the atomizer is overloaded with liquid, the mean particle size may be much larger than that predicted by the frequency equations. On the other hand, if the fluid supply is below the capacity of the atomizer, the distribution of the particles will shift to the smaller average particle sizes.

Various equations for predicting droplet size in an ultrasonically produced aerosol have been proposed. Lang (48) proposes the capillary-wave theory of droplet formation from the surface of an ultrasonic horn and has found the number-mean particle diameter, d, of the droplets in the aerosol to be

$$d = 0.34\lambda_c \tag{12.1}$$

where λ_c is wavelength of the capillary waves (cm) and is given by

$$\lambda_c = (8\pi T/\rho f^2)^{1/3} \tag{12.2}$$

where T is surface tension,

ρ is the density of the liquid, and

f is the ultrasonic frequency (Hz).

Lang found good agreement between experimental results obtained with either oil or water and results calculated from Eq. (12.2).

Popov and Goncharenko (49) propose the following equation for predicting size of droplets formed when using atomizing whistles, operating at frequencies within the range 6.8 and 32 kHz, in spraying materials including water, transformer oil, molten salts, waxes, and other organic materials:

$$d = (C/\xi)(3Q\eta T/\pi D\rho^2 f^2 g \cos \alpha)^{1/3} \tag{12.3}$$

where d is the average droplet size (microns),

C is a constant varying from 0.15 for d = 0.1 to 0.3 for d = 4.3,

ξ is amplitude of vibration (meters),

Q is the volume flow-rate of the liquid (m^3/sec),

η is the viscosity of the liquid [(kg-sec)/meter],

D is the nozzle diameter (meter),

g is acceleration of gravity (meters/sec^2), and

α is the angle between the spray cone and the vertical.

An equation for determining the optimum frequency for controlled droplet formation from liquid jets is:

$$f = u_j/4.058d_j \tag{12.4}$$

where u_j is the velocity of the jet and d_j is the diameter of the jet. When a liquid jet is disturbed at moderate intensities and at the frequency given by Eq. (12.4), the jet will break up into droplets of uniform size. Occasionally minor satellite droplets form on the axis of the jet. Uniform droplets are also formed by discrete frequencies within a range such that $3.5\ d_j < \lambda < 7d_j$. Dabora shows that $d = 1.89\ d_j$ and $\lambda = 2.38\ d$ at the optimum frequency.

Several examples of applications of ultrasonic atomization have been reported. Modified Hartmann whistles have been used to atomize fuels in turbines and combustion engines. Brockington (50) proposes ultrasonic atomization of diesel fuels to improve efficiency of partial-aspiration engines and eliminate the combustion-delay period by achieving good air-fuel mixing without wetting the manifold walls. He claims that the ultrasonic system enables the engine to run more economically at high loads, reduces smoke, and gives quieter operation.

Mebes (51) proposes using an ultrasonic Hartmann whistle in a reaction chamber located in the exhaust stream to complete the combustion of unburned constituents. He also suggests the use of catalysts in the reaction chamber, but claims good efficiencies for the ultrasonically induced reactions even without the catalysts. His recommended frequency range of 60 to 120 kHz is probably an order of magnitude too high for a Hartmann whistle of practical size.

Charpenet (52) has found a definite improvement in efficiency of burning both liquids and solids in an ultrasonic burner consisting of a steel combustion chamber and an external annular siren. Frequencies used were between 5 and 22 kHz. When the acoustic field was turned on, the flame temperature immediately rose by approximately 80°C and the exhaust CO_2 increased by approximately 0.8%. Fuels used were fuel oil and pulverized coal.

In commercial liquid-fuel burners with low consumption rates high pressure nozzles are used to atomize the fuel. The small orifices are easily plugged with dirt, which makes them either inoperative or lowers their efficiency. In addition, the compressor is usually noisy. Small, fairly inexpensive ultrasonic atomizers are an improvement over such compressor/nozzle combinations in several ways. For example, relatively large-diameter orifices can be used, which are self-cleaning under the influence of ultrasound. Also, the efficiency of the burner is higher than that of equivalent conventional nozzle-type burners. Small atomizers for use in portable, thermoelectric generators using a single transistor oscillator have been developed (53,54). These atomizers consume approximately 5 W under liquid load and approximately 2 W without the load. The unit operates from a 12-V dc supply. Feedback from the transducer to the oscillator controls the driving frequency so that it is maintained at the optimum frequency of the transducer. Fuel is fed by gravity. Thermal output is approximately 20,000 Btu/hr. Similar burners could be used in residential fuel-oil burners. Figure 12.13 shows the burner in operation. These burners are described by Hazard (53) and Hunter (54).

In a similar application in which the burner is recommended for use in oil-firing installations, Pohlmann and Lierke (55) use the pumping action of the ultrasonic transducer to provide oil to the atomizing surface. The oil reservoir is located beneath the transducer and one end of the transducer extends beneath the liquid surface. Throughput is 0 to 4 l/hr but higher rates are possible by increasing the size of the unit.

Molten metals with melting points up to 700°C have been atomized at frequencies up to 0.8 MHz to produce metal powders (56). Lead, tin, zinc, bismuth, and cadmium were successfully atomized by using a stepped-half-wave-velocity transformer fixed to a magnetostrictive transducer. Aluminum was atomized by using a noncorrosive transformer of composition Ti318A with an aluminum oxide coating. The results appear to confirm the capillary-wave theory of ultrasonic atomization.

Kirsten and Bertilsson (57) claim that using an ultrasonic atomizer in flame photometry and flame absorption spectrophotometry gives a tenfold increase in sensitivity compared with pneumatic installations. Solutions of sodium, calcium, and magnesium were used.

V. AGGLOMERATION AND FLOCCULATION

Although much as been published on the use of ultrasound to agglomerate smoke, several attempts to commercialize the technique have met with failure.

Moderately high-intensity ultrasound applied to relatively dense suspensions of paper pulp in water have been observed to promote flocculation of the pulp and thus clarify the liquid. There appears to be no general promotion commercially of this phenomenon.

VI. DRYING

Acoustic energy enhances drying. While it is an expensive method, it does have a practical use in drying heat-sensitive materials, especially those that degrade in a moisture environment, and chemicals that have a long drying cycle. Production of pharmaceutical materials is one area that has been proposed in which acoustic drying could be a benefit.

Although the effect appears to be primarily related to external mechanisms involving gas turbulence above the liquid interface and the production of reduced low-temperature boiling off of liquid vapors during the low-pressure phase of each cycle, there is some argument for internal effects, as well, which involve enhancement of moisture migration to the surface of the material.

Air-jet types of whistles have been the source of energy in most experiments with acoustic drying to date. Boucher (58) gives the optimum conditions for acoustic drying as (1) intensities above 145 db, (2) layers of less than 1 to 2 in. thick to be dried, and (3) frequencies within the range 6 to 10 kHz. Obviously, the operator must be protected against the sound energy at these intensities and frequencies. The safest protection is to acoustically isolate the sound from the surrounding environment by locating the sound source in acoustically treated chambers.

VII. AGRICULTURAL APPLICATIONS

VII.A. TOMATO POLLINATION

Tomato plants grown commercially in greenhouses are pollinated by hand. Workers pass daily along the rows of blossoming plants, contacting each cluster of blossoms with a low-frequency vibrator. Contact is made daily to insure that the blossoms are pollinated at the proper time, i.e., at maturity.

Tomatoes may be pollinated sonically or ultrasonically without contact (59). When a mass of fine dust particles is subjected to ultrasonic energy, the particles tend to disperse into a cloud. In pollinating tomatoes, the ultrasound produces a cloud of pollen. The cloud is uniformly dispersed within the blossom, resulting in uniform pollination. The sound energy does not rupture the sacs but merely scatters the available pollen within the blossoms. In fact, intensities that are high enough to rupture the sacs prematurely probably would damage the plant.

Sound intensities greater than 145 db are necessary for good pollination. Since moisture affects the adhesive character of the pollen particles, humidity affects the ability to raise a dust cloud of pollen with a corresponding effect on the pollinating ability of ultrasound. This has been proven by experiments conducted by DeTar, Haugh, and Hamilton (60), who showed that very high humidity increases the intensity required to cause pollen to fall and increases the number of blooms which fail to respond.

Research on tomato pollination was conducted at Battelle using modified Hartmann-type whistles operating at 10 kHz and 15 kHz. The work of DeTar, Haugh, and Hamilton was conducted using initially, a 20-W tweeter speaker with a cone mounted to concentrate the energy at the bloom. An industrial 40-W driver mounted with a 1 x 7 in. tube to carry the sound to the bloom proved to be more effective.

VII.B. GERMINATION OF SEEDS

Germination of seeds may be enhanced ultrasonically. The mechanisms involved are heat, transfer of moisture into the seed, and, in some cases, rupture of the seed coat. These mechanisms are variously active depending upon the seed. There is an optimum range of intensities for effective germination which also depends upon the delicacy of the seed. Very high intensities will damage many seeds while low intensities will have no effect.

The application of ultrasound to seed germination might be considered relatively inexpensive. A 1-gal cleaning tank could process several pounds of seeds per hour.

Wild rice, which grows in the lakes and marshes of northern Canada, is an annual and its seeds lie dormant during the winter. Freshly harvested seeds will not germinate until after they have been stored in nearly freezing water for up to 6 months. These seeds, however, can be artificially stimulated to germinate.

Halstead and Vicario (61) ultrasonically treated wild rice which had been stored for 90 days under water at 1 to 3°C. At the time of treatment the seeds were still dormant. The seeds were treated at 70 kHz for 10 min in water at 50 to 70°C, conditions which these experimenters found to be optimum. Under these conditions, 74 percent germination occurred. Longer exposure times resulted in fewer germinations. The method of germination did not affect the growth. The effectiveness was attributed to cavitation which changes the permeability of the seed coat and, perhaps, alters the response of the embryo to its environment. There was a marked temperature dependence.

VIII. PEST CONTROL

The use of sonic techniques of controlling pests, including birds, rats, and insects, has often been proposed. Possible methods include (1) using lethal intensities, (2) using intensities or frequencies that prove to be annoying to the pests, (3) duplicating the sounds of distress calls, and (4) duplicating the sounds of natural enemies.

Lethal intensities are impractical for several reasons. One is the need to entice the pest into a localized area where it can be exposed effectively. Another is the possible danger or annoyances to other individuals within the vicinity of the sound generator, especially at audible frequencies.

The possibility of using annoying sounds to rid an area of pests has not received much consideration. One possible source of annoyance in insects might be the resonance of antennae or other projections on the body which would produce discomfort.

Duplicating distress signals has sometimes proven to be effective for a short period of time. However, the animals often seem to develop an ability to distinguish between artificial or recorded calls and real distress signals, and the method loses its effectiveness. Busnel (62) has found that distress signals from a species in one locality may bring no response from the same species in another locality.

A report by the South Carolina Agricultural Research Service (63) claims that the bollworm (Heliothis Zea) responds to simulated bat calls in the same manner as it does to real bat cries. When insects, such as moths, hear the cries of bats, they immediately take evasive action.

Frings and Frings (64,65) have studied insect and animal sounds extensively from the viewpoint of pest control. Their conclusions are that their results allow them only to make suggestions for possible acoustical methods of pest control.

IX. CONTROL OF FOAMS

High-intensity ultrasonic energy in air can rapidly dissipate foams produced during various manufacturing processes. Bubbles are well matched acoustically to the surrounding atmosphere. The stresses developed in a high intensity field rupture the bubbles and release the internal gas. Only a small part of the bubble need be stressed to rupture. In a standing wave field, the high stress regions appear periodically and most of the bubbles will experience stresses that are of rupturing levels at some position on their surface.

Air-jet whistles have been used for defoaming purposes. They are usually located in a cylindrical chamber above the area where the reaction occurs. They may be located in branches of a pipe and directed upstream. The intensities of the sound must exceed 145 db. The frequency most often used is within the range 6 to 15 kHz. Acoustic treatment or isolation is essential to prevent exposure to damaging levels of intensity. Such treatment is not particularly difficult.

X. COATING MATERIALS AND PARTICLES

Ultrasonics has been used effectively to enhance uniform

coating of materials and to produce improved impregnation of porous or fibrous materials with resins, paints, enamels, and similar materials. Individual particles can be coated without agglomeration in air by means of high-intensity whistles. A stream of thin resin carrying particles to be coated is directed into the active zone but outside the main air stream of the whistle. The two constituents may be introduced separately and the particles will be coated individually. They must then be maintained in suspension until the solvent evaporates to avoid sticking to one another or to the walls of the container on contact.

Coating of fine particles can be done in liquid suspensions. The materials may be dispersed ultrasonically by placing a high-intensity probe in a beaker containing the materials to be coated and the coating resin. Sometimes particles are coated effectively in a three-phase system in which the materials to be coated and the coating material are mixed in a liquid suspension at a temperature above the melting point of the coating material. Application of ultrasonics continues as the temperature is dropped below the melting point. The particles are then generally uniformly and individually coated.

Coating of glass fibers with resins or rubber compounds can be improved under the influence of ultrasound (66). The irradiation removes the air from between the fibers and permits complete penetration of the resin between the fibers. To demonstrate the impregnating effects produced ultrasonically, bundles of glass fibers were passed through a solution, in parts per volume, of resorcinol 2.03, 37% HCHO 2.63, 10% NaOH 2.00, a 41% butadiene-styrene-vinylpyridine terpolymer latex 9.77, a 39% butadiene-styrene copolymer rubber 10.25, a 60% natural rubber latex 13.35, and water 59.97. After treatment in the solution at 25 kHz, the bundle was dried and showed complete impregnation and coating. Without ultrasonics, the bundle was not completely coated.

XI. PREPARATION OF CARBON SPHERES

Wettig and Doerre (67) claim to form round, highly sorbent carbon particles by ultrasonically irradiating for 20 min 30 mg of carbon suspended in 14 ml benzene to which 0.8 mg pyrene in 1 ml benzene has been added. Following the ultrasonic treatment, the suspension is mixed for 50 min. The process is said to produce carbon spheres 100 μ in diameter.

XII. GLASSWARE TESTING

A glassware tester has been developed that is capable of detecting flaws in glass that are otherwise hard to detect, such as fire checks, abrasions, and cracks (see Fig. 12.14). The glassware is moved along a conveyer belt between an air nozzle and a velocity transformer that is ultrasonically resonated. The air blast impacts the ware against the velocity transformer at a predetermined and controlled rate. If the glass contains a defect, it ruptures, the extent of the rupture depending upon the impact force of the glass against the transformer. Inspection rates are at least 180 units per minute.

XIII. DISPERSIONS

When particulate matter in a liquid suspension or two immiscible liquids are subjected to ultrasonic stresses at intensities above the cavitation level, the particles are accelerated omnidirectionally through the liquid. This phenomenon has been applied in producing uniform dispersions of various kinds.

XIII.A. DYES AND PIGMENTS

Liquid whistles may be used to disperse pigments in enamel preparations. Best results are obtained with low-viscosity fluids. Kotlyarskii and others (68) have shown good results with chrome, helio blue, zinc white, Hansa yellow, and azure pigments.

Lifshits (69) conducted comparative studies of the influence of quartz transducers, magnetostrictive transducers, liquid whistles, and mechanical mixers on the production of aqueous dispersions of dyes. The dispersions were evaluated on the bais of settling rates. In every case, ultrasonic dispersion was superior to mechanical dispersion. Lifshits recommends the liquid whistle for producing suspension dyes and for homogenizing dyes that form solutions. He claims that ultrasonically dispersed dyes give high-quality color with low dye consumption.

XIII.B. PREPARATION OF SPECIMENS FOR STUDY UNDER ELECTRON MICROSCOPES

Ultrasonic energy is ideally suited to preparing the evenly dispersed deposits necessary for study under the electron microscope. Aleinikov (70) used the method to prepare silicate specimens 200 to 400 A thick. Very fine glass powders in a 5% ethanol suspension were treated ultrasonically for 15 to 30 min, after which they were ready for mounting on the copper screen.

Belyaeva and Tuchkova (71) claim that dispersion of particles is the most important and the most difficult operation in preparing certain mineralogical materials for study under the electron microscope. They state that ultrasonic dispersion of the particles seems to be more effective than any other method.

Jacke and Kolb (72) recommend the use of ultrasonics not only for dispersing the specimen in the solvent but also for applying the specimen to the grid of the microscope. The latter is accomplished by placing a drop of the dispersed mixture on the tip of a stepped mechanical amplifier and placing the microscope grid near the tip as the ultrasonic energy is turned on. The ultrasonic energy produces a fog of the dispersion which settles on the grid to provide the necessary deposit.

XIII.C. PREPARATION OF SOIL SAMPLES FOR ANALYSIS

Oxidizing agents, acids, and peptizing agents are used conventionally for promoting dispersions of soil samples for analysis.

Some of the soil chemicals, such as ammonium, may be lost in producing dispersions by the chemical method. Edwards and Bremner (73) have shown that ultrasonics produces more stable suspensions and produces them faster than does the conventional method. Also the ultrasonic method causes no significant change in pH and electrical conductivity and no dissolution of organic or inorganic material. The time required to ultrasonically disperse various types of soils varies with the soil. The major problem is to decide how long to treat the specimen. Treatment past that required for complete dispersion may reduce the size of primary particles, which leads to analytical errors. For example, sand-size minerals may be reduced to clay-size particles by excessive treatment.

Transducers equipped with stepped mechanical amplifiers are recommended for this type of soil analysis, although any system capable of producing the necessary high intensity should be satisfactory.

XIII.D. DISPERSIONS OF CLAY SUSPENSIONS

The liquid whistle has been used to disperse dry-clay-body scrap during the manufacture of ceramic products (74). The whistle does the work of conventional mixing equipment at production rates, i.e., 50 gpm. For this application, the tungsten carbide parts are used where this is practical. The housing is lined with rubber to resist abrasion.

Ultrasonic treatment of clay suspensions in water at cavitational intensities has an immediate effect of dispersing the clay particles uniformly throughout the water, thus increasing the rheological quantities (75-85). These quantities reach an optimum value within a few minutes, after which the particles coagulate. The coagulated particles are so stable that further ultrasonic treatment does not destroy them. Kruglitskii and co-workers (82) obtained mechanical, X-ray, and electron-microscope data on clays before and after treatment. They conclude that ultrasonic energy destroys the original structure of the aqueous clay dispersion, promotes an even distribution of water coating, disperses the clay mineral particles, and causes the formation of new mineral aggregates with higher strength.

The influence of ultrasonic energy on the properties of the suspension can be varied by adding reagents prior to ultrasonic treatment. Addition of colloids such as carboxymethyl cellulose and NaCl or $CaCl_2$ to the clay prior to treatment enhances the thixotropic properties and the stability of the dispersions (77,78,81).

Peptizing electrolytes and stabilizing chemicals produce various effects, depending upon the amount of chemicals added (83-85). Large amounts of stabilizers apparently protect or shield the particles from the effects of the ultrasonic energy. Moderate amounts with ultrasonics produce beneficial effects on the mechanical properties.

Variations from low concentration to high concentration of peptizing electrolytes, such as NaOH, produce corresponding effects of peptization, spontaneous dispersion, and coagulation. Therefore, the addition of peptizing electrolytes may result in an improvement in the properties of the dispersion or may produce an adverse effect, depending upon the quantity used.

XIII.E. DISPERSION OF CHLORINATED PESTICIDES AND OTHER SOLUTIONS

Riggs and Biggar (86) have produced ultrasonically aqueous suspensions of p,p'-DDT, aldrin, dieldrin, and endrin which greatly exceed the theoretical solubilities of these materials in water. No emulsifiers were required and the suspensions remained stable for several days after 10 to 15 min of ultrasonic irradiation at cavitation levels.

Producing concentrations of solutions that exceed the theoretical solubilities is not unusual with ultrasonics. Uniform suspensions in large volumes of liquid may be produced more effectively mechanically by first ultrasonically preparing a concentrated dispersion of the material and diluting to the desired concentration. For example, in the production of a nondrying paint used to mark defects located by nondestructive testing, zinc oxide was dispersed through mineral oil. Mechanical mixing produced poor dispersions when the zinc oxide was added to the bulk liquid in the proper amount. A concentrated paste of the oxide in mineral oil produced ultrasonically was easily and uniformly dispersed through larger volumes of oil by mechanical stirring.

A 5-minute treatment of CaF_2 in water at intensities above the cavitation level produces a dispersion of the fluoride which is four times the maximum concentration possible by other known means. The phenomenon has a possible application in water fluoridation. Calcium fluoride is a safe chemical to handle, in contrast to other chemicals used for water fluoridation. Its chief drawback is that it requires several days for a desired quantity to dissolve unaided in still water, so that storage is a problem in supplying the normal daily consumption of any city using calcium fluoride. Using ultrasonics, water is easily fluorinated with calcium fluoride to 28 times the recommended concentration. Therefore, it may be possible to ultrasonically fluorinate water in a side branch through which 1 of every 28 gal flows. The treated water could be mixed with the main stream of water, where it is dispersed, and stored for use.

XIII.F. EMULSIFICATION OF FLOTATION AGENTS

Emulsions for flotation purposes have been prepared ultrasonically (87-90). It is claimed that these preparations are superior to mechanically mixed materials and that they result in savings of 25 to 50% in the amounts of reagents used. Liquid whistles are recommended for this purpose.

XIII.G. DISPERSION OF SODIUM IN HYDROCARBONS

Fine-particle dispersions of sodium in hydrocarbons can be prepared ultrasonically (91). By choosing a saturated hydrocarbon with a boiling point well above the melting point of the sodium, a homogeneous dispersion can be produced with the sodium in a liquid state. High-boiling light petroleum (b.p. 100° to 120°C) and yellow petroleum jelly are suitable dispersing media.

These dispersions can be produced with a stepped-horn or other suitable transmission line attached to either a piezoelectric or magnetostrictive driver. The material is treated in a cylindrical Pyrex cell. The transmission line is sealed into the end of the cell by a Neoprene ring.

XIII.H. DISPERSION OF HETEROGENEOUS PHASES IN MOLTEN METALS

The use of ultrasound to disperse various materials in molten metals has been considered by several investigators. It has not been used in actual production, partly because materials used in the ultrasonic apparatus would have difficulty surviving the environments.

Seemann and Staats (92) dispersed various insoluble materials in molten aluminum at 740 and 800°C using a sintered alumina rod to transmit ultrasonic energy into the melt. Materials, including iron and titanium, were dispersed rapidly.

Both titanium and titanium carbide caused pronounced grain refinement in the aluminum. Addition of 1% aluminum oxide doubled the high-temperature strength.

FIG. 12.1. Typical commercial-type ultrasonic cleaning tank (courtesy of Crest Ultrasonics Corporation).

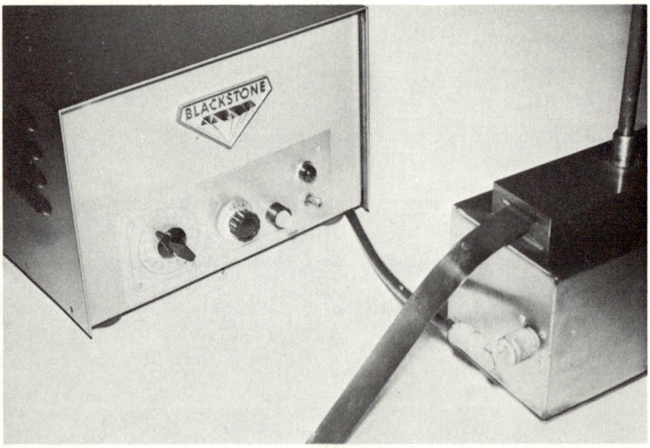

FIG. 12.2. Ultrasonic strip cleaner (courtesy of Blackstone Corporation).

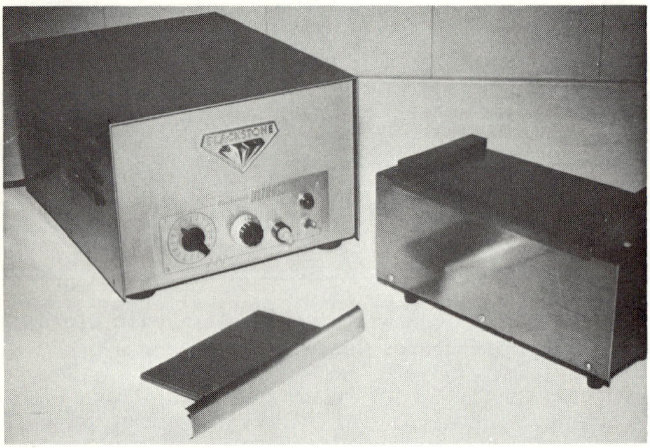

FIG. 12.3. Ultrasonic continuous dry cleaner (courtesy of Blackstone Corporation).

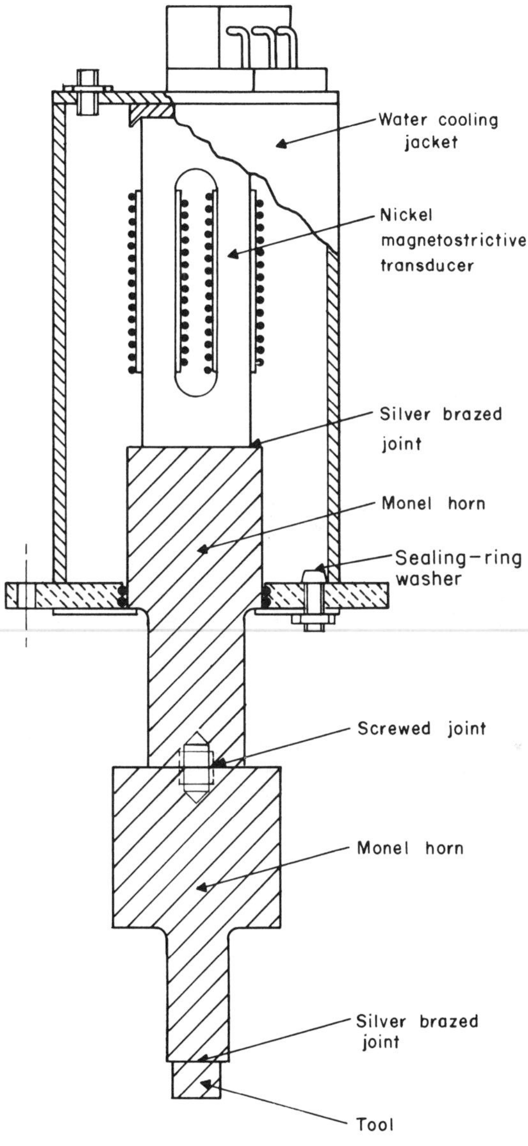

FIG. 12.4. Typical early model ultrasonic machining assembly consisting of three half-wavelength resonant sections.

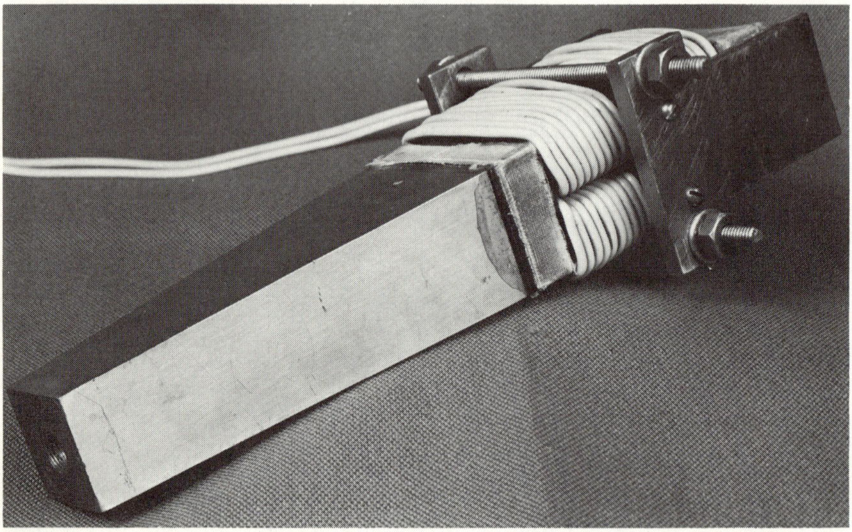

FIG. 12.5. Two-kilowatt magnetostrictive transducer.

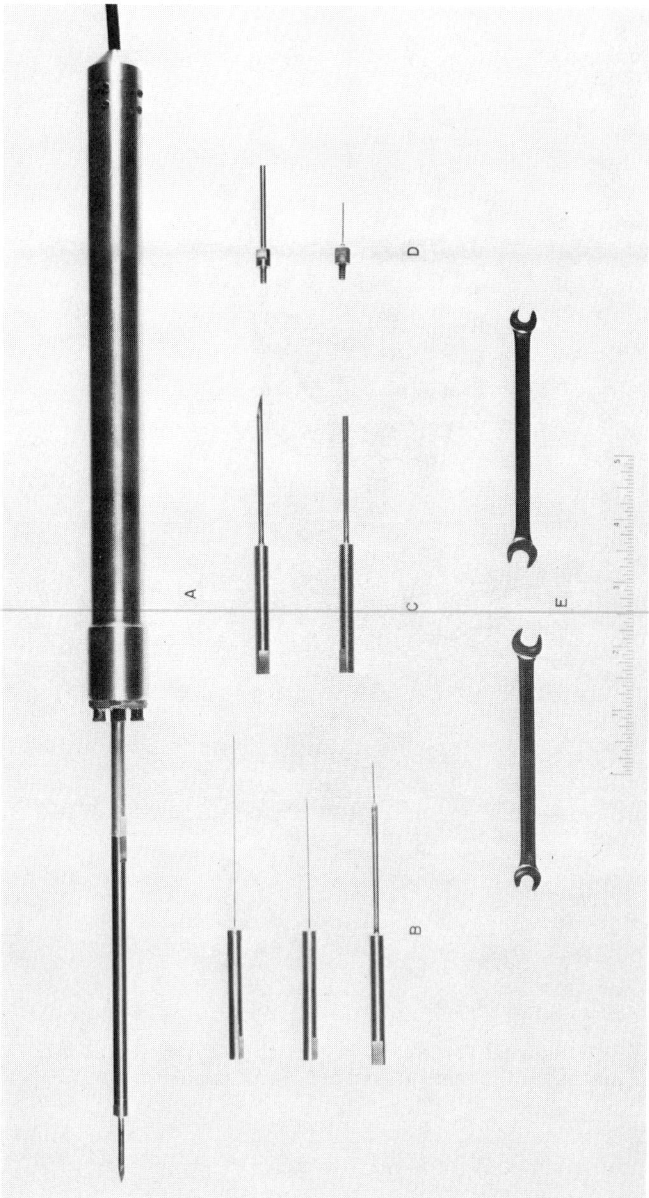

FIG. 12.6. Hand-held ultrasonic drill with tools, tool-holders, and extension for deep holes.

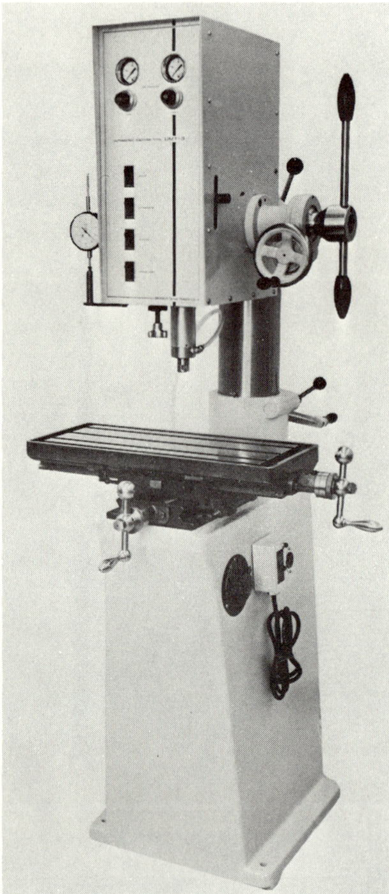

FIG. 12.7. Commercially available rotary-type ultrasonic drill press (courtesy of Branson Sonic Power Company).

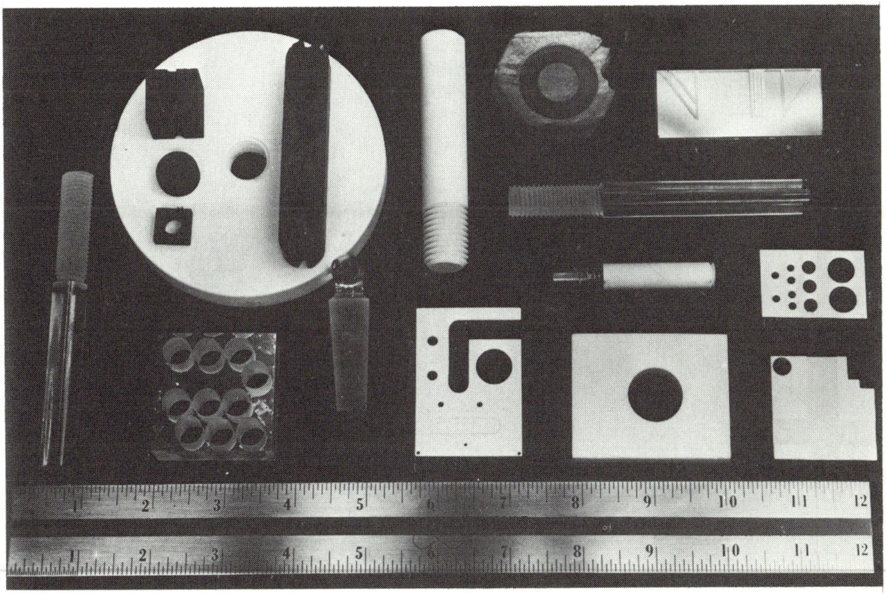

FIG. 12.8. Typical parts machined with a rotary-type ultrasonic drill press (courtesy of Branson Sonic Power Company).

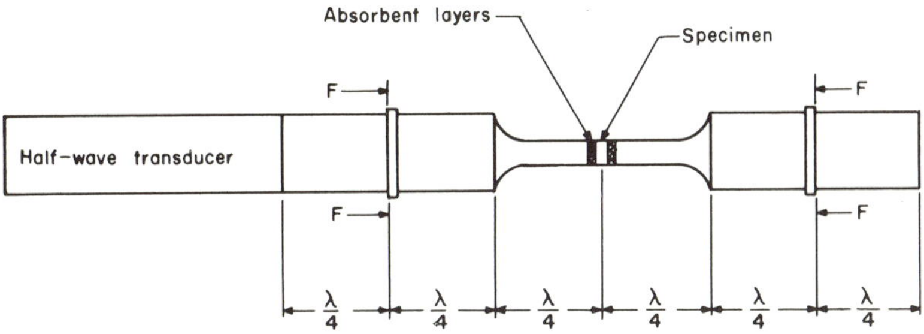

FIG. 12.9. Basic design of ultrasonic system for producing powder-metallurgy parts.

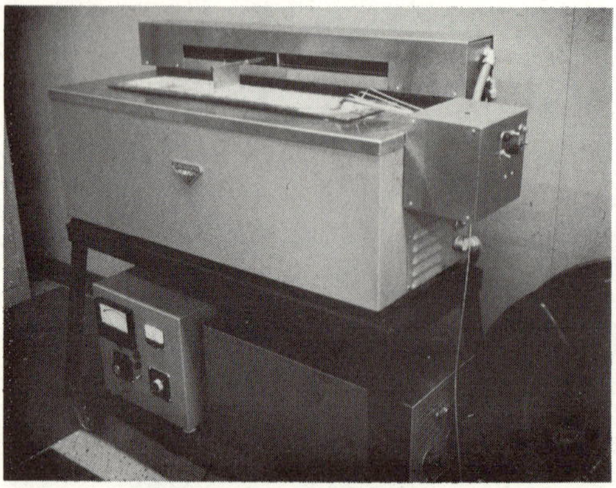

FIG. 12.10. Commercial soldering pot for fluxless soldering of aluminum sheets and tubes such as those used in the manufacture of aluminum heat exchangers (courtesy of Blackstone Corporation).

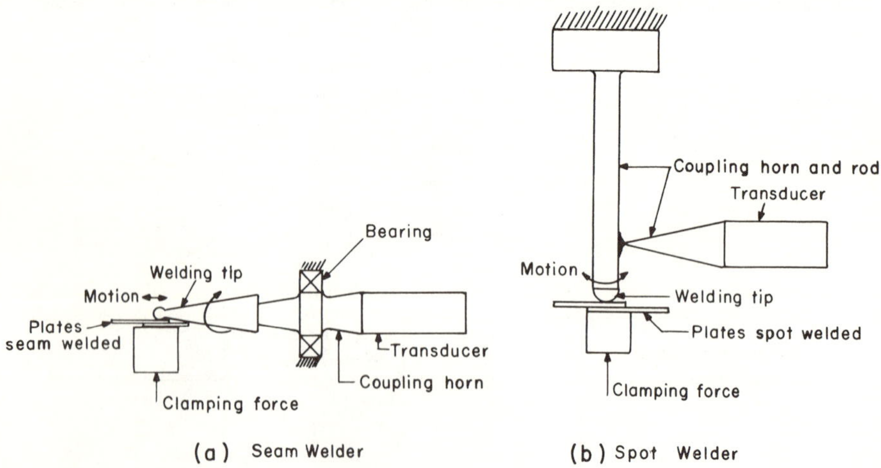

FIG. 12.11. Ultrasonic welder arrangements.

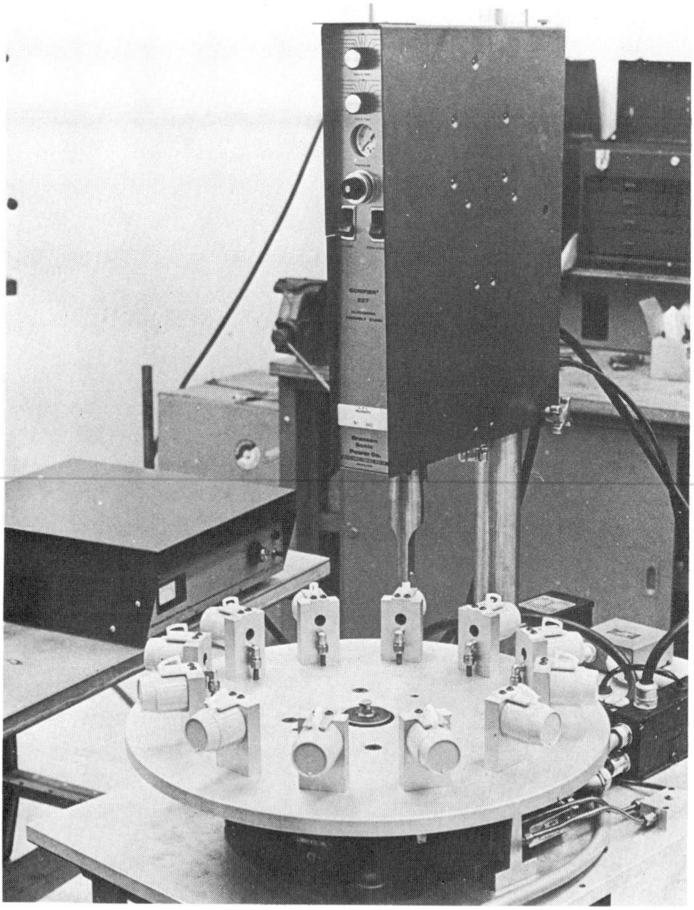

FIG. 12.12. Assembling thermoplastic products by ultrasonic bonding (courtesy of Branson Sonic Power Company).

FIG. 12.13. Small ultrasonic burner.

FIG. 12.14. Ultrasonic glassware tester (courtesy of Blackstone Corporation).

REFERENCES

1. Ultrasonic Manufacturers Association, Inc., Recommended Standard Cavitation Activity Measuring Procedure, 271 North Avenue, New Eochelle, N. Y. 10801 (Copyright 1964).

2. D. Ensminger, "Recommended Standard Cavitation Activity Measuring Procedures", paper delivered at the Fifth Annual Technical Meeting and Exhibit of the American Association for Contamination Control, Houston, Texas, March 29 - April 1, 1966.

3. S. E. Jacke, "Cavitation Activity Measurements", paper delivered at 1965 IEEE Ultrasonic Symposium, Boston, Massachusetts, December 2, 1965.

4. J. A. Ferris, ASTM Special Technical Publication, No. 342, October, 1962.

5. J. O. Farrer, British Pat. 602,801 (June 3, 1948).

6. L. Balamuth, U.S. Pat. 2,580,716 (January 1, 1952).

7. B. Langenecker, Paper B-2 delivered at 1969 IEEE Ultrasonics Symposium, St. Louis, Missouri, September 24-26, 1969.

8. L. Balamuth, SAE paper No. 650762 presented at National Aeronautic and Space Engineering and Manufacturing Meeting, Los Angeles, California, October 4-8, 1965.

9. B. Langenecker, IEEE Trans. Sonics and Ultrasonics, SU-13(1), 1-8 (1966).

10. E. Lehfeldt, VDI Z., 111(6), 359-363 (1969).

11. N. L. Pozen, V. N. Semirog-Orlik, and I. A. Troyan, Fiz. - Khim. Mekhan. Materialov., Akad. Nauk Ukr. SSR, 5(1), 112-113 (1969).

12. O. Izumi, K. Oyama, and Y. Suzuki, Trans. Japan Inst. Metals, 7(3), 162-167 (1966).

13. G. Kralik and B. Weiss, Z. Metallk., 58(7), 471-475 (1967).

14. B. Langenecker, U.S. Pat. 3,276,918 (October 4, 1966).

15. B. Ya. Pines and I. F. Omel'yanenko, Fiz. Metal. Metalloved., 28(1), 110-114 (1969).

16. V. N. Chachin and A. L. Skripnichenko, Izvest. Akad. Nauk Beloruss SSR, (Fiz.-Tekhn.), (3), 43-46 (1968).

17. E. Schmid, Proc. International Conference on Strength of Metals Alloys 1967, 9, 798-804 (Pub. 1968).

18. C. E. Winsper and D. H. Sansome, J. Inst. Metals, 97(9), 274-280 (1969).

19. R. Pohlman and E. Lehfeldt, Ultrasonics, 4, 178-185 (1966).

20. G. Kralik, Acta Phys. Austriaca, 20(1-4), 370-375 (1965).

21. V. P. Severdenko and V. V. Clubovich, Dokl. Akad. Nauk USSR, 7(2), 95 (1963).

22. M. Lorant, Tooling, 20(12), 51 (1966).

23. N. Maropis and J. C. Clement, ASM Technical Report No. C6-26.3, American Society for Metals, Metals Park, Ohio, 1966, 16 pp.

24. Anon.,Steel, 159(20), 38 (1966).

25. Anon., Machinery, 74(9), 88-89 (1968).

26. J. B. Jones, Metal Progr., 93(5), 103-107 (1968).

27. H. G. Dohmen, Aluminum, 42(9), 559-566 (1966).

28. L. Balamuth, Ultrasonics, 4, 125-130 (1966).

29. J. W. Cunningham and R. J. Lanyi, Study of the Feasibility of Applying Ultrasonic Energy to the Rolling Process for Sheet Metals, Final Report on Contract NOw64 029J, March 15, 1965.

30. E. Lehfeldt and R. Pohlman, Planseeber. Pulvermet., 16(4), 263-276 (1968).

31. V. R. Pokryshev and V. I. Marchenko, Poroshkovaya Met., Akad. Nauk Ukr. SSR, (2), 30-33 (1969).

32. B. Ya. Pines, I. F. Omel'yanenko, and A. F. Sirenko, Poroshkovaya Met., (8), 106-111 (1967).

33. R. Hoffmann, U.S. Pat. 3,338,294 (August 29, 1967).

34. L. Gross, R. Hoffmann, and R. Jefson, Ultrasonics, 7(4), 245-248 (1969).

35. N. E. Weare, J. N. Antonevich, and R. E. Monroe, Welding J. Research Suppl. (August, 1960), 11 pp.

36. S. K. Ginzburg, A. M. Mitskevich, and Yu. G. Nosov, Svarochn. Proizv., (5), 45-47 (1967).

37. Anon., Aeroprojects, Inc., Research Report No. 65-13, December, 1964, 28 pp.

38. J. B. Jones and C. F. DePrisco, U.S. Pat. 3,319,984 (May 16, 1967).

39. M. V. Bruk, Svarochn. Proizv., (10), 10-11 (1967).

40. S. Shin and H. T. Gencsoy, Welding J., 47(9), 398-s-403-s (1968).

41. I. Varga, Femip. Kut. Intez. Közlemeny., 8, 173-186 (1966).

42. P. F. Pfaelzer and J. Frisch, Ultrasonic Welding of Metals in Vacuum, Final Report on Contract No. W-7405-eng-48, California University, Berkely, California, College of Engineering, December, 1967, 65 pp.

43. V. L. Russo, Sudostroyeniye, (4), 37-41 (1958).

44. C. C. Libby, IEEE Trans. Sonics and Ultrasonics, SU-15(1), (1968).

45. D. J. Kolb, Machine Design, 39(7), 180-185 (1967).

46. Anon., Modern Packaging, 37(11), 18-19 (1964).

47. M. S. Akutin, V. A. Sorokin, and V. S. Osipchik, Plast. Massy, (12), 16-18 (1967).

48. R. J. Lang, J. Acoust. Soc. Am., 34(1), 6-8 (1962).

49. V. F. Popov and G. K. Goncharenko, Isv. Vysshikh. Uchebn. Zaved. Khim. i Khim. Tekhnol., 8(2), 331-337 (1965).

50. P. A. C. Brockington, Commercial Motor, 30(12), 33 (1966).

51. B. Mebes, Swiss Pat. 484,359 (February 27, 1970).

52. M. L. Charpenet, Riv. Combust., 18(12), 522-532 (1964).

53. H. R. Hazard, Proceedings Annual Power Sources Conference, 20, 152-155 (1966).

54. H. H. Hunter, Ultrasonics, 7(1), 63-64 (1969).

55. R. Pohlmann and E. G. Lierke, VDI Z., 108(34), 1669-1674 (1966).

56. E. G. Lierke and G. Griesshammer, Ultrasonics, 5, 224-228 (1967).

57. W. J. Kirsten and G. O. B. Bertilsson, Anal. Chem., 38(4), 648 (1966).

58. R. M. G. Boucher, Ultrasonic News, 3(2), 8-9, 14-16 (1959).

59. Battelle Memorial Institute News Release, December 5, 1963.

60. W. R. DeTar, C. G. Haugh, and J. F. Hamilton, "Acoustically Forced Vibration of Greenhouse Tomato Blossoms to Induce Pollination", Paper presented at the 1967 Winter Meeting American Society of Agricultural Engineers, Cobo Hall, Detroit, Michigan, December 12-15, 1967.

61. E. H. Halstead and B. T. Vicario, Can. J. Botany, 47(10), 1638-1640 (1969).

62. R. G. Busnel, J. Giban, Ph. Gramet, H. Frings, M. Frings, and J. Jumber, Paper HA6 presented at the Second ICA Congress in Boston, Massachusetts, 1956.

63. Anon., "Electronic Bat Sounds Drive Insects:, News release from South Carolina Agricultural Research Service, Clemson, South Carolina.

64. H. Frings and M. Frings, Sound, 1(6), 13-20 (1962).

65. H. Frings and M. Frings, Sound, 2(1), 39-45 (1963).

66. I. Iwami, T. Matsunaga, and K. Yoneyama, West German Pat. 1,934,349 (February 19, 1970).

67. K. Wettig and W. Doerre, East German Pat. 68,307 (August 5, 1969).

68. L. B. Kotlyarskii, A. V. Zelenaya, Z. M. Kishinevskaya, and A. G. Zelenyi, Lakokrasochnye Materialy i ikh Primenenie, (5), 51-57 (1963).

69. A. G. Lifshits, Nauchn.-Issled. Tr. Tsentr. Nauchn.--Issled. Inst. Khlopchatobumazhn. Prom., 25, 43-47 (1961).

70. F. K. Aleinikov, Zavodsk. Lab., 32(2), 215 (1966).

71. I. D. Belyaeva and G. A. Tuchkova, Eksperim.--Metod. Issled. Rudn. Mineralov. Akad. Nauk SSSR, Inst. Mineralog., Geokhim. i Kristallokhim. Redkikh Elementov, pp. 99-113 (1965).

72. S. E. Jacke and D. J. Kolb, Ultrasonic News, pp. 14-17 (Winter, 1961).

73. A. P. Edwards and J. M. Bremner, J. Soil Sci., 18, 47-63 (1967).

74. H. N. Walker and W. W. McCarthy, Am. Ceram. Soc. Bull., 46(2), 188-190 (1967).

75. N. N. Kruglitskii and V. V. Simurov, Ukr. Khim Zh., 30(8), 823-830 (1964).

76. N. N. Kruglitskii, V. V. Simurov, F. D. Ovcharenko, S. P. Nichiporenko, and A. B. Ostrovskaya, Ukr. Khim. Zh., 30(12), 1283-1289 (1964).

77. N. N. Kruglitskii, V. V. Simurov, F. D. Ovcharenko, and S. P. Nichiporenko, Ukr. Khim. Zh., 31(7), 669-673 (1965).

78. N. N. Kruglitskii, V. V. Simurov, F. D. Ovcharenko, and S. P. Nichiporenko, Ukr. Khim. Zh., 31(7), 674-678 (1965).

79. N. N. Kruglitskii, V. V. Simurov, F. D. Ovcharenko, and S. P. Nichiporenko, Dokl. Akad. Nauk SSSR, 159(6), 1367-1370 (1964).

80. N. N. Kruglitskii, V. V. Simurov, V. V. Minchenko, F. D. Ovcharenko, and S. P. Nichiporenko, Dokl. Akad. Nauk SSSR, 162(4) 861-864 (1965).

81. N. N. Kruglitskii, V. V. Simurov, F. D. Ovcharenko, and S. P. Nichiporenko, Dopovidi Akad. Nauk Ukr. RSR, (10), 1336-1339 (1965).

82. N. N. Kruglitskii, V. V. Simurov, F. D. Ovcharenko, and S. P. Nichiporenko, Fiz.-Khim. Mekhan. Dispernykh. Struktur, Akad. Nauk SSSR, Sb. Statei, pp. 151-157 (1966).

83. N. N. Kruglitskii, V. V. Simurov, F. D. Ovcharenko, S. P. Nichiporenko, S. V. Barshchevskaya, and E. V. Dzhus, Ukr. Khim. Zh., 32(9), 971-978 (1966).

84. F. D. Ovcharenko, N. N. Kruglitskii, V. V. Simurov, S. P. Nichiporenko, and S. V. Barshchevskaya, Dopovidi Akad. Nauk Ukr. RSR, (3), 367-369 (1966).

85. N. N. Kruglitskii and V. V. Simurov, Khim. Prom. Ukr., (1), 1-4 (1967).

86. R. L. Riggs and J. W. Biggar, Proc. of Soil Sci. Soc. Am., 29(5), 629 (1965).

87. M. A. Sokolov, V. G. Varmalov, and V. M. Ermolaev, Gorn. Zh., (12), 51-53 (1966).

88. D. S. Emel'yanov, N. I. Kamenev, and I. V. Trushlevich, Peredovoi Opyt. v Stroit. i Ekspluat. Shakhi, (2), 16-25 (1963).

89. A. M. Fedotov, USSR Pat. 1,453,098 (September 16, 1966).

90. M. Ghiani, Ind. Mineraria, 15, 509-515 (1964).

91. M. W. T. Pratt, R. Helsby, and H. M. Stanier, Chem. Soc., (London), Spec. Pub. No. 22, 284-289 (1967).

92. H. J. Seemann and H. Staats, Z. Metallk., 59(5), 347-356 (1968).

Chapter 13

APPLICATIONS OF ULTRASONICS BASED ON CHEMICAL EFFECTS

I. INTRODUCTION

The applications discussed in this chapter are those in which the chemical effects are of primary interest. The chemical effects are never divorced from the mechanical activity and often both factors are important.

The influence of ultrasonic energy on chemical activity may involve any or all of the following: production of heat, promotion of mixing, promotion of intimate contact between materials, dispersion of contaminated layers of chemicals, and production of free-chemical radicals.

This chapter is brief. Its brevity does not reflect the extent of the influence of chemical activity in applications of ultrasonics. However, its influence in many applications, such as those discussed in Chap. 12, may be secondary.

II. ACCELERATED ETCHING

Ultrasound applied during chemical or electrolytic etching often increases the rate of material removal by a factor. Dimensional control depends upon several factors, including intensity and directionality of the ultrasound waves, the flow characteristics of the etchant, and the selectivity of the etchant for one component material over another in a multicomponent system.

Both mechanical and chemical mechanisms appear to be operative. At low intensities, below cavitation levels, ultrasonic activity helps maintain close contact between the work surface and the etchant. The mechanism is primarily a stirring action with the direction of the force maintained normal to the surface to be etched. The etchant is forced through any gas barrier formed during the process and contaminated material is dispersed; thus fresh etchant is supplied to the metal surface.

Above the cavitation level, etching may proceed at a much accelerated rate. Flow of etchant becomes heterogeneous and the etching patterns may show the irregular patterns of the cavitation bubbles.

Ultrasonically assisted etching in each of the above categories, i.e., below and above cavitation levels, has been investigated. An example of the lower-intensity application is photoetching, a process in which a photographic image is developed in a coating of resist material laid over the surface of the metal to be etched. When the photograph is developed the resist is removed from the areas where etching is to occur. For photoetching, intensities must be kept low to prevent removing the photoresist by cavitation. The etched depression is rounded, and if etching proceeds to large depths, the resist is undercut. This type of etching has been used only for etching small surface areas, although it could be applied on a continuous basis. Ultrasound is introduced through plastic members attached to the transducers.

Etching core material from between layers of cladding during laboratory fabrication of nuclear fuel elements is an example of ultrasonic etching at intensities above cavitation levels. During one laboratory study, elements were fabricated in a large sheet and cut to size, exposing the core to the surrounding medium. The exposed core material was then removed to a predetermined depth and cladding material was welded into the recess. Aqua regia was used as etchant and this preferentially attacked the core material. A thin sheet attached to an ultrasonic machine tool was guided into the slot to be etched and oscillated slowly back and forth along the length of the slot. This technique produced a straight bottom in the slot. Complete removal was obtained to the desired depth. A problem in adapting this method to large-scale fabrication of fuel elements exists in locating tool materials that resist attack by aqua regia.

Electrochemical machining of carbon steel, which ordinarily may proceed at rates to 5 mm/min at current densities of 300 amp/cm^2 may be increased to 10 mm/min with the application of ultrasound (1).

Ultrasound is superior to conventional magnetic stirring for isolating precipitates and inclusions from steels for analytical purposes (2). About 5 g of common steel samples, 1 mm thick, can be dissolved ultrasonically at 60°C within 2 hr in a 14% I_2 - MeOH solution. The isolation ratios of oxide inclusions from 18-8 stainless steel and Fe-O system steels are higher than those obtained by magnetic stirring. Ultrasonics has also been used successfully to isolate carbides at room temperature from steels of the Fe-Mo-C, Fe-V-C, and Fe-Ti-C systems in acid.

Not all chemical etching processes are accelerated by ultrasound. Lavrent'eva and Pakhomov (3) found that ultrasound produced a severalfold increase in the rates of dissolution of gallium in HNO_3, H_2SO_4, and HCl, of indium in HNO_3 and HCl, and of arsenic in

HNO_3 at room temperature and at 16 to 20 kHz, but it produced no significant increase in dissolution rate for tin or cadmium in H_2SO_4.

In a patent, Weinberg (4) describes a system for etching metals rapidly and uniformly, using either chemical or electrolytic etching under the influence of ultrasound. Representative materials and applications claimed are chemical milling of aluminum, titanium, beryllium, boron, nickel, and tungsten. The method consists of immersing the material to be etched in a rigid brittle material, such as glass, suspended in an ultrasonically activated tank of tap water. An example was given of increasing the rate of etching of 6061 Al alloy masked by an acid resist in aqueous solutions of HNO_3, HF, and HCl from 0.0005 in/min without ultrasound to 0.0015 in/min with ultrasound and without lateral undercutting of the resist edges. The etched surfaces were claimed to be smoother and flatter than those etched without ultrasound. It was proposed that ultrasound dispersed the surface sludge and hydrogen bubbles which retard the attack during static etching. Electrolytic etching in the same solutions increased the rate of metal removal at 50 A/ft^2 current density from 0.00015 in/min without ultrasound to 0.0045 in./min with ultrasound. The gain was attributed to ultrasonic vibrations breaking the insulating ion barrier that forms at the metal/liquid interface during electrolytic etching. Etching a beryllium sheet with ultrasound in an etchant of 450 ml H_3PO_4, 26.5 ml H_2SO_4, and 53 g CrO_3 proceeded at 0.0009 in./min, producing a surface finish rated at 8 to 16 μ in. roughness; without ultrasound the rate was 0.0002 in./min and the roughness was 63 to 80 μ in. Tungsten sheet could be etched electrolytically in a 10% NaOH solution to a depth of 0.035 in. in 10 min with ultrasound, but without ultrasound no metal was removed.

III. TREATING BEVERAGES, JUICES, AND ESSENTIAL OILS

Ultrasonics applied to newly fermented alcoholic beverages produces effects similar to those resulting from a long period of aging (5). Singleton (6) ultrasonically treated red and white table wine, red and white port wine, and sherry-base wine at intensities above the cavitation level in the presence, separately, of nitrogen, hydrogen, carbon dioxide, air, and oxygen. Some compositional changes were effected, accompanied by a readily detected odor and flavor difference. However, these changes did not necessarily improve the quality ratings of the wines. In fact, the conclusion of this author was that overall quality scores were usually decreased by application of ultrasound. This is in contrast with the conclusion of Shakhsuvaryan (7) that ultrasound may be used for the improvement of wine quality and appearance. He found that ultrasonically treating white wine affected with "AcOH disease" caused a reduction of nitrogen compounds and tannin contents and stopped further formation of volatile acids. These effects were accompanied by an increase in clarity.

Removal of sulfur dioxide from pear juices and wines by means of intense ultrasound has been claimed (8).

Berishvili (9) found that a higher percentage of juice could be extracted from grapes treated ultrasonically at 19 to 21 kHz for 5 to 45 min. The amount of the increase of juice yield ranged from 1 to 6.2%, depending upon the variety and maturity of the grapes.

Berk and Mizrahi (10) conducted laboratory experiments on the use of ultrasound in producing low-viscosity concentrated orange juices. The irradiation was combined with evaporation techniques. The treatment produced unpleasant, garlic-like odors that had to be eliminated by flash evaporation; however, the ascorbic acid content was not affected. Ultrasound did improve the concentration process, but the process is limited by the size and cost of equipment that would be required for industrial processing.

Gilyus (11) found that ultrasound could reduce the processing time for removing sugar from sugar beets by a factor of two. The effectiveness of ultrasound decreased with increase in temperature.

Carpenter (12) has patented a process utilizing ultrasound to extract fatty substances from vegetable-oil-yielding cells. The ultrasound helps rupture the cells, increases diffusion through the walls of unruptured cells, and emulsifies the oils in the suspending liquid medium. A similar patent was issued to Romagnan (13) for extracting aromatic oils from vanilla beans.

Ultrasound also has been used to change the characteristics of essential oils and perfumes. The effects produced depend upon the intensity of the sound, the duration, and the atmosphere in which the treatment occurs. As an example of the effects produced, the "weedy" odor of fresh geraniol is removed by a 1-min treatment at 1 MHz at cavitating intensities. The "weedy" odor is similar to that of newly mowed hay or dry grass.

IV. TREATMENT OF SEWAGE

There are two categories of sewage treatment: (a) treatment of industrial wastes involving discharged chemicals and other debris from industrial processes and (b) treatment of community, or residential, wastes.

The use of ultrasonics in at least three different ways has been proposed for sewage treatment. One method involves enhancing or promoting chemical reactions such as oxidation of aromatic compounds and depolymerization of organic compounds in industrial wastes (14). A second method involves the use of ultrasonics to clean or strip masses of material from the upstream side of a filter and to concentrate the material in an area from which it can be easily removed (15). The third method that has been proposed for treating sewage involves the effect of increasing the propagation rate of bacterial growth under the influence of ultrasound. All of these methods have shown feasibility on a small scale but are limited commercially where large volumes are involved. Supply of sufficient energy to promote the activity associated with each method does not appear to be economically feasible at the present time.

V. EXTRACTION PROCESSES

Extraction processes actually involve several effects, including emulsification, cleaning, certain chemical effects, and diffusion. Ultrasonic energy is used to disrupt cells and to remove certain essential oils and materials that are otherwise difficult to remove from porous media, producing generally higher yields.

The beneficial effects must be weighed against the economics of the applications. Ultrasonics usually is limited to exotic, small-batch extractions. The possible chemical reactions also must be considered, i.e., whether the chemical to be removed will be adversely affected by the ultrasound.

Adamski and Socha (16) showed that ultrasound had about the same effect as heat in removing capsaicine from the fruit of Capsicum annuum, and did not cause any decomposition of capsaicine.

Aimukhamedova and Korneva (17) used ultrasound at 22 kHz, 300 kHz, and 750 kHz to disintegrate cells of poppies to extract morphine. Similar amounts were extracted at all three frequencies. Ultrasound did cause decomposition of the morphine with time when the process was carried out in water, so the optimum extraction time was short (10 min or less). However, the yield increased when an alcohol-$CHCl_3$-NH_3 mixture, aqueous HCl, $Ba(OH)_2$, or sulfamic acid solutions were used as extractants instead of pure water. When aqueous $Ba(OH)_2$ at a concentration of 50 g/l was used, more alkaloids were removed than were removed at lower concentrations.

Extraction of tanning materials can be improved by ultrasound. The amount of tanning materials extracted ultrasonically at 800 kHz in 45 min is equivalent to that extracted in 8 hr by usual extraction procedures. The effectiveness is attributable both to mechanical action of the ultrasound and to heat due to absorption.

Chapman and Bradbury (18) make use of the capability of ultrasound for disrupting wool fibers in HCO_2H to separate orthocortex and paracortex from wool for analysis.

Chen and Chon (19) have found that increased extraction efficiency results from applying ultrasonics at intensities above the cavitation threshold during continuous liquid-liquid extraction processes. They attribute the increased effectiveness to increased turbulence, which may also increase the interfacial area of contact between the phases, thus enhancing the mass transfer. The increase in extraction efficiency is nearly linear with increased intensity of ultrasound.

Chen and Fairbanks (20) have shown that ultrasonic energy at 20 kHz can increase the rate of flow of oil through sandstone.

Yields of alkaloids increase under the influence of ultrasound. Periods of maceration are 30 to 60 min. Also, yields obtained during continuous solvent extraction are greater than those obtained by conventional means (21). A French patent has been issued (22) which claims that higher yields of sugar juice are obtained from sugar beets when the beets are subjected to an initial compression and sudden decompression followed by an ultrasonic disruption of the cells. Ultrasonics at low frequencies can increase the yield of bitter substances from hops in aqueous suspension (23,24). Increased yield of cottonseed oil (25) and of adrenaline from ground adrenals (26) have been claimed to result from the application of ultrasound. Vaisman (27) claims a 10% increase in the amount of extracted alkaloids and glycosides from belladonna leaves and lily of the valley when ultrasonics at 80 kHz and 2.5 W/cm^2 were applied for 1 hr, as compared with the extracts prepared by the percolation method.

Whiervaara (28) studied the effects of ultrasound on the extraction of dehydrogenases from yeast cells. Using a titanium probe operating at 18 to 20 kHz to generate an intensity of 35 W/cm^2, he found that maximum enzyme activities were liberated after 5 min of treatment, although the amount of protein liberated increased linearly during the first 15 min. Enzyme liberation increased with increased power and decreased with increased salt concentration.

Woodford and Morrison (29) found that ultrasonic irradiation increased the extraction rate of rhein glycosides, free anthraquinones, and total water-soluble solids from whole senna pericarps. The ratio of rhein glycosides to total water-soluble solids was higher in extracts prepared using ultrasound than in preparations obtained by simple maceration. The ratio was slightly lower at higher intensities than at lower intensities.

Ultrasonic processes for extracting crude drugs should be avoided where molecules that are subject to destruction by intense ultrasonics are involved.

Ultrasonics has been found to enhance flotation of mercury and antimony (30). Ultrasound inhibits flotation of galena pulps in aqueous suspensions of the ore with small additions of ethylxanthate as a collector (31). The ultrasound apparently affects the hydrophobic properties of the galena surface.

VI. DEMULSIFICATION OF CRUDE PETROLEUM

The idea of demulsifying oil and water is in opposition to the emulsifying effects usually expected of ultrasonic energy. However, the same principle that applies in preparing dispersions can be used to promote demulsification. This is done very effectively by ultrasonically dispersing a demulsifying agent through the emulsion. Skripnik and his co-workers (32) have shown that ultrasonic dispersion of the demulsifier in removing salt and water from crude oil imparts considerable improvement over conventional methods.

VII. MISCELLANEOUS CHEMICAL EFFECTS AND APPLICATIONS

The application of ultrasonics during the cracking of solar oil distillates has been shown to increase the yield of gasoline, with the yield increasing with intensity and temperature (33). At the same time, the paraffin content decreased with intensity, but the amount of naphthenic hydrocarbons did not change. Carbene and carboid contents increased with intensity and temperature. The yield of cracking gas also increased, with considerable decrease in the amount of hydrogen in the gas at higher intensities and temperatures. The effects of raising either the temperature or the ultrasonic intensity were similar.

Ultrasound has been found to increase the dissolution of hemicellulose during the beating of softwood sulfite and sulfate pulp (34). Studies were performed at 25 kHz and 400 kHz, with the higher frequency being more effective.

The chlorination of unbleached beechwood kraft pulp in the presence of intense ultrasound decreased chlorination rate and lowered consumption of chlorine, but gave higher brightness and more complete delignification (35).

The application of ultrasound to the desalination of seawater has been considered on several occasions. In purification by zone freezing, impurities show a preference for the liquid phase. As the freezing progresses, a contaminated zone builds up near the liquid/solid interface. A percentage of the contaminate is trapped in the ice, requiring one or more additional cycles before the desired level of purification is reached. Ultrasound applied at the surface disperses the debris and reduces the number of passes required. Mechanical stirring produces the same effect and is less expensive.

Ultrasound has been applied to desalination of seawater in conjunction with distillation (36). The ultrasound produces a mist which increases the surface area in the still and increases the rate of distillation.

Another method of applying ultrasound to the desalination of seawater involves the effects of ultrasound on electrochemical processes (37). This approach has not been used commercially.

A Czechoslovakian patent (38) claims that slaking lime is improved by activation in an ultrasonic field. The activation is accelerated by effectively increasing the access of water to lime grains. By this process, mortars are produced by simultaneously activating the lime and adding the sand.

The dissolution of steel in hydrochloric acid (39) and in phosphoric acid (40) is speeded up under the influence of ultrasound. However, the effect has been attributed to ultrasonic generation of heat. Heating is, therefore, recommended over ultrasonics for this purpose.

VIII. ELECTROLYSIS AND ELECTROPLATING

Electrolysis is the production of chemical changes by the passage of an electric current through an electrolyte. Electroplating is plating by electrodeposition, i.e., by depositing the plating on a surface by means of electrolysis.

Several studies have been conducted relative to the effects of ultrasonic energy during electrolysis and electroplating operations. Brown (41) reports that, at intensities exceeding 1 W/cm^2, metal particles appear in suspension, probably as a result of cavitation erosion of the plating. Ultrasonics within the range 20 to 40 kHz permits the use of higher current densities and results in purer, harder, and stronger coatings. Brown recommends its use in plating with nickel and chromium.

Bystrov (42) confirms Brown's claims of the use of higher current densities and the production of harder coatings plus smaller grain sizes. He claims that ultrasound at 0.02 to 7 W/cm^2 and 27 kHz had no effect on electrodeposition of nickel at current densities of 0.5 to 1 A/dm^2, but at 5 A/dm^2, ultrasound at 0.6 to 1 W/cm^2 sharply increased electrodeposition by factors of 3 to 6 times. Increasing the intensity to 10 W/cm^2 had practically no effect. At low intensities, the effect is similar to that of chemical mixing. Similar results have been observed in plating zinc.

Kenahan and Schlain (43) report that at all current densities ultrasonics produces finergrained deposits of electrolytic manganese than those obtained in normal cell operation, but it resulted in less adherent deposits at high current densities. The ultrasound did not affect cell voltage or cathode current efficiency.

According to Bondarenko (44), movement of gas bubbles and

cavitation in an ultrasonic field can affect the acceleration of the process of electrocrystallization, but they are not determining factors in acceleration of the process. He concludes that the chief factor accelerating electrocrystallization is the steady-state microeddies that arise in the electrolyte at the surface of the cathode in the ultrasonic field.

Ultrasound reduces the cathodic and anodic polarization during electrodeposition, permitting an increase in the rate of metal deposition. With copper deposited on a helical steel wire from pyrophosphate baths, Domnikov (45) found that deposition rate was 4 to 4.5 times that obtained by intensive mechanical agitation.

Electrodeposition of chromium shows similar beneficial effects of ultrasonic irradiation (46).

Simpson (47) produced striped electroplating, which is the deposition of metals in closely spaced parallel lines, by applying ultrasound at 2 MHz during electrodeposition of copper. The lines were caused by standing waves in the solution. The lines disappeared as the thickness of the metal deposit increased.

REFERENCES

1. N. I. Ivanov, V. P. Rassakazov, and F. V. Sedykin, Fiz. Khim. Obrabot. Mat., (1), 38-42 (1969).

2. O. Kammori, I. Taguchi, K. Takimoto, and A. Ono, Nippon Kinzoku Gakkaishi, 3(4), 493-497 (1969).

3. V. G. Lavrent'eva and D. A. Pakhomov, Prom. Khim. Reaktivov Osobo Chist. Veshchestv, (10), 5-8 (1967).

4. H. P. Weinberg, U.S. Pat. 3,411,999 (November 19, 1968).

5. J. A. Bachmann, U.S. Pat. 2,086,891 (July 13, 1937).

6. V. L. Singleton, Wines and Vines, 34, 31-34 (1963).

7. A. V. Shakhsuvaryan, M. G. Gulyamov, K. I. Rasulova, and V. Lapina, Tr. Tashkentsk. Politekhn. Inst., (22), 335-340 (1963).

8. E. S. Drboglav, V. S. Maiorov, and L. M. Lipovich, Tr. Tsentr. Nauchn.-Issled. Inst. Pivovarennoi Bezalkogol'noi Vinnoi Prom., (11), 59-60 (1963).

9. L. I. Berishvili, Novye Fiz. Metody Obrabotki Pishchevykh Produktov, Kiev, Sb., pp. 248-251 (1963).

10. Z. Berk and S. Mizrahi, Conserve Deriv. Agrumari (Palermo), 14(2), 55-59 (1965).

11. I. P. Gilyus, Akust. Ul'trazukovaya Tekh., 3, 40-46 (1968).

12. J. Carpenter, U.S. Pat. 2,717,768 (September 13, 1955).

13. L. Romagnan, U.S. Pat. 2,601,635 (June 24, 1952).

14. Yu. Yu. Lure, P. F. Kandzas, and A. A. Mokina, Nauchn. Soobsch. Vses. Nauchn.-Issled. Inst. Vodosnabzh. Kanaliz., Gidrotekhn. Sooruzhen i Inzh. Gidrogeol., Ochistka Prom. Stochn. Vod., Moscow, pp. 7-9 (1963).

15. R. Davidson, W. G. Palmer, and T. H. Forrest, U.S. Pat. 3,489,679 (January 13, 1970).

16. R. Adamski and A. Socha, Farm. Pol., 23(5-6), 435-436 (1967).

17. G. B. Aimukhamedova and G. M. Korneva, Izv. Akad. Nauk Kirg. SSR, Ser. Estestv. i Tekhn. Nauk, 4(6), 17-24 (1962).

18. G. V. Chapman and J. H. Bradbury, Arch. Biochem. Biophys., 127(1-3), 157-163 (1968).

19. E. C. Chen and W. V. Chon, Chem. Engr. Progr., Symp. Serv., 63(77), 44-48 (1967).

20. W. I. Chen and H. V. Fairbanks, IEEE Trans. Sonics and Ultrasonics, SU-15(1), Abstract B8 (1968).

21. A. E. DeMaggio and J. A. Lott, J. Pharm. Sci., 53(8), 945-949 (1964).

22. J. O. Roeder, French Pat. 1,368,693 (July 31, 1964).

23. Z. I. Emel'yanova and V. A. Leoninck, Ferment. Spirt. Prom., 34(7), 16-18 (1968).

24. J. Hoggan, Ultrasonics, 6(4), 217-219 (1968).

25. D. S. Iskanderova, N. U. Rizaev, R. M. Mirzakarimov, A. Sultanov, and M. I. Niyazov, Dokl. Akad. Nauk Uzb. SSR, 24(8), 30-31 (1967).

26. L. S. Kazarnovskii, V. N. Solon'ko, and L. A. Shinyanskii, Med. Prom. SSSR, 18(8), 29-30 (1964).

27. G. A. Vaisman, M. I. Gurevich, E. S. Skvirskaya, and V. Ya. Gorodinskaya, Farmatsevt. Zh. (Kiev), 18(4), 61-65 (1963).

28. K. Whiervaara, Acta Chem. Scand., 21(1), 295-297 (1967).

29. R. Woodford and J. C. Morrison, J. Pharm. Pharmacol., 21(9), 595-601 (1969).

30. V. A. Glembotskii, P. M. Solozhenkin, and L. L. Ogneva, Izv. Akad. Nauk Tadzh. SSR, Otd. Fiz. Tekh. Khim. Nauk, (1), 58-62 (1966).

31. A. E. Kolchemanova, Intensifikatsiya Flotation. Protsessa, Akad. Nauk SSSR, Gos. Kom. po Toplion. Prom. pri Gosplane SSSR, Inst. Gorn. Dela, pp. 39-43 (1964).

32. E. I. Skripnik, V. I. Dolganov, A. Z. Simileiskii, and V. G. Dyrin, Neft. Khoz., 41(7), 51-56 (1963).

33. G. A. Balakishiev, I. G. Ismailov, M. I. Korneev, and E. B. Mezhebovskii, Izv. Vysshikh. Ucheb. Zaved. Neft i Gaz, 8(12) 71-74 (1965).

34. T. Iwasaki, J. Nakano, M. Usuda, and N. Migita, Kami-Pa Gikyoshi, 21(10), 557-563 (1967).

35. L. Boskova, R. Borisek, and G. Katuscakova, Sb. Vyskum. Prac. Odboru Celulozy Papiera, 13, 63-69 (1968).

36. K. D. Brown, U.S. Pat. 3,317,405 (May 2, 1967).

37. E. Yaeger, National Academy of Science--National Research Council, Publication No. 942, 229-254 (1963).

38. M. Dlabaja, Czech. Pat. 108,767 (October 15, 1963).

39. Khr. Petrov, Godishnik Mashinno--Elektrotekhn. Inst., 10(1), 163-168 (1961).

40. Khr. Petrov and Iv. Petkov, Godishnik Mashinno--Elektrotekhn. Inst., 10(1), 153-162 (1961).

41. R. Brown, Machinery (London), 99, 415-416 (1961).

42. Yu. M. Bystrov, Prom. Primenenie Ul'trazvuka, Kuibyshevsk. Aviats. Inst., Kuibyshev, Sb., pp. 196-202 (1961).

43. C. B. Kenahan and D. Schlain, U.S. Bur. Mines, Rept. Invest., No. 6073, 1962, 26 pp.

44. A. V. Bondarenko, Tr. Novocherk. Politekhn. Inst., 133, 59-78 (1962).

45. L. Domnikov, Metal Finishing, 67(4), 59-61 (1969).

46. A. M. Ginberg, V. N. Nud'ga, and Yu. N. Petrov, Elektrokhimiya, 3(6), 789-790 (1967).

47. R. J. Simpson, Electroplating Metal Finishing, 18(9), 305-306 (1965).

Chapter 14

MEDICAL APPLICATIONS OF ULTRASONIC ENERGY

I. INTRODUCTION

The brevity of this chapter is in no way indicative of the importance of ultrasonics in the field of medicine. Hundreds of published articles and several books (1-4) on the subject are available. Ultrasonics has found usage in all aspects of the medical field, including diagnostic, therapeutical, and surgical applications. Sterilization and cleaning of surgical instruments, preparation of emulsions, and preparation of pharmaceutical materials are proven benefits of ultrasonics to clinical practice. Thus, medical applications of ultrasonics include both low-intensity applications and high-intensity applications.

Diagnostic medical applications are based on the same principles that are applied in inspecting materials and in various other types of sensing and control. The merits of ultrasonic diagnosis are its safety, convenience, and ability to detect phenomena to which X-rays and other means of diagnosis are insensitive.

The principles and mechanisms involved in the industrial high-intensity applications of ultrasonics also have medical importance. The production of heat in the body has been recognized for its therapeutic value. Mechanical effects may include embolism destruction or increased diffusion through cell walls and better dispersion of injections. Chemical effects may be beneficial in relation to certain natural chemical reactions that occur in the body or to the activity of various injected materials. Ultrasonic therapy also includes atomization of liquids, including water, used for medicinal purposes.

Focused high-intensity ultrasonic energy can produce permanent changes deep within tissue without adversely affecting intervening intact structures. Effects produced can be selective. Body tissues are affected on either a macroscale or microscale in any number of ways, depending both upon the characteristics of the tissue and upon the intensity of the ultrasound. Production of lesions in the brains of experimental animals for anatomic, physiologic, and psychologic studies is a routine procedure with appropriate instrumentation (5). Normal structures other than the brain, including the pituitary body and the liver, have been ultrasonically irradiated under relatively precisely controlled conditions of exposure. Carcinogenic tissues have also been irradiated. Intense ultrasound has been used at several centers for the permanent modification of tissue in the human but not on a routine basis.

In this chapter, the emphasis is on applications of ultrasonic energy to medical practice rather than on the equipment used.

II. POWER MEASUREMENTS AND DOSAGES

The intensity of the ultrasonic energy used in medical applications can be measured calorimetrically, by absorption probes, or by radiation pressure gauges (or balances) calibrated in watts. Power measurements are essential in laboratory work. Clinical instruments available on the market are designed to produce intensities at or below the accepted limit. For general therapeutic applications, this limit is approximately 3 W/cm^2.

There are two aspects of dosage: intensity and time. Certain phenomena occur at intensities above certain threshold intensities that do not occur at intensities below these thresholds. Thus, if a certain beneficial effect is obtained at one intensity, doubling the intensity will not necessarily double the rate at which the beneficial effect occurs. In fact, it could prove to be harmful. The dosages should be determined for each type of application and must be carefully controlled.

III. BASIC MECHANISMS AND PRINCIPLES

III.A. MECHANISMS

All methods of medical diagnosis using ultrasonics are based upon either of two principles: (a) reflections from interfaces or (b) Doppler effects produced by motions within the body. Both of these are low-intensity methods. In the reflection method, the same principles are used that are employed in nondestructive testing. Intermediate intensities are used in ultrasonic therapy. Although not all mechanisms associated with the results obtained through ultrasonic therapy are known, there are certain possibilities which certainly must be considered in order that this form of energy may be used most beneficially and harmful effects avoided.

The generation of heat is associated with absorption, and many physicians think of ultrasound as a means of producing heat well within the body for therapeutical purposes. There undoubtedly are some beneficial thermal effects derived from absorption of ultrasound, but generation of heat cannot explain all phenomena observed under the influence of ultrasound in biological tissues. Absorption within tissues varies with the type and condition of the tissue.

Ultrasound causes increased vascular and fluid circulation. Neurogenic effects result in changes in skeletal muscle tone. Associated with ultrasonic irradiation therapy is general relaxation and sedation. Diffusion of chemicals and ions through cell membranes increases, leading to improved organic exchanges as in metabolism and osmosis or to better and more effective utilization of injections. A 3.6-fold increase in the penetration of iodine from a hydrophilic ointment through the skin into the bodies of dogs, rabbits, and humans using therapeutic doses of ultrasound at 800 kHz for 10 min has been reported (6). Ultrasound applied to a dialyzer designed for extracting specific fractions of urine increased the extraction rates by several orders of magnitude (7). Killing efficiency of chemical agents in an ultrasonic sterilizer increased owing to increased transfer rates under the influence of ultrasound.

Rupture of cells and cell structures can occur because of the high stresses generated by extremely high intensities, but such intensities are not used in routine therapeutical practice. Also, at these high intensities, biochemical and functional changes occur which can be related to the action of ultrasound on intracellular molecular complexes. Molecular complexes in tissues and cell structures can be broken loose, resulting in liberation or increased activity of cell enzymes. In common laboratory practice, results of this type are usually associated with cavitation; however, cavitation probably does not occur within the body at therapeutic intensities. Researchers at the University of Illinois (8) have shown that cavitation is not an important factor in the mechanism of production of paralysis of the hind legs of frogs. The same results were obtained even at pressures that were sufficiently high to suppress cavitation.

Ultrasonic energy inhibits the biological activity of fibrinogen and produces an anticoagulation factor out of its molecule (9). If a high-intensity ultrasonic probe is brought into contact with a clot in a blood vessel, the clot will dissolve rapidly. Figure 14.1 shows a commercial instrument for dispersing blood clots and ureteral stones. The wire is passed through a catheter to the site requiring dispersion.

III.B. EFFECT ON HUMAN BLOOD

Extremely high intensities can cause various changes in blood. However, at noncavitating intensity levels corresponding to therapeutical intensities in the human body, there appears to be no danger to blood. Howkins and Weinstock (10) have shown that focused ultrasound below cavitation intensities can hemolyze blood if the focus is at the surface of the liquid and exposed to air. However, no damage is caused at the same intensity if the focus occurs beneath the surface.

III.C. EFFECT ON TISSUE REGENERATION

The use of ultrasonic therapy in the epiphyseal area of growing bone has been viewed with caution because high-intensity ultrasound was believed to cause damage to bone cortex and bone marrow. Vaughen and Bender (11) studied the effects of clinical doses (1 W/cm^2) of ultrasound upon the epiphyses of growing rabbits. The left-knee area of the animals was treated under water for 5 min daily from the age of 3 months until there was evidence (revealed radiographically) of epiphyseal closure, at 6 and 8 months. The untreated hind leg was used as a control. Their findings were that there was no difference in bone length, microscopic appearance, or rate or manner of epiphyseal closure between the treated and the control leg.

Regeneration of amputated tails of tritons (marine snails) has been shown to be stimulated by exposure to ultrasound (12). The blastemal RNA in treated tritons reached levels in 12 days following amputation that corresponded to levels reached in 21 days in the controls.

Some significant work has been accomplished by Dyson, Pond and co-workers (13) in the anatomy department of Guy's Hospital Medical School, London. They have shown accelerated tissue regeneration in adult rabbits under the influence of ultrasound. A small, 1-cm^2 hole is excised through each ear of the rabbit. One of these is used as a control and the other is treated ultrasonically.

The transducer used consisted of a lead-zirconate-titanate disk, 1 cm in diameter, operated at its third harmonic to obtain a frequency of 3 MHz. The area of the ear to be insonated is clamped lightly between the transducer holder and an absorption chamber. At the optimum intensity, around 0.5 W/cm^2, the holes closed in about 34 days in male rabbits and in about 46 days in female rabbits, while the corresponding controls closed in 50 and 70 days, respectively. Higher intensities (8 W/cm^2) caused damage.

These investigators were able to show that the effects were not due to heat. The ultrasound probably accelerates migration of cells to the site of injury and probably also increases the rate of protein synthesis.

Ultrasonically induced regenerated tissue contains a smaller proportion of collagen than normal tissue and, therefore, may have lower tensile strength.

IV. DIAGNOSIS

Ultrasonics is much safer than X-rays for use in medical examination. It is also sensitive to certain abnormalities not detectable by X-rays without resorting to certain expedients. For example, radiographic detection of aneurysms (ballooning of a blood vessel in a weakened portion) is possible if an injection of materials that cause a contrast in X-ray absorption is made into the blood-stream prior to taking the picture. The picture must be taken within a short period of time following the injection. The aneurysm is indicated by a characteristic change in blood-flow pattern. Ultrasonics, on the other hand, makes possible B-scan photographs of aneurysms without injecting anything into the blood stream and without the time limitation of the radiographic method. The actual size of the aneurysm can be measured by this means. Figure 14.2 is a B-scan indication of an aneurysm of the aorta obtained by scanning the patient in a lateral direction.

IV.A. PRINCIPLES

The principle of reflections of acoustic energy from a boundary between two media exhibiting different acoustic impedances is the basis of one type of diagnostic procedure. Equipment used and procedures are similar to those used for nondestructive testing by the pulse-echo technique.

A second diagnostic technique is based upon the Doppler effect. When an ultrasonic wave is reflected from a moving target, the reflected wave is shifted in frequency from that of the incident wave by an amount that depends upon the velocity and direction of the motion. This phenomenon is finding wide usage in the medical field for measuring blood-flow rate and the movement of certain components of organs such as the heart or heart valves.

IV.B. EQUIPMENT

A list of companies that manufacture ultrasonic diagnostic equipment has been compiled by Crawford (14).

In reflection-type, often referred to as pulse-echo, equipment, A-scan, B-scan, or combinations of both of these methods are used for presenting data on the screen of an oscilloscope or for permanent recordings. The A-scan presents echo amplitude and distance. Its primary use is in echoencephalography where it has been especially useful in detecting midline shifts due to concussion or to tumors. It has also been used in obstetrics, gynecology, and ophthalmology in conjunction with B-scan techniques.

With B-scan equipment, radar and sonar techniques of data processing are used to synthesize the received signals into a pattern on an oscilloscope that corresponds to a cross section of the region scanned lying in a plane parallel to the direction of beam propagation. The position of the probe is synchronized with the sweep of one of the axes of the oscilloscope, and the echo amplitude appears as a spot of varying intensity at a position on the screen corresponding to the position of the interface causing the echo.

The transmitted pulse must be as short as possible to resolve small structural details. To do this, the transducer is highly damped just as transducers for obtaining high-resolution in nondestructive testing are highly damped.

As is true in nondestructive-testing applications, coupling between the transducer and the body is important. For direct-contact procedures, coupling is through a film of oil or grease which must be free of air bubbles. For immersion methods, coupling is through a bath of liquid, usually water. The acoustical match between soft tissue and water is good and little energy is lost in irradiating soft tissue. The match is poor between water and bone, and also the attenuation in bone is high. For this reason, considerable energy is lost in echoencephalography, in which contact is made almost immediately with the skull bone, and this may prevent the observation of certain abnormalities that would be observable if the bone had good transmission properties.

Instruments based upon Doppler effect are less expensive and more easily adapted by the practicing physician than are the echo-type units. These features make this type of instrument easier to market. The output from these units can be reproduced through headphones, recorded on strip chart, or recorded on a tape recorder. Some units are focused so that an indication of a blood-flow profile may be obtained using a pulse-Doppler principle. Applications range from detecting fetal heart-beat to indicating blood pressure.

IV.C. DIAGNOSIS BY REFLECTION METHODS

IV.C.1. Cardiac

Ultrasonic diagnostic procedures have generated considerable interest and application in the study of cardiac functions. Major causes of deaths following heart transplants have been acute rejection and infection. A-scan techniques have been used at the Stanford University Medical Center (15) in an effort to detect early signs of rejection. When rejection is indicated, dosages of immunosuppressive agents are lowered and active therapy of the rejection is begun. The method consists of measuring the dimensions of the heart as it fills with fluid. As the heart fills at the onset of rejection, the muscle walls swell and stiffen. Measurements are made using a 0.75-in. diameter, 2.25-MHz transducer at a pulse repetition rate of 1000/sec. Echo indications from the anterior wall and the posterior wall surfaces provide the measurement information, with the distance from the anterior wall to the posterior wall indicating overall heart size.

At the same center, echographic methods (methods using reflections of ultrasonic pulses from interfaces) have been used during cardiac catheterization to estimate the severity of valvular regurgitation in heart patients (16). The estimations are based upon calculations of end-diastolic and end-systolic volumes and stroke volumes based upon ultrasonic measurements. The method has been used on several patients with the conclusion that echographic determinations of stroke volume are atraumatic, safe, and acceptable in patients who do not have valvular regurgitation and that ultrasound echocardiography does provide a reliable means of estimating the severity of valvular regurgitation in patients who have it.

Echocardiography can be used at a patient's bedside to diagnose pericardial effusion (escape of fluid through a rupture in the pericardium (17). In such cases, the echo received from the posterior wall of the heart is split when the transducer is located on the anterior chest surface.

Echocardiography is also of value in evaluating the degree of stenosis (narrowing of the opening) in mitral-valve disease (18). The transducer is aimed at the anterior mitral leaflet and the echo signal is recorded on a strip chart so that an upward movement of the recorder pen corresponds to the movement toward the ultrasonic transducer and a down-stroke represents movement away from the transducer. Thus the slope of the tracing is an indication of the velocity of motion. The degree of stenosis affects the rate at which the blood flows through the opening. The speed of the diastolic downstroke during the period of ventricular filling is affected, slowing down with increased stenosis. This slowing down is readily discernible by the corresponding slope of the ultrasound cardiogram strip-chart recording. Rigidity or calcification of the valve is indicated by a decrease in the total amplitude of the movement of the anterior mitral leaflet between the closed position during ventricular systole and the position of maximum opening in early diastole.

Intracardiac scanning has been used to obtain plan-position-indication (C-scan) displays of the interior of the heart (19,20). This method is used to detect and measure atrial septal defects. A miniature probe is inserted into the right atrium through the external jugular vein or the femoral vein. The probe is located in the desired position with the aid of a fluoroscope. The probe may consist of a single directional element with scanning being accomplished by moving and rotating the probe. Or the probe may consist of multiple elements, with scanning being accomplished by sector techniques in which the positions of points on the recorded image are correlated with the beam direction. One such catheter-borne unit consists of four transducers spaced 90 degrees apart in a plane normal to the axis of the catheter (21). The transducers are pulsed sequentially at the rate of 1000/sec. The data are collected over a period of about 8 sec, and the cardiac cycle is divided into 24 equal increments which depict the contour of the left ventricle at various stages of the cycle. The data are sorted and stored by computer and displayed after the data acquisition is completed. A-scan is used with the C-scan to provide three-dimensional information. The same technique is used in the inferior vena cava of the liver, thus avoiding some of the problems attributed to air in lungs and gas in intestines which are encountered when scanning the liver externally.

In order to obtain stationary patterns without the use of a computer during ultrasonic tomography of the heart, the picture must be completed within one-eight of the cardiac cycle when the tomographs are for systolic phase (when the movement of every part of the heart is maximum) and within one-quarter of the cardiac cycle when the tomographs are for the diastolic phase (when the movement of the heart is slowest).

Focusing transducers have been used to obtain high-resolution pictures of the heart. Movie films of the active heart using this method have been produced at the rate of 7 frames/sec (22), a rate which is fast enough for examination of the movements of the internal structures of the heart.

IV.C.2. Liver

Air in the lungs and gases in the intestines plus certain structural characteristics of the liver present certain difficulties in ultrasonic diagnosis. Nevertheless, certain investigators have reported encouraging results in the use of ultrasonics for diagnosing liver disorders. The method of tomography used by

Kimoto and fellow workers (19) is promising for liver visualization. Also Kelly, Okuyama, and Fry (67) have shown some promising results with cat, hog, and beef liver by using a 1-in.-diam focusing, lead-zirconate-titanate transducer operating at 1 MHz. Reflections from the biological tissue are rereflected and focused on a 1/8-in.-diam lithium sulfate transducer, thus increasing the sensitivity of the system. Small, intact animals (cats and young pigs) are mounted in a chair-type device and immersed to the top of the chest in degassed distilled water. With the technique described and with sufficient amplifier sensitivity, the liver structure of both cats and young pigs can be detected with approximately equal clarity. Ultrasound methods are capable of detecting lesions in the liver not found by X-ray. Holmes (24) has obtained ultrasonic indications from abscesses of the liver, cirrhosis of the liver, and carcinoma of the liver.

IV.C.3. Ophthalmology

The use of ultrasonics in ophthalmology makes possible accurate diagnosis of conditions existing in the soft tissues of the orbit in the light-opaque portions of the eye. A typical frequency is 15 MHz, with focused transducers. The method has the capability of outlining tumors and detached retinas, measuring the length of the axis of the eye, or detecting foreign bodies close to the posterior eye wall. An instrument that combines a diagnostic transducer for locating foreign bodies with a surgical instrument for removing the object permits a rapid removal of foreign bodies from the eye by directing the surgical tool to the body with least damage to the eye.

One method of scanning is the immersion method. A dam or open-ended tube containing a liquid bath is located about the eye; the ultrasonic instrument is immersed in this bath and performs the necessary search with A-scan or B-scan display of the information. A second method is to anesthetize the eye and place the transducer in direct contact with the cornea using distilled water, methyl cellulose, or contact-lens wetting agents for coupling fluids (23).

Greguss (25) indicates certain errors that can result from the fact that the fluids in the eye are not homogeneous. He claims that errors of up to 4 mm in the measured positions of interfaces can result if the heterodispersive nature of these fluids and the components of the eye are not taken into account. These errors can be calculated and fairly accurate measurements are possible with the present state of the art.

Greguss (26) also claims a holographic technique for use in ophthalmology. The ultrasonic wave propagates down a liquid-filled tube into the eye. A sonosensitive plate is placed between the transducer and the eye. The initial wave passing through the plate is the reference beam, while the wave reflected from the eye is the information-bearing wave.

IV.C.4. Neurology

Ultrasonic echoencephalography provides a rapid means of detecting lateral shifts in the midline septum caused by tumors or concussion. Every accident ambulance in Japan carries echoencephalographic equipment in order to identify patients with possible subdural hemorrhage so that these patients may be taken directly to special neurological units for treatment (27). Posttraumatic intercranial hemorrhage or skull and brain trauma are quickly diagnosed and the lesions are located rapidly without discomfort to the patient (28). Ultrasonic pulses are transmitted through the temples. On an A-scan presentation, echo indications from the midline and the opposite temple areas of the skull are presented on the oscilloscope screen. A shift in the midline is easily discernible.

B-scan reproductions are being used in an attempt to improve the information capability of the ultrasonic method. High acoustic absorption in the skull, interference of the hair with coupling, contour of the skull, complexity of the brain, and complicated shape of the ventricular system are problems that interfere with obtaining high resolution. High frequencies required for good resolution are attenuated in the skull and small echoes from interfaces between slight differences in acoustic impedance are lost by the same means. B-scan can be used effectively with small children and babies because their skulls are softer and have attenuation coefficients lower than those of adults (29). Makow and Real (30) developed a 360-degree compound scanning device for examining the brain using an immersion method. In using this device, they observed certain phenomena which led them to believe ventricular pulsations could be used to identify the origin of some echoes. This was not an entirely new discovery but it is an observation that now appears to be clinically significant. According to de Vlieger (29) echo pulsations provide significant clinical information:

1. Echo-pulsation characteristics depend upon the regions of the brain from which the echoes are obtained.
2. Respiration affects the shape of the echo-pulsation curve. Usually, expiration lengthens the rise time of the curve; however, an inverse effect may appear in certain cases, for example with patients having a predominantly thoracical type of respiration.
3. Pulsations in the pressure of the cerebro-spinal fluid appears to influence echo pulsations.
4. Arterio-venous aneurysms cause abnormal dicrotic pulsation curves.
5. Some heart diseases cause abnormal echo-pulsation curves corresponding to abnormal cerebral pulsations.

During an epileptic seizure, the ventricle echo oscillates delicately and rapidly in the prodromal stadium of seizure, but the oscillation stops at the conclusion of the seizure and the echo begins a rhythmical pulsation.

In addition to those disorders previously mentioned, echoencephalography is claimed to be useful for diagnosing hydrocephalus, brain abscess, cerebro-vascular accident, hypertensive encephalography, and cerbral atrophy (31).

IV.C.5. Abdomen and Uterus

The ultrasonic B-scan technique for examining the abdomen has been used for many years. It is useful for diagnosing pelvic tumors, hydatidiform moles, cysts, and fibroids. It can be used to diagnose pregnancy at 6 weeks (counted from the first day of the last menstrual period) and after, and permits the obstetrician to follow the development of the fetus throughout pregnancy, including size and maturity and the location of the placenta. Also, the presence of twins or multiple pregnancy is revealed. Use of ultrasound eliminates the need for X-rays and hence the exposure of the mother and child to X-ray irradiation. Fetal death can be confirmed much earlier by ultrasound than it can by radiography (32).

Available to the diagnostician are two methods of coupling energy from the transducer to the patient: immersion or direct coupling.

A-scan is used with B-scan, particularly to obtain specific measurements, such as the biparietal diameter of the fetal head.

Equipment for automatically scanning the abdomen is available. The commonly used frequency is 2.5 MHz.

A hydatidiform mole produces a characteristic ultrasonic pattern that is easily recognized (32). A radiating sunburst-type of pattern of multiple echoes fills the area corresponding to the uterine cavity and no evidence of fetal structures can be seen. By taking somagrams (ultrasonic pictures) over a period of several days, it is possible to differentiate between the hydatidiform mole and a missed abortion, which sometimes is confused with the mole by its somagram.

The surface of both cystic and solid tumors can be outlined readily. A homogeneous mass such as a cystic or a solid tumor presents no internal reflections because of the absence of internal interfaces. However, the attenuation in the liquid of a cyst is lower than it is in solid tumors, and this permits a larger echo from the posterior wall of the cystic tumor than is possible from a solid tumor for a given instrument sensitivity. In addition, the cystic-tumor indication shows a well-outlined, thin, uniform, and smooth surface, while echoes from the solid tumor are more irregular and often more dense and thickened.

Free fluid in the abdominal cavity produces no echo but can be identified by its effect on abdominal structures such as a floating bowel.

Papillations seen within a cyst and echoes from dense adhesions and metastatic nodules projecting from a tumor to surrounding organs, such as the intestines, or to the cavity wall are indications suggestive of a malignant condition.

Single cysts or polycystic kidney disease can be identified by somagram.

Early pregnancy diagnosis is performed while the patient has a full urinary bladder. The fluid in the bladder provides a useful transmission path of low attenuation and also forces the bowel out of the way for more direct access to the uterus. An embryo first appears as a small circle usually situated in the upper part of the uterus (33). The fetus goes through a stage of rapid growth from the sixth to tenth week and this growth is easily followed by ultrasonic means. After this period, the gestation-sac indication begins to fragment, and this is followed by a period of about 4 weeks during which echoes are rather random and difficult to interpret. However, at about the fourteenth week the fetal head begins to appear and from then on it can be found and measured.

Following an abortion, ultrasonic diagnosis can reveal retained products of conception such as placenta, membranes, or dead fetus requiring surgical attention.

A pregnancy complicated by hydramnios is indicated by an entirely echo-free zone between the fetus and the wall of the uterus. This is due to the fact that attenuation in amniotic fluid is low and it contains no reflecting interfaces (34). Acranial monsters are usually found in connection with hydramnios. In these cases, the normal ultrasound contour of the fetal skull is partly or completely missing, depending upon how much of the cranial vault is missing. In such cases, the uterus is thoroughly scanned, both longitudinally and transversely, in various planes in order to verify the condition of the skull.

Prenatal hydrocephalus can be determined by oversized skull dimensions and low echo density corresponding to thin skull bones.

IV.C.6 Miscellaneous

The potential of ultrasonic diagnosis by echo methods extends to all parts of the body. Since air and other gas has a much lower acoustic impedance than either liquid or solids, air cavities produce distinct echoes. Thus stationary air embolisms can be identified by distinct echoes. Their identification is possible because of the fact that they are located in areas usually filled with fluid. Posterior-wall reflections disappear in the presence of embolisms. Moving embolisms have been located by Doppler methods (35), as discussed in the following section. The particular interest in circulating bubbles has been in relation to decompression sickness caused by gases coming out of solution from the blood when the pressure on the body decreases too rapidly. Pulmonary embolisms can be detected and identified by echo methods. Gallstones have an acoustic impedance that is equivalent to bone and thus provide good ultrasonic echoes.

By using an aluminum transmission line, Lees and Barber (36) have been able to obtain discernible echoes from the enamel-dentin junction of teeth as well as from the dentin-pulp interface. Kossoff and Sharpe (37) investigated the use of ultrasonics to detect degenerative pulpitis, a condition characterized by a degeneration of pulp tissue in which the pulp is replaced by air. It cannot be detected by radiography. Kossoff and Sharpe used excised teeth for their studies and their results appear promising.

Certain rheumatic manifestations, especially arthritic conditions, are easily diagnosed ultrasonically. However, these can also be observed radiographically and with better resolution.

The ultrasonic B-scan technique has been proven useful in diagnosing disorders of the thyroid (38,39). Typical disorders located ultrasonically are (a) neoplastic lesions including cystic nodules, solid adenoma, and carcinoma and (b) nonneoplastic lesions such as Graves disease, subacute thyroiditis, and chronic thyroiditis. A heavily spotted pattern is usually referred to as a "malignant pattern" and should be considered a warning sign whether it is found in the thyroid area or in any other soft-tissue area.

IV.D. DIAGNOSIS BY DOPPLER METHODS

The Doppler principle refers to a change in frequency between the initial transmitted wave and a received wave due to relative motion between a transmitter, a target, and a receiver. Fluid moving at a continuous rate does not cause a Doppler shift in frequency but it does affect the magnitude of the shift caused by relative motion between the three components: transmitter, target, and receiver. Also, variations in fluid motion which cause the beam of ultrasound to be deflected and thus give the appearance of motion cause a Doppler shift in frequency.

The application of the Doppler principle to medical diagnosis is widely accepted. One particularly exciting application is for detecting fetal blood flow and heart beat. A mother can be reassured that her unborn baby is still alive by listening to its heart beat through a set of headphones used with such equipment.

Blood-flow rate can be measured through intact blood vessels using the Doppler principle. The carrier frequency ranges from 2 to 10 MHz, at which frequency the wavelength is short enough that significant echoes from blood particles (erythrocytes) are obtained to produce a Doppler shift. The Doppler frequency signal is calibrated in terms of velocity or detected by earphones, as mentioned previously, and the velocity is rated from low to high according to the pitch of the Doppler signal. The method is suited for FM/FM telemetering.

Baker (40) has described a pulsed ultrasonic Doppler flowmeter capable of obtaining a flow-velocity profile across a blood vessel. Ultrasonic pulses of 5 MHz and 1-μ sec in duration are projected into the blood stream under study. The Doppler shift of the back-scattered signals is sensed in a phase detector. The flow profile is obtained and mapped by sequential sampling at selected intervals over one period of pulsatile flow, using a comb-type gate.

The value of the Doppler principle is not limited to blood-flow measurements or indications. It is useful in detecting and monitoring motions of internal-organ components, such as heart valves and heart muscle, where the sounds generated by these motions cannot be heard. A single unit can be synchronized to record on a multichannel recorder the sequence of heart-muscle and heart-valve movements. The Doppler signal can be separated from other sound from the different parts of an organ by filtering, since these signals are pitched much lower than the carrier frequency of the ultrasonic Doppler system. Thus it has been possible to use Doppler units for measuring mitral-valve and ventricle-wall velocities (41), providing a means of detecting mitral stenosis. Opening and closing of aortic valves also can be measured (42). The processed Doppler signals are recorded on magnetic tape simultaneously with the frequency-modulated electrocardiogram or with measurements of intracardiac pressures.

Doppler-type diagnostic units have found use in determining the severity of atherosclerosis, of locating congenital heart defects, and of continuously monitoring fetal heart beat during birth.

Wells (43) discusses the principles and uses of Doppler methods of diagnosis and includes a good general presentation of physical principles of ultrasonic diagnosis.

V. THERAPY

Therapeutical applications of ultrasound may be divided into two categories: (a) those involving inhalation therapy in which the ultrasound is used to nebulize medicine and (b) those involving direct irradiation of body tissues. The first of these is used in hospital care for treating patients with respiratory ailments such as cystic fibrosis. With high-frequency nebulizers, the droplets are small enough to be drawn into the alveoli and absorbed there.

Many of the identified mechanisms of ultrasonic-irradiation therapy have been discussed previously. A large number of maladies have been treated ultrasonically and, in some cases, with a degree of success not obtained by any other treatment.

V.A. EQUIPMENT

Therapeutical equipment introduces considerably more energy into the body than do the diagnostic instruments. Such equipment has been available for many years. Crawford lists manufacturers of these instruments (14). Except in rare cases of specialized local treatment applications, commercially available therapeutic equipment is limited in power output to approximately 3 W/cm^2. Coupling is usually through a heavy film of oil or grease, but immersion methods are sometimes used. Good coverage of couplant over the area to be treated is essential. Air-bubbles inhibit energy transfer and overlapping an area not covered with couplant can produce a painful burning sensation.

Figures 14.3 through 14.5 illustrate a second type of ultrasonic therapeutical apparatus. Ultrasonic nebulization of pharmaceuticals occurs without producing destructive temperature levels. Units have been made for atomizing very small quantities of material for inhalation therapy. The particles of liquid enter the body at room temperature, which is desirable both from the standpoint of comfort to the patient and from the standpoint of nondecomposition of the pharmaceutical material being inhaled.

V.B. METHODS AND APPLICATIONS

V.B.1. Laryngeal Papillomatosis

Laryngeal papillomatosis is a rare juvenile disease involving wartlike growths on the larynx. The disease may strike in infancy before the age of one year and persist until adolescence. Rarely does it disappear on its own before adolescence. Frequent surgery is required to keep the child alive. Until ultrasonic treatments were adopted, there was no known cure.

The ultrasonic treatment of laryngeal papillomatosis was an outgrowth of a search by Birck (44) for a more effective means of treating a patient who had repeatedly been treated surgically but in whom the disease showed signs of spreading into the tracheobronchial tree. Reports of successful therapy of plantar warts (other warts are also treated successfully by means of ultrasound) was the incentive for using ultrasound on the patient. Treatment with a commercially available therapeutical unit in the throat area daily for 1 week brought favorable results, as it did in several succeeding cases. The objectionable aspect of this treatment was the amount of anesthesia given the patient. For this reason, Birck went to a high-intensity probe-type transducer, with the transducer located in the tip of the probe in order to bring the source into direct contact with the diseased tissue. The application is now administered immediately following surgical removal of the diseased tissue. The relatively large number of patients who appear to have been cured of the disease following this procedure attests to the efficacy of the technique. Others also are treating this disease with ultrasound with apparent success (45-47). The use of frequencies above 1 MHz helps localize the intensity by absorption, and surrounding tissues suffer no adverse effects.

V.B.2. Cancer

Some past research on the use of ultrasound in the treatment of cancer has been reported and some of the reported results have been encouraging. Only extremely high intensities were used in the previous work, an approach that is open to question. The high intensities may produce metastases. The problems of completely locating, identifying, and thoroughly irradiating the diseased tissue are always present.

Typical high-intensity studies of the use of ultrasound in treating cancers were conducted by Soviet scientists (48,49). They report treating 400 rabbits no earlier than 10 days following inoculation with Brown-Pierce type of tumor by which time the tumor is very large. They claim a minimum traumatic effect, while attaining a selective effect in the tumor. There is some edema following irradiation, sometimes accompanied by slight skin maceration. Their results indicate that the controlled application of such high-level sound (on the order of 500 W/cm^2) constitutes a means of destroying at least some types of malignant neoplasms in humans. The irradiation caused the "liquidation (resolution)" of the entire malignant tumor, according to the reports. Their histological investigations reportedly showed a picture of a gradual growth of dystrophic processes in the tumor tissue of the irradiated nodule and of nonirradiated metastases and a rapid reduction in the ability of tumor cells to reproduce. The treated animal was not only healed but became immune, causing the authors to conclude they were on the track of a serum for cancer. They claim they definitely found antibodies to Brown-Pierce tumors in irradiated specimens. The exposure time and frequency of application were not reported. Transplantation of the same type of tumor to the experimental animal could not be accomplished after irradiation. This implies that a type of immunity to cancer has been produced.

According to these same Soviet reports, nine human patients with melanoblastoma were treated. An irradiated tumor, with pronounced metastases in the first stage, in one patient was irradiated ultrasonically and consequently liquidated. Discussion covering the remaining eight patients was somewhat apologetic, explaining the difficulties of irradiating this type of tumor.

Much further work is necessary in this field.

V.B.3. Meniere's Disease

Meniere's disease is a disease of the inner ear producing disabling attacks of giddiness and vomiting. These symptoms have been attributed to an imbalance of pressure in the two fluid-filled cavities of the semicircular canals. In a large percentage of cases, the disease can be controlled by control of diet and medication. Formerly, medication was used to bring relief to the patient with an acute case, but patients suffered repeated attacks of vertigo and the frequency and severity of these attacks were unpredictable. Medical treatment would often lose its effectiveness, with surgery being the only alternative. However, surgery involved the destruction of both the auditory and the vestibular portions of the inner ear which produced permanent deafness. For this reason there was a reluctance to operate on patients with good hearing in the affected ear, and patients in whom both ears were affected were offered no chance of surgical cure.

It is claimed that Meniere's disease offered the first opportunity for surgical application of ultrasound (27). The treatment involved irradiation of the semicircular balance canals. The equipment used in the initial surgical applications was designed and manufactured in Italy by the Federicci Company of Milan (50). The equipment used in this early work operated at approximately 1 MHz and was capable of a maximum acoustical output of 15 W/cm^2. It was equipped with a dial supposedly indicating the irradiation intensity.

Ultrasonic treatment in the early history involved giving the patient light sedation and local anesthesia in order to obtain patient cooperation in identifying the treatment stages, particularly the end point (50). Access to the horizontal semicircular canal was obtained by means of a postauricular simple mastoidectomy. The canal was skeletonized, its surface flattened, and its wall thinned until the blue line was visible. Frenzel glasses were then placed over the patient's eyes so that an assistant could monitor the patient for induced nystagmus (rapid, involuntary oscillation of the eyeballs). The tip of the applicator was placed on the posterior half of the flattened horizontal semicircular canal, and the beam (adjusted by dial for 6 to 9 W/cm^2) was directed into the labyrinth in such a direction as to avoid irradiating the facial nerve. (Irradiation of this nerve could result in either temporary or permanent facial paralysis of the affected side.) An irritative nystagmus directed toward the operated ear would develop within a few minutes of irradiation. After about 15 to 30 min of irradiation, the nystagmus would subside with no recurrence. Irradiation was continued until a paralytic nystagmus directed toward the opposite ear was noted, at which time the intensity, according to the dial on the intensity meter, was raised to 9 to 12 W/cm^2. The end point was indicated if no further irritative nystagmus was induced and if the paralytic nystagmus remained. A fair percentage of success was reported for this procedure, and failures were attributed to insufficient radiation of the canal during surgery, which was perhaps occasionally due to insufficient thinning of the horizontal canal but was more probably the result of improper tuning.

Dial readings in electronic circuitry cannot be relied upon for indicating radiation intensities. Newell reports that other investigators have measured the output of the Federicci equipment calorimetrically and found that the output could vary from 6 to 0 W with as many as six maxima and minima occurring over a frequency change of 10% (51).

Improvements in the technique of treating Meniere's disease include the use of 3 MHz instead of the 1 MHz used with the original Federicci apparatus. The higher frequency is more directional, and therefore makes irradiation of the facial nerve easier to avoid. (In those few cases in which paralysis of this nerve has been caused by ultrasonic irradiation, the effect usually is temporary.)

The Commonwealth Acoustic Laboratories in Australia has developed an ultrasonic unit for the treatment of Meniere's disease that has received general approval by those who have had opportunity to evaluate it (52). The tip of the applicator contains a flat 3.5-MHz transducer, 1.5 mm in diameter, mounted at an angle of 20 degrees to the axis of the 1-mm-diam stainless steel tube to which it is attached. It is capable of output intensities variable up to 25 W/cm^2 (70 mW). With the slender construction of the applicator, the transducer can be inserted into the middle ear and the energy directed in a cylindrical beam directly into the vestibule through the round-window membrane (53). The beam is then directed through the oval window to the saccule, the utricle, and the ampulla of the lateral and the superior canals. The cochlear duct is at an angle and receives energy only after multiple reflections and after attenuation of the ultrasound has occurred. A large area of the membranous labyrinth is exposed, and since ultrasound increases the permeability of membranes, the result is prevention of the accumulation of hydrops. The niche beside the sound window is filled with saline for coupling energy from the transducer into the window. During treatment, the ultrasound is turned off every 5 min and 1 or 2 drops of saline are added to the niche. During this period, at 25 W/cm^2, the temperature rise is 6°C in the facial nerve. The rise in the niche in 2 min is 15°C.

With the CAL unit, selective treatment of the membranous labyrinth is possible with minimal side effects. The treatment requires only reflection of the tympanic membrane--a surgical procedure that inflicts little trauma. The patient is discharged in 2 or 3 days.

With this treatment, irritative horizontal nystagmus develops within the first 3 min of treatment time. As the treatment continues, vertical nystagmus develops and either alternates or combines with the horizontal, producing a rotary nystagmus. The intensity of the motion decreases with continued exposure, becoming latent after 5 to 10 min of irradiation. Following this period, both types of nystagmus reverse their direction, with the horizontal one becoming paralytic before the vertical one reverses. Irradiation continues for 3 min following reversal of the vertical nystagmus.

A device for ultrasonically treating Meniere's disease in much the same manner as the CAL unit has been developed in Sweden (54).

This unit employs a concave focused transducer which uses a hollow water-filled cone as a transmission line. The liquid in the cone is continuously recirculated to maintain cooling. The tip of the cone is at an angle to the axis of the cone for convenient irradiation of the labyrinth. Only a small drop of coupling liquid is used between the tip and the labyrinth which permits operation with a completely dry cavity.

In its present state, ultrasonic treatment may totally replace destructive surgical treatment of Meniere's disease.

V.B.4. Rheumatic Diseases and Related Disorders

Although ultrasound has been used for the treatment of rheumatic diseases, few recent articles describing techniques and results are available. Certain types apparently are benefited by such treatment, while in others, such as rheumatoid arthritis, the treatment produces only a transitory analgesic effect. Diseases reported to have been favorably influenced are osteoarthritis, chondrosis, and spondylosis (55). Satisfactory results have been achieved with rheumatoid spondylitis but other inflammatory forms of rheumatic diseases show little or no response.

Kaclova, Kolar, Babicky, and Kacl (56) have conducted some laboratory experiments that may have clinical significance with reference to treating certain forms of arthritis. These investigators exposed the right tibias of male Wistar rats to ultrasound at therapeutic intensities for 5 min. Each rat was given an intraperitoneal injection of $^{45}CaCl_2$ 48 hr before it was killed. They found that the per cent of ^{45}Ca retained in tibias and scapulas was lower in the treated animals killed 4 to 120 days following exposure to ultrasound in comparison with nonexposed animals used for controls. The amount of ^{45}Ca retained in the incisors was the same in the treated animals and controls. However, when the incisors were irradiated directly the intake of ^{45}Ca was less in the exposed animals than it was in the controls, as revealed by examinations 8 and 28 days following exposure. These investigators claim that changes in intake of ^{45}Ca were not restricted to the exposed organ but occurred in all calcified tissues.

Ultrasonic therapy is effective in treating muscle spasm. When the nerve segment with which the affected limb is associated is ultrasonically irradiated, the result is relaxation of tension, significant loosening, and improved circulation in the limb. The pain is relieved, which leads to a decrease in hypersensitivity, and the improved circulation leads to regenerative processes in the limb and joint.

Ultrasound is widely used in treating athletic injuries ranging from sore and pulled muscles to "charley horse."

V.B.5. Diffusion of Subcutaneous Injections

The increased permeability of membranes and cell walls under the influence of ultrasound has been used to good advantage. Physicians report better results from cortisone injection when the injection is followed by ultrasonic therapy about the area of the injection. This is true also with many other types of injections.

VI. SURGERY

VI.A. PRINCIPLES

In addition to its use in treating Meniere's disease, which in a sense is a surgical procedure since the labyrinth is modified during the application, ultrasonic energy has been applied in a few other types of surgery. Ultrasonic surgery is based simply on the destructive nature of ultrasound (its ability to rupture tissues) and its ability to promote erosion of hard tissues (similar to ultrasonic machining of brittle materials).

VI.B. EQUIPMENT

In comparison with diagnostic and therapeutic equipment, there is very little equipment on the market for surgical use. Small hand-held transducers are available for removing dental calculus. A similar instrument can be obtained for use in cataract surgery for emulsifying and removing the fluid in the lens.

Equipment for altering tissue structure (or forming lesions in the brain) is locally fabricated. Use of such equipment requires special medical and operational training.

VI.C. METHODS AND APPLICATIONS

VI.C.1. Dentistry

The use of ultrasound to prepare cavities in teeth prior to filling received considerable publicity during the 1950's and early 1960's. Ultrasonic dental units were small (about the size of a large fountain pen) magnetostrictive transducers operating within the range of 20 to 30 kHz in a longitudinal mode. The drilling method was patterned after that with the industrial ultrasonic machine tools which utilized an abrasive slurry for cutting. The slurry used with ultrasonic dentistry consisted of a fine-grit aluminum oxide in water flooding the tooth. The slurry was recirculated by means of a small pump.

Tools were blunt and shaped according to the pattern of the cavity to be prepared. Only light pressure was applied and there was no danger of damaging soft tissue. Heavy pressures could cause excessive heating and pain. Not only are these high pressures undesirable physiologically but they also have an adverse effect on the operation of the tool and the drilling rate. Proper light pressure should permit good drilling and have no adverse effect on the tissues.

The ultrasonic method of dental-cavity preparation could not compete with more modern developments of conventional dental equipment. The ultrasonic equipment developed has not gone to waste, however. It is finding wide acceptance as a clinical tool for dental hygiene. Removing scale from teeth, both supragingival and subgingival, requires considerable effort on the part of the dentist using conventional scrapers. The scaling implement gouges the gums, causing bleeding, and the overall operation is uncomfortable to the patient. The ultrasonic dental drill has been adapted for use in removing dental calculus. The tips used to irradiate the teeth are interchangeable and water irrigated. The water couples energy to the teeth and at the same time cools the tooth and tool. The cavitation and light chiselling action combine to remove the calculus. The method also provides a good means of treating the gums. Patients sometimes complain of sore gums following ultrasonic treatment. The exact cause of this pain is not certain; however, if light pressure only is applied and movement of the tip is maintained during the operation, such pain should be minimal or nonexistent.

Some Czechoslovakian scientists (57) have investigated the possibility of making use of the increased permeability of tissues under the influence of ultrasonic energy for the application of local anesthetics in dentistry. A mixture of procaine, adrenaline, synderman, adeps lanae, and oleum olivae was applied to the skin or gingiva of 50 patients prior to tooth extraction and ultrasound applied to the area to promote diffusion. In most cases the anesthesia was ineffective and in difficult cases, it had to be supplemented by a local injection.

Aluminum wires reportedly have been successfully bonded to excised teeth imbedded in a plastic block (58). These bonds were produced using a magnetostrictive transducer operating at 60 kHz which applied an ultrasonic shear stress to the interface between the dentine and the wire. Pressure was applied across the wire and tooth in a direction normal to the surface of the tooth as in conventional ultrasonic welding. The energy was applied for a brief period of time (usually less than 2 seconds). A good bond was claimed for the process, with the wire breaking before the bond. Whether the break occurred at the point where the wire deformation started as a result of the application of ultrasound and, hence, at a point of stress concentration was not reported. The method has not evolved to clinical practicality.

VI.C.2. Stapedectomy

Otosclerosis is a common cause of deafness. A spongy bone growing in the inner ear gradually obstructs the vestibular or cochlear window or both and causes progressively increasing deafness. In recent years, stapes surgery has become fairly common and has been successful in restoring at least partial hearing to patients. Curettes, hooks, and microdrills are generally used for this type of surgery. These instruments produce trauma to the inner ear, with drills causing the greatest damage because of heat, lesions from dust particles thrown into the vestibule, and microfractures in the petrous bone. At one time, the ultrasonic dental drill was suggested as a possible aid to stapes surgery. Early attempts were disappointing in that they resulted in the destruction of the organ of Corti and other neurosensorial elements in the inner ear. This should not have been entirely unexpected from an instrument that produces the amplitude of vibration that is characteristic of the dental drill, considering the relative sensitivity of these components of hearing to damage by loud audible sounds. Much of the failure in using these instruments has been attributed to the low frequency (which was only slightly above the hearing range), high amplitude, and heat resulting from high absorption in the bone. A persistent, continuous application would certainly be destructive to the hearing organs. However, a procedure is possible that avoids the severe acoustic trauma of these early experiments.

Arslan, one of the pioneers in the use of ultrasound for treating Meniere's disease, has had considerable success in stapes surgery using a specially constructed probe operating at 40 kHz (59). The amplitude of vibration is lower than it is at lower frequencies for a given intensity. Tests with laboratory animals showed that no injury was caused when the probe was used on the edges of the window of the periosteum but injury to the inner ear was caused by applying the probe to the promontory or apical turns. Arslan has used the probe clinically to disintegrate the footplate down to the vestibule endosteum, the most difficult and dangerous step in stapes surgery, and to make intercrural section of the plate. Arslan applies the tool for intervals of four seconds at a time and permits a five-second cooling period between applications. As in dentistry, the heat generated increases rapidly with increased pressure. For such a delicate operation, one would expect to apply very light pressure and the effectiveness would be expected to be better than the effectiveness of curettes, hooks, and drills. Arslan's results show no injury to cochlear receptors, and none of his patients show greater hearing loss than those operated on with conventional stapedectomy techniques.

VI.C.3. Neurosonic Surgery

Very high intensity ultrasound produces, under properly controlled dosage conditions, selective changes in the central nervous system. Pioneering research at the Bioacoustics Laboratory of the University of Illinois has led to applications in humans, such as the alleviation of tremors due to Parkinson's disease. These researchers conducted thorough investigations into mechanisms and dosages and developed electronic generators, transducers, and devices for proper orientation of the patient's head with respect to the location of the transducers. With the University of Illinois apparatus, the skull cap is removed and the skin is attached to a special metallic pan which is then filled with degassed physiological saline. The transducers are then precisely located in the saline bath so that the energy will be focused in the desired part of the brain.

There are other, definitely less expensive, methods of treating Parkinson's disease, which include the use of probes that pierce the region to be treated and generate heat at the point to be treated or that freeze the tissue in the region. These methods damage intervening tissue and also present the risk of rupturing blood vessels. The advantage of the ultrasonic method is in the focusing so that tissue between the focal region and the transducer suffers no damage and no instrument needs to penetrate the brain. The main disadvantages of the method are the requirements for removal of the skull cap and specialized equipment to perform the operation.

Fry (60) considers the selective action of ultrasound and its possible application in fundamental neurological research, in neurosurgery, and in gaining an understanding of basic cellular mechanisms to be among the benefits to be derived from the use of focused ultrasound in forming lesions in the central nervous system. Fry points out, as does Lele (61), that white matter (fiber tracts) is more easily disrupted (i.e., lower intensities are required) than gray matter. Fry also states that the most desirable condition of application is to use an irradiation procedure which combines the choice of dosage conditions for selective action and focusing, rather than focusing irrespective of selectivity.

Lele (61) claims that focused ultrasound is used successfully in several laboratories of basic neural sciences without the continual assistance of physicists and electronics technicians. Applications of focused ultrasound bringing satisfactory results at the Massachusetts General Hospital include transdural irradiation of superficial cortical epileptogenic foci in cases of posttraumatic epilepsy, performing commissural myelotomy of the spinal chord for intractable pain and for anterior horn neuronolysis in a case of multiple sclerosis, and irradiation of painful subcutaneous neuromata (61).

Focusing of ultrasound for neurosurgery may be accomplished either by shaping the piezoelectric transducer in the form of a bowl or by using focusing lenses with either a single transducer element or an array of elements. Focusing lenses of Lucite have proven to be satisfactory for this purpose in spite of high attenuation within the lens material.

The size of the smallest lesion that can be made is inversely related to frequency, since the size of the focal region depends upon the wavelength of the ultrasound. At a frequency of 2.7 MHz, Lele claims that the smallest lesion that could be made with certainty was 1.2 mm in length and 0.4 mm in diameter. The lesions that could be made consistently with single pulses were 11 mm long, while approximately 17 mm was the maximum length of lesions that could be produced consistently using multiple pulses.

In related research, Fry, Heimburger, Gibbons, and Eggleton (62) have been able to visualize ultrasonically essentially all soft-tissue fluid-filled space interfaces and, in some instances, interfaces between gray and white matter by using an acoustically transparent window in the skull for transferring the ultrasound. By using this diagnostic technique, lesions produced ultrasonically or by any other technique can be visualized immediately after placement.

VI.C.4. Selective Hypophysectomy

Hypophysectomy refers to the surgical removal of the pituitary body. It is prescribed for patients with advanced cancer of the breast and prostate with bone metastases, diabetic retinitis, malignant exophthalmus (abnormal protrusion of the eyeball), Cushing's syndrome, and acromegaly (a chronic disease characterized by permanent enlargement of the head, thorax, hands, and feet).

Arslan (63) has reported favorable results using ultrasound for selective hypophysectomy with patients having cancer of the breast and prostate, Cushing's syndrome, acromegaly, and retinitis diabetica. The operation consists of a surgical procedure to expose the dural sheath of the pituitary to firm contact by the ultrasonic probe via the sphenoid sinuses, and irradiating the pituitary at 3 MHz and an ultrasonic intensity of 2.5 W/cm^2 for 20 min or more. Following irradiation, the anterior septum is replaced in its original position (i.e., its position before displacement in preparing access to the pituitary) and the columellar incision, necessary for access, is sutured. The operative cavity is loosely packed. The irradiation may be repeated, in a completely bloodless manner, by introducing the probe through one nasal fossa, in a month or more following the initial operation.

VI.C.5. Pulverization of Ureteral Stones

It is possible to pulverize ureteral stones by ultrasonic means by placing these stones in a beaker of water and irradiating them with high-intensity ultrasound. The disintegration is caused by cavitation-induced stresses and relative motion and impact between particles and between particles and the beaker walls. Similar mechanisms are not free to operate within a body so that intensities that are effective externally would not be effective in the body under the restraints imposed by the ureteral tracts. An approach being investigated involves the insertion of probes into the ureteral tract through a catheter. The stones are disintegrated by impact by the probe tip. Extreme care is necessary in this operation to avoid damaging the surrounding tissues.

VII. ANCILLARY APPLICATIONS OF ULTRASONICS IN MEDICAL PRACTICE

Hospital and clinical use of ultrasonic cleaning tanks for cleaning surgical instruments is generally accepted. Other ancillary uses include preparation of emulsions, cell disruptions and analytical purposes.

Andrianova (64) reports using ultrasound for the preparation of 10% emulsions of corn oil in sterilized 5% glucose solution for intravenous injection.

A large number of high-intensity ultrasonic probes are on the market for use in cell disruption for use in the preparation of cell-free extracts. These have varied uses, particularly in the study of certain blood diseases. The probes consist of a mechanical amplifier (tapered solid horn usually made of an alloy of titanium) driven by a high-powered transducer and capable of irradiating the suspension of cells at very high intensities.

Ultrasound is useful for other analytical purposes. For instance, exposure of seminal stains to ultrasonics facilitates the separation of spematozoa from the material. Grochowska (65) has used ultrasound to improve the extraction of poisons from human tissue for analytical purposes. Poston (66) has used ultrasound to treat sputum for cytologic diagnosis of cancer of the lung.

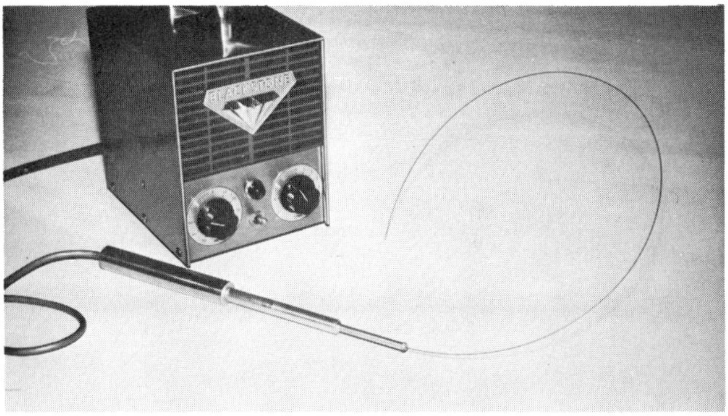

FIG. 14.1. Instrument for the dispersion of blood clots and ureteral stones (courtesy of Blackstone Corporation).

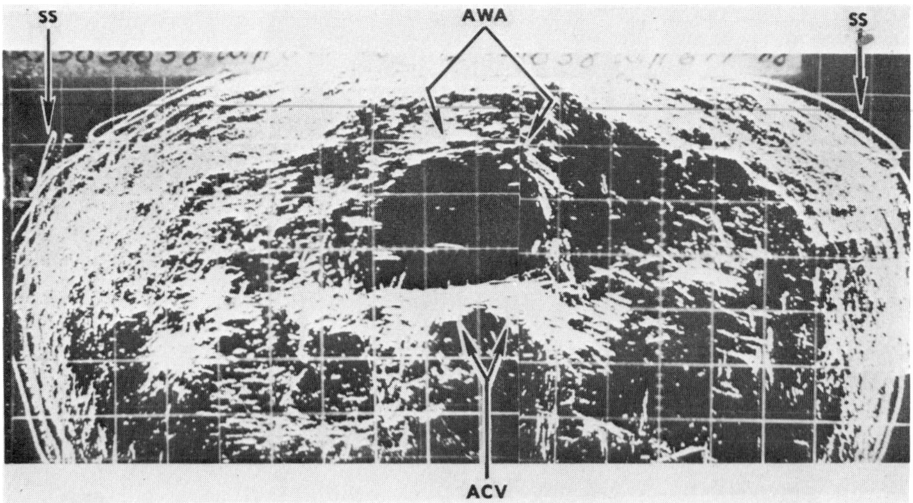

FIG. 14.2. A B-scan transverse section of an aneurysm (ballooning out) in the wall of the aorta (large artery) in the abdomen. SS = skin surface; AWA = anterior wall of abdominal aortic aneurysm; ACV = anterior contour of vertebral bodies. This pattern was made 3 cm above the level of the upper border (crest) of the ilium (Hip bone). (Courtesy of A. K. Freimanis, M.D., of The Ohio State University College of Medicine.)

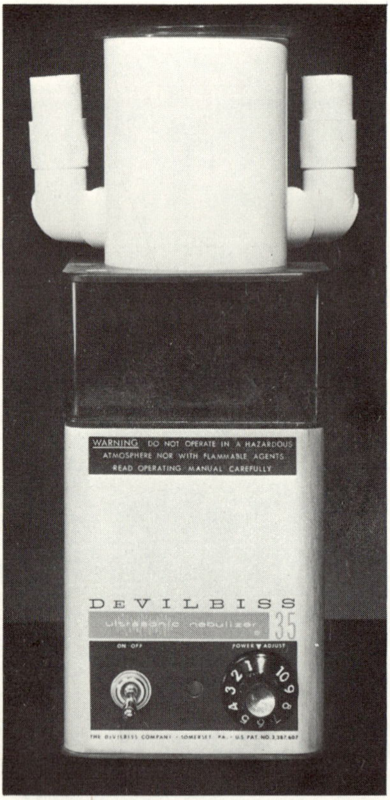

FIG. 14.3. Ultrasonic nebulizer for clinical use (courtesy of DeVilbiss).

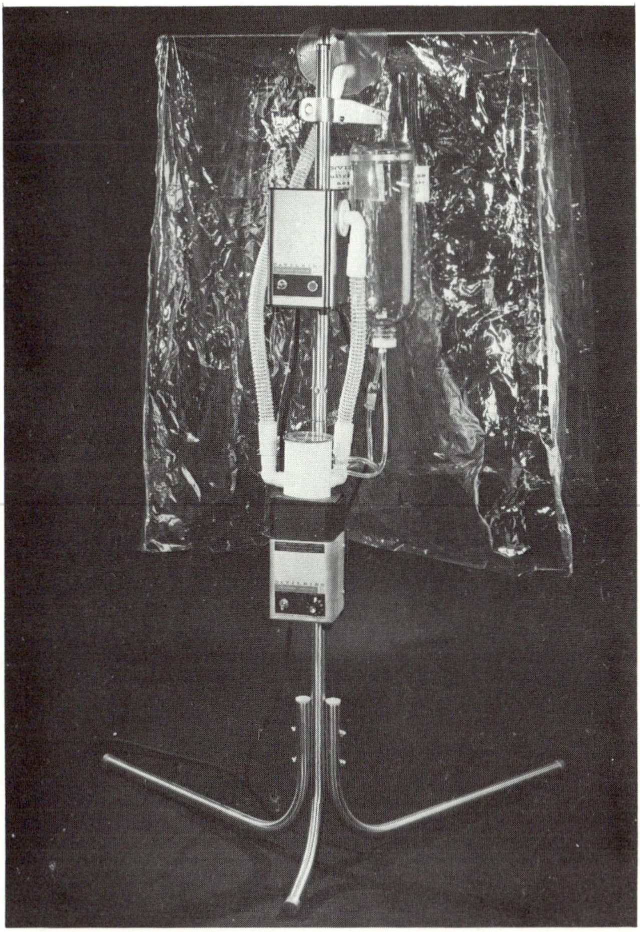

FIG. 14.4. Ultrasonic nebulizer for clinical use equipped with floor stand and canopy (courtesy of DeVilbiss).

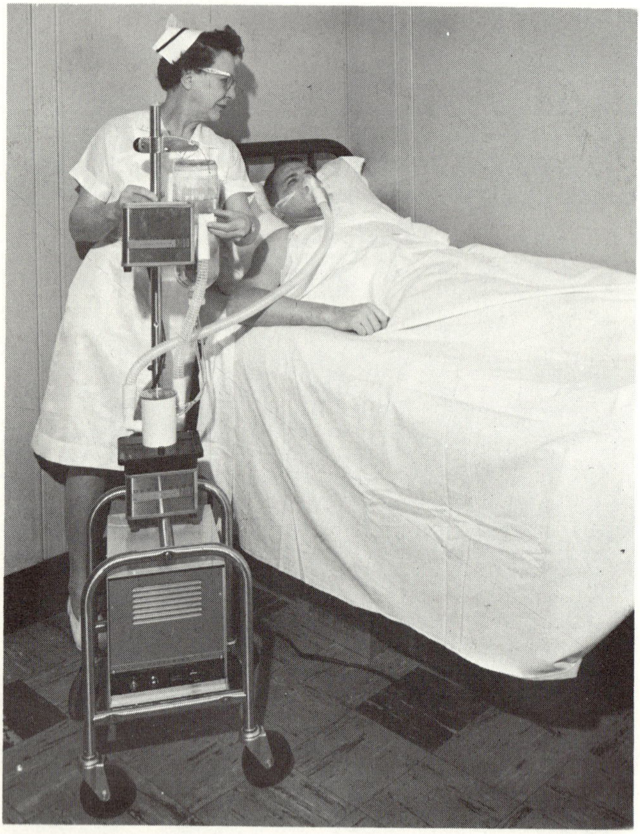

FIG. 14.5. Ultrasonic nebulizer in clinical use (courtesy of DeVilbiss).

REFERENCES

1. B. Brown and D. Gordon, Ultrasonic Techniques in Biology and Medicine, Charles C. Thomas, Publisher, Bannerstone House, Springfield, Illinois, 1967.

2. D. Gordon, Ultrasound as a Diagnostic and Surgical Tool, Williams and Wilkins, Baltimore, Maryland, 1964.

3. J. M. Van Went, Ultrasonic and Ultrashort Waves in Medicine, Elsevier, Amsterdam, Houston, London, New York, 1954.

4. E. Wiedau and O. Röhrer, Ultraschall in der Medizin, Steinkopf, Dresden and Leipzig, 1963.

5. W. J. Fry and F. J. Fry, Paper C-1 presented at the 1965 IEEE Ultrasonics Symposium, Boston, Massachusetts, December 1-4, 1965.

6. J. Dohnalek, I. Hrazdira, J. Cecava, F. Novak, and J. Svoboda, Cesk. Dermatol., 40(3), 173-176 (1965).

7. P. Lasner, I. Embreckt, Dr. Miyake, and H. H. Zinsser, Paper J-6 presented at 1966 IEEE Ultrasonics Symposium, October 12-14, 1966.

8. W. J. Fry, D. Tucker, F. J. Fry, and V. J. Wulff, J. Acoust. Soc. Am., 23(3), 364-368 (1951).

9. B. Bilkova, Scripta Med., 42(1), 11-16 (1969).

10. S. D. Howkins and A. Weinstock, Ultrasonics, 8(3), 174-176 (1970).

11. J. L. Vaughen and L. F. Bender, Paper K9 presented at the 54th Meeting of the Acoustical Society of America, University of Michigan, Ann Arbor, Michigan, October 24-26, 1957.

12. V. Preda, C. Craciun, M. Cristea, M. Gocan, and M. Georgescu, Stud. Cercetari Embriol. Citol., Ser. Citol., 1(2), 155-160 (1965).

13. M. Dyson, J. B. Pond, J. Joseph, and R. Warwick, paper presented at IEEE Ultrasonics Symposium, St. Louis, Missouri, September 24-26, 1969.

14. A. H. Crawford, Ultrasonics, 8(2), 105-111 (1970).

15. Anon., Ultrasonics, 7(4), 217-218 (1969).

16. R. L. Popp and D. C. Harrison, Circulation, 41, 493-502 (1970).

17. R. L. Soulen, M. S. Lapayowker, and J. L. Gimenez, Radiology, 86(6), 1047-1051 (1966).

18. I. Edler, Ultrasonics, 5, 72-79 (1967).

19. S. Kimoto, R. Omoto, M. Tsunemoto, T. Muroi, K. Atsumi, and R. Uchida, Ultrasonics, 2, 82-86 (1964).

20. R. Omoto, Ultrasonics, 5, 80-83 (1967).

21. R. C. Eggleton, C. Townsend, J. Herrick, G. Templeton, and J. H. Mitchell, IEEE Trans. Sonics and Ultrasonics, 17(3), 143-153 (1970).

22. A. Asberg, Ultrasonics, 5, 113-117 (1967).

23. G. A. Leary, Ultrasonics, 5, 84-87 (1967).

24. J. H. Holmes, Digest, pp. 27-34 (1964).

25. P. Greguss, Ultrasonics, 2(3), 134-136 (1964).

26. P. Greguss, "Ultrasonic Holography in Ophthalmology," Optics Technol. 1(1), 40-41 (1968).

27. C. N. Smyth, Ultrasonics, 5, 59 (1967).

28. F. Lahoda, Wehrmedizinische Monatsschrift, 13, 8-11 (1969).

29. M. de Vlieger, Ultrasonics, 5, 91-97 (1967).

30. D. M. Makow and R. R. Real, Ultrasonics, 3, 75-80 (1965).

31. R. E. Schlagenhauff, J. A. Mazurowskii, and M. S. Megahed, IEEE Trans. Sonics and Ultrasonics, SU-15(3), 173-178 (1968).

32. H. E. Thompson, J. Acoust. Soc. Am., 44(5), 1365-1372 (1968).

33. I. Donald and U. Abdulla, Ultrasonics, 5, 8-12 (1967).

34. B. Sundén, Ultrasonics, 5, 67-71 (1967).

35. A. Evans and D. N. Walder, Ultrasonics, 8(4), 216-217 (1970).

36. S. Lees and F. E. Barber, Science, 161(3840), 477-488 (1968).

37. G. Kossoff and C. J. Sharpe, Ultrasonics, 4, 77-83 (1966).

38. Y. Fujimoto, A. Oka, R. Omoto, and M. Hirose, Ultrasonics, 5, 177-180 (1967).

39. B. Damascelli, N. Cascinelli, T. Livraghi, and U. Veronesi, Ultrasonics, 6, 242-243 (1968).

40. D. W. Baker, Paper J6 presented at the 1969 IEEE Ultrasonics Symposium, St. Louis, Missouri, September 24-26, 1969.

41. Y. Yoshitoshi, K. Machi, H. Sekiguchi, Y. Mishina, S. Ohta, Y. Hanoaka, Y. Kohashi, S. Shimizu, and H. Kuno, Ultrasonics, 4, 27-28 (1966).

42. J. B. Kostis, Circulation, 40, 197-207 (1969).

43. P. N. T. Wells, Physical Principles of Ultrasonic Diagnosis, Academic, London and New York, 1969.

44. H. G. Birck and H. E. Manhart, Arch. Otolaryngol., 77, 603-608 (1963).

45. I. C. Jenkins, J. Laryng. Otol., 81, 385-390 (1967).

46. R. Preibisch-Effenberger, Arch. Klin. Exp. Ohr. Nas. Kehlkopf-heilk, 186, 146-152 (1966).

47. P. Schläppi, Pract. Oto-Rhino-Laryngol. (Basel), 29, 365-370 (1967).

48. A. K. Burov, Dokl. Akad. Nauk USSR, 106, 239 (1956).

49. A. K. Burov and G. D. Andreewskaya, Dokl. Akad. Nauk USSR, 106, 445 (1956).

50. R. J. Wollfson, E. E. N. T. Digest, pp. 73-74, 76-78, 80-81 (December, 1963).

51. J. A. Newell, Phys. Med. Biol., 8(3), 241-264 (1963).

52. G. Kossoff, IEEE Trans. Sonics and Ultrasonics, SU-11(2), 95 (1964).

53. Anon., Ultrasonics, 6(1), 3 (1968).

54. S. J. Johnson, Ultrasonics, 5, 173-176 (1967).

55. F. Tschannen, Ultrasonic News, 1(4), 25 (1957).

56. J. Kaclova, J. Kolar, A. Babicky, and J. Kacl, Cesk. Stomatol., 65(6), 437-442 (1965).

57. F. Novak, J. Cecava, J. Dohnalek, and I. Hrazdira, Cesk. Stomatol., 66(4), 271-274 (1966).

58. Anon., Ultrasonics, 5, 201 (1967).

59. Anon., Medical World News, p. 33 (January 27, 1967).

60. W. J. Fry, J. Acoust. Soc. Am., 28(4), 719-721 (1956).

61. P. P. Lele, Ultrasonics, 5, 105-112 (1967).

62. F. J. Fry, R. F. Heimburger, L. V. Gibbons, and R. C. Eggleton, IEEE Trans. Sonics and Ultrasonics, 17(3), 165-169 (1970).

63. M. Arslan, Ultrasonics, 5, 98-101 (1967).

64. I. G. Andrianova, Khim.-Farm. Zh., 1(7), 55-56 (1967).

65. Z. Growchowska, Mikrochim. Acta., (1), 1-6 (1967).

66. F. Poston, Am. J. Med. Technol., 31, 279-284 (1965).

67. E. Kelly, F. Fry, and D. Okuyama, "Ultrasonic Differentiation of Normal Liver Structure as a Function of Age and Species." Proc. 6th International Congress on Acoustics, Tokyo, Japan, Paper M-1-1, Pp. M-1-4 (Aug. 21-28, 1968).

SUBJECT INDEX

A

D

R

S

U

V

W

Y